UNIVERSITY OF HERTFORDSHIRE
The School of Pharmacy

Applications of LC-MS in Toxicology

Applications of LC-MS in Toxicology

Edited by

Aldo Polettini

BSc, PhD
Associate Professor
Laboratory of Analytical Toxicology
Department of Legal Medicine & Public Health
University of Pavia, Italy

London • Chicago Pharmaceutical Press

Published by the Pharmaceutical Press
An imprint of RPS Publishing

1 Lambeth High Street, London SE1 7JN, UK
100 South Atkinson Road, Suite 206, Grayslake, IL 60030-7820, USA

© Pharmaceutical Press 2006

(PhP) is a trade mark of RPS Publishing

RPS Publishing is the wholly-owned publishing organisation of the Royal Pharmaceutical Society of Great Britain

First published 2006

Typeset by Type Study, Scarborough, North Yorkshire
Printed in Great Britain by The Bath Press, Bath

ISBN-10 0 85369 629 2
ISBN-13 978 0 85369 629 2

All rights reserved. No part of this publication may be reproduced, stored in a retrieval system, or transmitted in any form or by any means, without the prior written permission of the copyright holder.

The publisher makes no representation, express or implied, with regard to the accuracy of the information contained in this book and cannot accept any legal responsibility or liability for any errors or omissions that may be made.

A catalogue record for this book is available from the British Library

Contents

Preface ix
About the editor xi
Contributors xiii
Abbreviations xv

1   Ionisation, ion separation and ion detection in LC-MS   1
Lucia Politi, Angelo Groppi and Aldo Polettini

Introduction   1
Ionisation   1
Ion separation   6
Ion detection   11
Tandem mass spectrometry   12
Acquiring and processing LC-MS data   16
Conclusions   19
References   19

2   Method development and optimisation of LC-MS   23
Paul J. Taylor

Introduction   23
General   25
Mass spectrometry   27
Chromatography   33
Sample preparation   36
Preliminary method validation   38
Conclusion   39
References   39

3 Quantification using LC-MS 43
Robert Kronstrand and Martin Josefsson

Introduction 43
Samples and matrices 44
Chromatography 51
Ionisation 52
Fragmentation and mass analysis 54
Determining the concentration 57
Concluding remarks 64
References 64

4 Method validation using LC-MS 71
Frank T. Peters

Introduction 71
Guidelines and literature on validation 71
Validation parameters 72
Experimental design for method validation 88
Conclusions 90
References 90

5 Identification and confirmation criteria for LC-MS 97
Laurent Rivier

Introduction 97
Identification 98
Confirmation of identity 100
Value and role of mass spectral library searching in the identification of organic compounds 100
Recommendations of regulatory bodies 102
Specific difficulties in constructing an in-house LC-MS system 104
Nature of databases for identification in LC-MS 104
Requirement for pure standards 105
Decision limits 105
Final considerations 107
Conclusions 108
Acknowledgements 108
References 108

6 Systematic toxicological analysis with LC-MS 111
Pierre Marquet

Introduction   111
Molecular identification using LC-MS   112
Chromatographic separation   124
Sample preparation   125
Practicality and efficiency of comprehensive LC-MS(-MS) systematic toxicology analysis procedures   126
Conclusion   127
References   127

7 Analysis of therapeutic drugs with LC-MS 131
Hans H. Maurer and Frank T. Peters

Introduction   131
Sample preparation for LC-MS(-MS) analysis of therapeutic drugs   132
Separation systems for LC-MS(-MS) analysis of therapeutic drugs   138
Ionisation and detection of therapeutic drugs in LC-MS(-MS) analysis   139
Validation of LC-MS(-MS) procedures for the analysis of therapeutic drugs   142
Conclusions and perspectives   144
References   144

8 LC-MS of drugs of abuse and related compounds 149
Maciej J. Bogusz

Introduction   149
General methodological considerations   150
Illicit drugs   151
Therapeutic drugs with abuse potential   170
LC-MS screening for drugs of abuse   176
Future perspectives   179
References   180

9 LC-MS in doping control 193
Detlef Thieme

Introduction   193
  Small molecules   195
  Large molecules   206
  Summary   212
  References   212

10   Pesticide analysis using LC-MS   217
Félix Hernández, Juan V. Sancho and Oscar J. Pozo

General overview   217
Sample matrices and analytes investigated   223
Sample preparation   225
LC-MS   227
Quantification and confirmation of analytes   232
Selected applications   237
Concluding remarks   238
References   239

11   Peptide analysis using LC-MS   243
Isabel Riba Garcia and Simon J. Gaskell

Analysis of peptides   243
Analysis of proteins   251
Summary   253
References   253

12   LC-MS in forensic chemistry   257
Jehuda Yinon

Introduction   257
Lacrimators   257
Inks   259
Dyes in textile fibres   260
Chemical warfare agents   260
Explosives   263
References   268

Index   271

Preface

THE SUCCESS OF any analytical technique depends not only on the informing power it provides, but also on the reliability and robustness of its application. In this regard, liquid chromatography–mass spectrometry (LC-MS) is no exception. In fact, LC-MS possesses very attractive features for analytical toxicologists, e.g. the direct analysis of polar, high-molecular-weight and thermally labile compounds. None the less, this technique roamed through a relatively long pioneering age of moving belt, particle beam and thermospray interfaces, and only began to spread in the laboratories in the 1990s, much later than gas chromatography (GC)-MS. This happened thanks to the appearance of atmospheric pressure ionisation (API) on the market, which brought about ruggedness in LC-MS analysis.

A quick search of PubMed (www.ncbi.nlm.nih.gov/entrez) shows that the number of publications with the term 'LC-MS' or synonyms in the title increased exponentially starting from 1990, approaching the number of 'GC-MS' publications per year in the mid-1990s and almost doubling it from 2003 on. These data unequivocally attest to the success of LC-MS.

Having reached maturity, the potential of LC-MS is none the less far from being exhaustively explored. Actually, LC-MS is still undergoing huge technical developments. In the last few years, the well-established electrospray ionisation (ESI) and atmospheric pressure chemical ionisation (APCI) techniques have been flanked by new types of ion sources such as atmospheric pressure photo-ionisation (APPI), atmospheric pressure laser ionisation (APLI) and sonic spray ionisation (SSI). The scenario of mass analysers is even more differentiated. Together with the traditional quadrupoles (Q) and quadrupole ion traps (ITs), new or re-designed mass analysers have become available to routine laboratories, such as time-of-flight (TOF), ion cyclotron resonance (ICR) and linear ITs (LITs), also widening the range of hybrid configurations (QTOF, QLIT, LIT-ICR). This does not include software developments, such as the so-called information (or data)-dependent acquisition, and software tools for metabolite identification and protein identification. This picture has both a positive and a negative side. The positive side is that there is a lot to expect from this technique in the future in terms of analytical solutions. The negative side is that LC-MS, with this choice of technical alternatives, will probably never reach the degree of standardisation of a one-button technique.

The aims of this book are (i) to provide a guide for the successful selection and operation of LC-MS equipment for analytical toxicology applications, and (ii) to present the current applications of LC-MS in the different fields of analytical toxicology and related areas.

Chapter 1 presents an overview of the different instrumental alternatives available on the market in relation to ionisation, ion separation and ion detection, together with two further sections dedicated to tandem mass spectrometry and data acquisition/processing, respectively.

A systematic, step-by-step approach to developing and optimising a successful quantitative LC-MS method is illustrated in Chapter 2, whereas Chapter 3 provides a more comprehensive view of the problems encountered and viable solutions in LC-MS quantification, and Chapter 4 discusses the aspects involved in validation of a quantitative LC-MS method.

Chapter 5 tackles the complex and long-debated issue of identification and confirmation criteria for LC-MS, and Chapter 6 reviews the potential of LC-MS in the screening and identification of exogenous compounds in biological samples (so-called systematic toxicological analysis [STA]).

Chapters 7–11 are dedicated to specific application fields of LC-MS, i.e. therapeutic drug analysis (Chapter 7), drug of abuse analysis (Chapter 8), doping control (Chapter 9), pesticide analysis (Chapter 10) and peptide analysis (Chapter 11). Finally, Chapter 12, although outside of the main scope of the textbook, presents an overview of non-toxicological LC-MS applications in forensic chemistry.

This book would have never been written without the effort of all the contributing authors who agreed to share their knowledge on LC-MS and had the patience to fulfil the editor's requests.

This book also owes a great deal to Lucia Politi, Angelo Groppi (who both co-authored the first chapter with me) and Luca Morini for their invaluable help in revising the manuscript, and to the Pharmaceutical Press editor Laurent Galichet for his continuous encouragement and help.

Finally, my deepest gratitude goes to my wife Cristina, for her immense patience and support, and for her dedication to family, despite her work, that allowed me to draw enough time to take care of this book.

Aldo Polettini
March 2006

About the editor

Aldo Polettini graduated in Biological Sciences at the University of Milan (1980–1985) and obtained his PhD in Forensic Toxicology at the University of Pavia (1987–1993). He worked as a Post-doctoral Researcher in many organisations including: the Joint Research Centre of the European Union, Ispra, Italy (1986), the Municipal Institute for Medical Research, Barcelona, Spain (1992) during the Olympic Games, the National Institute of Public Health and Environmental Protection, Bilthoven, The Netherlands (1994), and the National Institute on Drug Abuse, Baltimore, MD, USA (2000). He is currently Associate Professor of Forensic Toxicology and responsible for the Laboratory of Analytical Toxicology of the Department of Legal Medicine and Public Health of the University of Pavia.

He is the author of over 60 peer-reviewed papers, a member of the Editorial Board of the *Journal of Analytical Toxicology* and a member of the Systematic Toxicological Analysis (STA) and Guidelines Committee of The International Association of Forensic Toxicologists (TIAFT).

His main research interests are the development of new analytical methods for trace organic compounds in biosamples using hyphenated chromatographic and mass spectroscometric techniques, STA by gas chromatography–mass spectrometry (GC-MS) and liquid chromatography (LC)-MS, doping control, and drugs of abuse in hair.

Contributors

Maciej J. Bogusz MD, DrSc (Toxicology)
Department of Pathology and Laboratory Medicine,
King Faisal Specialist Hospital and Research Centre,
Riyadh,
Saudi Arabia

Simon J. Gaskell PhD
Michael Barber Centre for Mass Spectrometry,
School of Chemistry,
University of Manchester,
Manchester,
UK

Angelo Groppi MSc
Department of Legal Medicine and Public Health,
University of Pavia,
Pavia,
Italy

Félix Hernández PhD
Research Institute for Pesticides and Water,
Universitat Jaume I,
Castellón,
Spain

Martin Josefsson PhD
National Board of Forensic Medicine,
Department of Forensic Genetics and Forensic Chemistry,
University Hospital,
Linköping,
Sweden

Robert Kronstrand PhD
National Board of Forensic Medicine,
Department of Forensic Genetics and Forensic Chemistry,
University Hospital,
Linköping,
Sweden

Pierre Marquet MD, PhD
Department of Pharmacology–Toxicology,
University Hospital,
Limoges,
France

Hans H. Maurer PhD
Department of Experimental and Clinical Toxicology,
University of Saarland,
Homburg (Saar),
Germany

Frank T. Peters PhD
Department of Experimental and Clinical Toxicology,
University of Saarland,
Homburg (Saar),
Germany

Aldo Polettini PhD
Department of Legal Medicine and Public Health,
University of Pavia,
Pavia,
Italy

Lucia Politi PhD
Department of Legal Medicine and Public Health,
University of Pavia,
Pavia,
Italy

Oscar J. Pozo PhD
Research Institute for Pesticides and Water,
Universitat Jaume I,
Castellón,
Spain

Isabel Riba Garcia PhD
Michael Barber Centre for Mass Spectrometry,
School of Chemistry,
University of Manchester,
Manchester,
UK

Laurent Rivier PhD
Laurent Rivier Scientific Consulting,
Place de l'Europe 7,
Lausanne,
Switzerland

Juan V. Sancho PhD
Research Institute for Pesticides and Water,
Universitat Jaume I,
Castellón,
Spain

Paul J. Taylor
Department of Clinical Pharmacology,
Princess Alexandra Hospital,
Woolloongabba,
Queensland,
Australia

Detlef Thieme PhD
Institut für Rechtsmedizin,
Universität München,
München,
Germany

Jehuda Yinon PhD
Weizmann Institute of Science,
Rehovot,
Israel

Abbreviations

1,4-BD	1,4-butanediol	CMHC	3-chloro-4-methyl-7-hydroxycoumarin
2,4,5-T	2,4,5-trichlorophenoxyacetic acid	CN	2-chloroacetophenone
2,4-D	2,4-dichlorophenoxyacetic acid	CS	o-chlorobenzylidene malononitrile
3-PBA	3-phenoxybenzoic acid	CSF	cerebrospinal fluid
4-F-3-PBA	4-fluoro-3-phenoxybenzoic acid	CV	coefficient of variation
6-AM	6-acetylmorphine	CZE	capillary zone electrophoresis
11-OH-THC	11-hydroxy-Δ^9-tetrahydrocannabinol	DAD	(photo) diode array detector/detection
AACC	American Association of Clinical Chemistry	DAM	diacetylmorphine (heroin)
AAFS	American Academy of Forensic Sciences	DAP	dialkylphosphate
		DBCA	3-(2,2-dibromovinyl)-2,2-dimethylcyclopropane-1-carboxylic acid
AC	acetylcodeine		
ACN	acetonitrile		
AIDS	acquired immune deficiency syndrome	DC	direct current
AMP	amphetamine	DCCA	3-(2,2-dichlorovinyl)-2,2-dimethylcyclopropane-1-carboxylic acid
ANOVA	analysis of variance		
AOAC	Association of Official Analytical Chemists		
		DDA	data-dependent acquisition
AORC	Association of Official Racing Chemists	DEA	2,6-diethylaniline
		DEAMPY	2-diethylamino-6-methyl pyrimidin-4-ol
APCI	atmospheric pressure chemical ionisation	DEDTP	diethyl dithiophosphate
		DEET	N,N-diethyl-m-toluamide
API	atmospheric pressure ionisation	DEP	diethyl phosphate
APPI	atmospheric pressure photo-ionisation	DETP	diethyl thiophosphate
BMA	benzphetamine	DMDTP	dimethyl dithiophosphate
BP	buprenorphine	DM-MA	desmethyl malathion
BPG	buprenorphine glucuronide	DMP	dimethyl phosphate
BTA	1,2,3-benzotriazin-4-one	DMT	N,N-dimethyltryptamine
BZE	benzoylecgonine	DMTP	dimethyl thiophosphate
C6G	codeine-6-glucuronide	EDDP	2-ethylidene-1,5-dimethyl-3,3-diphenylpyrrolidine
C_8	octasilane		
C_{18}	octadecasilane	EGDN	ethylene glycol dinitrate
CBD	cannabidiol	EI	electron ionisation
CBN	cannabinol	EMDP	2-ethyl-5-methyl-3,3-diphenyl-1-pyrroline
CDEPA	2-chloro-N-[2,6-diethylphenyl]acetamide		
		EME	ecgonine methyl ester
CE	collision energy	EMIT	enzyme multiplied immuno technique
CE	capillary electrophoresis (Chapter 11)		
CEM	channel electron multiplier	EOF	electro-osmotic flow
CID	collision-induced dissociation	EPO	erythropoietin
CIT	5-chloro-1,2-dihydro-1-isopropyl-[3H]-1,2,4-triazol-3-one	ESI	electrospray ionisation
		EtG	ethyl glucuronide
CL-20	2,4,6,8,10,12-hexanitro-2,4,6,8,10,12-hexaazaisowurtzitane	EtS	ethyl sulphate
		ETU	ethylenethiourea

FAB	fast atom bombardment	IT-MS	ion trap–mass spectrometry
FD	field desorption	IUPAC	International Union of Pure and Applied Chemistry
FDA	US Food and Drug Administration		
FIA	flow injection analysis	LAMPA	lysergic acid methylpropylamide
FLD	fluorimetric detection	LC	liquid chromatography
FLN	flunitrazepam	LC-DAD	liquid chromatography–(photo) diode array detector/detection
FTICR	Fourier-transform ion cyclotron resonance		
		LC-LC	coupled-column liquid chromatography
FWHM	full width at half maximum		
GA	O-ethyl-N,N-dimethylphosphoramidocyanidate (or tabun)	LC-MS(-MS)	liquid chromatography–(tandem) mass spectrometry
GB	isopropyl methylphosphonofluoridate (or sarin)	LH	luteinising hormone
		LIT	linear ion trap
GBL	γ-butyrolactone	LLE	liquid–liquid (solvent) extraction
GC	gas chromatography	LLOQ	lower limit of quantification
GC-MS	gas chromatography–mass spectrometry	LOD	limit of detection
		LOI	limit of identification
GD	pinacolyl methylphosphonofluoridate or soman	LOQ	limit of quantification
		LSD	lysergic acid diethylamide
GF	cyclohexyl methylphosphonofluoridate	m/z	mass-to-charge ratio
		M3G	morphine-3-glucuronide
GFTCh	Gesellschaft für Toxikologische und Forensiche Chemie	M6G	morphine-6-glucuronide
		MALDI	matrix-assisted laser desorption ionisation
GHB	γ-hydroxybutyrate		
GUS	general unknown screening	MAMP	methamphetamine
HBOC	haemoglobin-based oxygen carrier	MBTFA	N-methyl-bis-trifluoroacetamide
hCG	human chorionic gonadotrophin	MCA	malathion monocarboxylic acid
HF	human haemofiltrate	MCPA	2-methyl-4-chlorophenoxyacetic acid
hGH	human growth hormone	MDA	3,4-methylenedioxymethamphetamine
HILIC	hydrophilic interaction liquid chromatography		
		MDCA	malathion dicarboxylic acid
HIV	human immunodeficiency virus	MDEA	3,4-methylenedioxyethylamphetamine
HMTD	hexamethylenetriperoxidediamine		
HMX	1,3,5,7-tetranitro-1,3,5,7-tetrazacyclooctane	MDMA	3,4-methylenedioxyamphetamine
		ME	matrix effects
HNE	4-hydroxy-2-nonenal	MEKC	micellar electrokinetic chromatography
HPLC	high-performance liquid chromatography		
		MeOH	methanol
HR	high resolution	MO	malathion oxon
IA	immunoassay	MPNP	3-methyl-4-nitrophenol
ICAT	isotope-coded affinity tag	M_r	molecular weight/mass
ICH	International Conference on Harmonization of Technical Requirements for Registration of Pharmaceuticals for Human Use	MRM	multiple-reaction monitoring
		MRPL	minimum required performance level
		MS	mass spectrometry
		MS-MS	tandem mass spectrometry
ICR	ion cyclotron resonance	MSTFA	N-methyl-N-trimethylsilyltrifluoroacetamide
i.d.	internal diameter		
IDA	information-dependent acquisition	MTBE	tert-methylbutyl ether
IEC	ion-exchange chromatography	NBP	norbuprenorphine
IGF-1	insulin-like growth factor-1	NBPG	norbuprenorphine glucuronide
IMPY	2-isopropyl-6-methyl-pyrimidin-4-ol	NCCLS	National Committee for Clinical Laboratory Standards
IOC	International Olympic Committee		
IP	identification point	NG	nitroglycerin
IS	internal standard	NMT	N-methyltryptamine
ISO	International Organization for Standardization	NPD	nitrogen phosphorous-selective detection
IT	ion trap	OC	oleoresin capsicum

OCP	organochlorine pesticide	SI	similarity index
ODS	octadecylsilica	SIM	selected-ion monitoring
OPP	organophosphorus pesticide	SIR	selected-ion retrieval
PAGE	polyacrylamide gel electrophoresis	SOFT	Society of Forensic Toxicology
PBI	particle beam ionisation	SOP	standard operating procedure
PBM	probability-based matching algorithm	SPD	substance P derivative
PCB	polychlorinated biphenyl	SPE	solid-phase extraction
PE	process efficiency	SPME	solid-phase microextraction
PEEK	polyether-ether ketone	SRM	selected-reaction monitoring
PETN	pentaerythritol tetranitrate	SSI	sonic spray ionisation
PGC	porous graphitic carbon	STA	systematic toxicological analysis
PMA	*p*-methoxyamphetamine	TATP	triacetone triperoxide
PMF	peptide mass fingerprint	TBA	tetrabutylammonium
PNP	*p*-nitrophenol	TCPY	3,5,6-trichloro-2-pyridinol
PP	protein precipitation	TDM	therapeutic drug monitoring
PSD	post-source decay	tetryl	2,4,6,*N*-tetranitro-*N*-methylaniline
Q	single quadrupole	TFA	trifluoroacetic acid
QqQ	triple quadrupole	THC	Δ^9-tetrahydrocannabinol
QTOF	quadrupole/time of flight	THC-COOH	11-nor-9-carboxy-Δ^9-tetrahydrocannabinol
QTOFMS	quadrupole/time-of-flight mass spectrometry	THC-COOH-G	11-nor-9-carboxy-Δ^9-tetrahydrocannabinol glucuronide
RDX	1,3,5-trinitro-1,3,5-triazacyclohexane	THG	tetrahydrogestrinone
RE	extraction efficiency	TIAFT	The International Association of Forensic Toxicologists
RF	radiofrequency		
RFD	radiofrequency-only daughter-ion scan	TIC	total ion current (chromatogram)
rhEPO	recombinant human erythropoietin	TMS	trimethylsilyl
RIC	reconstructed ion current	TNT	2,4,6-trinitrotoluene
RP	reversed phase	TOF	time of flight
RSD	relative standard deviation	TOFMS	time-of-flight mass spectrometry
S/N	signal-to-noise ratio	TSP	thermospray
SCLC	small cell lung cancer	ULOQ	upper limit of quantification
SCX	strong cation exchange	WADA	World Anti-Doping Agency
SD	standard deviation		
SDS	sodium dodecylsulphate		

1

Ionisation, ion separation and ion detection in LC-MS

Lucia Politi, Angelo Groppi and Aldo Polettini

Introduction

This chapter focuses on the techniques of ionisation, ion separation and ion detection that are available for liquid chromatography–mass spectrometry (LC-MS) instruments. For each technique, a description of the principles of operation is given, together with examples of applications and/or hints for the successful operation of that technique.

A separate section is dedicated to the different operating modes accessible with tandem MS (MS-MS). In the last section, a description of the most important software tools for acquiring and processing LC-MS data is provided.

Sample preparation and chromatographic separation are, despite the notable selectivity of detection of MS and MS-MS, crucial steps for successful analysis, and their importance is discussed in Chapters 2 and 3.

Ionisation

A LC-MS interface has the double task of eliminating the solvent from the LC eluent and producing gas-phase ions from the analyte [1]. This can basically be accomplished in two ways:

- the two processes are separated in space, i.e. the analyte is sampled into the vacuum region of the mass spectrometer and then ionised at reduced pressure
- ions are created at atmospheric pressure (hence the name atmospheric pressure ionisation [API]) and then sampled into the vacuum region.

The second strategy has been proved to confer much more robustness and reliability to the ionisation process, and, for to this reason, API sources have currently almost completely displaced other ion sources that were routinely used previously, e.g. thermospray ionisation (TSI).

The family of API sources, originally consisting of electrospray ionisation (ESI) and atmospheric pressure chemical ionisation (APCI), has expanded in recent years to include atmospheric pressure photo-ionisation (APPI), atmospheric pressure laser ionisation (APLI) and sonic spray ionisation (SSI). These have all further increased the analytical potential of LC-MS in one way or another. These ion sources are presented and discussed in the following sections. Other ion sources, i.e. TSI, particle beam (PBI), fast atom bombardment (FAB) and matrix-assisted laser desorption ionisation (MALDI) will not be treated here because of their obsolescence and/or their limited application in LC-MS equipment routinely used in analytical toxicology laboratories.

However, a description of the principles of MALDI can be found in the chapter dedicated to peptide analysis (*see* Chapter 11).

Electrospray ionisation

ESI was the first to be developed in the family of API sources and, at the time of writing, is the most frequently used ionisation mode in the field of analytical toxicology [2, 3]. An ESI source consists of a capillary tube, the ESI probe, to which a voltage (typically 3–5 kV) is applied. At flow rates of a few microlitres per minute, the difference in potential between the ESI probe and the orifice opening into the vacuum region of the mass spectrometer is sufficient to nebulise the liquid flowing into the capillary. As a result of the limitation in the flow rate applied, this device can be directly coupled to a nano- or micro-LC apparatus, whereas coupling to a normal LC requires splitting of the eluent. In order to handle higher flow rates, a gas (nitrogen) flowing coaxially to the ESI probe is necessary to maintain a stable spray (pneumatically assisted ESI).

ESI requires the analyte to be ionised in solution. Depending on the voltage polarity, nebulised droplets trapping the ionised analyte will be positively or negatively charged. The reduction in size caused by solvent evaporation accounts for the increase in charge density in the droplet, ultimately leading to its explosion when repulsive forces between charges exceed the cohesive forces of the droplet. This process occurs repeatedly until gas-phase ions are produced by direct emission from the microdroplets [4, 5].

If a compound can be charged at multiple sites in solution, it will carry multiple charges under ESI conditions. This allows the use of mass analysers with limited mass-to-charge (m/z) range (e.g. quadrupoles [Q]) to achieve the analysis of high-molecular-weight compounds such as peptides. For example, a compound of molecular weight 12 000 carrying 20 charges will appear at m/z 600. The combination of this feature with the ionisation mechanism allowing the formation of gas-phase ions from ions in solution makes ESI the best solution for the LC-MS analysis of peptides.

Typical ions produced under ESI conditions are:

- in the positive mode: protonated molecular ions [M + H]$^+$, sodium or potassium adducts, or solvent adducts
- in the negative mode: deprotonated molecular ions [M − H]$^-$, or formate or acetate adducts [6].

Cluster ions may sometimes be formed under ESI or other soft ionisation API techniques [7]. Adduct formation should be avoided whenever possible, particularly for quantification purposes, as it is often irreproducible. This is not always easy to accomplish, as will be illustrated in Chapters 3 and 10. Some quantitative applications, however, exploit adduct formation [8].

Due to the low amount of energy transferred to the molecule during the ionisation process, ESI is a mild ionisation technique producing very little fragmentation. As will be discussed later in this chapter, this has a positive impact on the sensitivity of MS-MS analysis as most of the total ion current (TIC) pertains to one ion (the precursor ion in MS-MS mode). The flip side is that ESI-MS does not provide as much structural information. However, fragmentation can be promoted by imparting kinetic energy to the ions of the analyte within the intermediate vacuum region placed between the ion source and the mass analyser. By applying a difference in potential at the two ends of this region (this parameter is called the orifice, fragmentor, skimmer or capillary voltage, depending on the manufacturer), ions are accelerated and are forced to collide with other molecular species (gas molecules, residual solvent, co-eluting compounds, etc.) and the energy gained as a result of these collisions is dissipated by fragmentation [9]. The higher the voltage and the length of the path in the intermediate vacuum region, the higher the fragmentation rate. This technique is called in-source collision-induced dissociation (in-source CID) in order to differentiate it from CID obtained under MS-MS conditions (described in the MS-MS section of this chapter). The application of in-source CID to obtain full-scan mass spectra with structural information (fragmentation), which is therefore useful for identification purposes, is described in Chapter 6.

In order to correctly select/develop chromatographic separation before ESI-MS, it is important to remember the following fact – signal intensity obtained by ESI depends on the analyte concentration more than on the analyte mass flow (amount per unit time) into the ion source, inferring that better signal-to-noise (S/N) ratios are obtained at lower flow rates [10]. This is said to depend on the solvent evaporation process, which is more efficient at lower flow rates. This improves the transmission of gas-phase ions into the vacuum region.

This explains why a drying gas (nitrogen) and a heating device are also included in currently marketed ESI sources to assist droplet formation and solvent evaporation. Nevertheless, higher flow rates will certainly result in more frequent servicing of the interface.

In addition to flow rate, the composition of

the mobile phase also affects ESI [11, 12]. As a general rule, because ESI requires the analyte to exist in solution as an ion, the pH of the LC eluent should be selected accordingly. Should this not match with the pH of the mobile phase required for optimal chromatographic separation, it is always possible to modify the latter by post-column addition. However, the rule of thumb 'basic analyte, acidic mobile phase and vice versa' has more than one exception, as discussed in Chapter 2. It should be kept in mind that the chemical environment in the electrospray plume may change dramatically during the spray process [13, 14]. For example, the production of abundant protonated ions from solutions where their equilibrium concentration is low may be due to a lower than expected pH as a result of solvent oxidation, charge enrichment, uneven droplet subdivision or solvent evaporation.

Buffers are obviously important in order to obtain and control the right pH of the LC eluent. Their selection should be made in the light of two practical considerations:

(i) non-volatile buffers should be avoided as they deposit into the ion source during the vaporisation/ionisation process
(ii) their concentration, as well as the concentration of any other ionised species in the LC eluent, should be kept as low as possible; in fact, an ionised species present at a much higher concentration than the analyte will cover the surface of droplets and prevent the analyte passing into the gas phase, thus reducing its response.

This competition between the analyte and other ions present in the LC eluent is called the matrix effect (ME) – a well-known phenomenon that is particularly evident with ESI [15–17]. Interferents present in the sample extract and co-eluting with the analyte may cause 'ion suppression' (and sometimes 'ion enhancement' [18]). If not properly addressed by means of careful development and optimisation of sample preparation and chromatographic separation, ME may cause a variable response of the analyte and possibly affect the limit of detection (LOD) of the procedure as well as its quantitative performance, as will be extensively illustrated in the following chapters.

Finally, the percentage of organic modifier in the LC eluent should be kept as high as possible as it improves the efficiency of solvent evaporation and, as a result, the yield of ions transmitted to the vacuum region. This may be a problem in the case of polar analytes separated under reversed-phase (RP) chromatography. Glucuronide conjugates, for example, require a low percentage of organic modifier to be sufficiently retained in an RP column. Again, post-column addition helps, provided that the increase in the S/N ratio due to the increase in the organic modifier is not counterbalanced by an increase in flow rate [19]. Figure 1.1 shows the gain in the S/N ratio obtained for the analysis of two ethanol metabolites, ethyl glucuronide and ethyl sulphate, by means of a post-column additional flow of acetonitrile (ACN).

Alternatively, normal-phase chromatography with an aqueous/organic mobile phase enables the retention of polar species with a higher percentage of organic modifier, resulting in higher ionisation efficiency when compared with RP separation [20].

Figure 1.2 shows the range of application of LC-ESI-MS in terms of analyte polarity and molecular weight compared with gas chromatography (GC)-MS analysis.

Sonic spray ionisation

Developed by Hirabayashi at the beginning of the 1990s [21, 22], this ion source has only recently become commercially available [23]. SSI is a modified ESI source where charged droplets are formed without applying an electrical field. If the flow of the nebuliser gas (nitrogen) is fast enough, charge separation at the surface of the droplets occurs. The formation of charged droplets, and subsequently of gas phase ions, is therefore a result of the non-uniform charge distribution during droplet formation. The phenomenon is proportional to the speed of the nebuliser flow and ion abundances increase to a maximum when reaching sonic speed (Mach 1), hence the name of this ion source [24, 25].

A major practical advantage of SSI is that ionisation does not need high voltages or high temperatures to take place, thus extending the

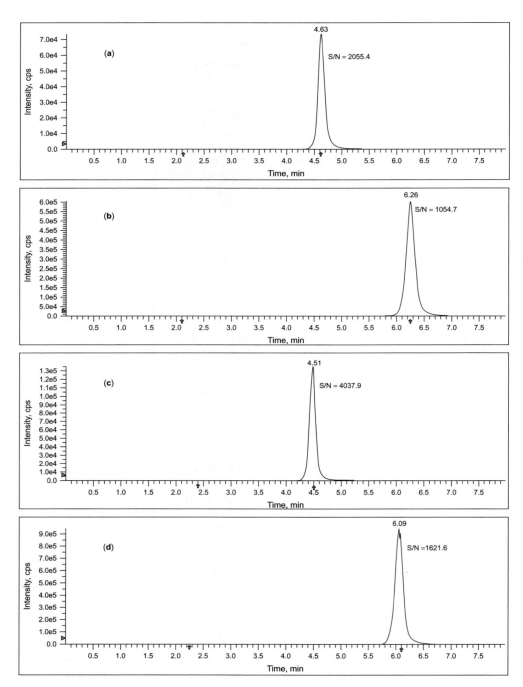

Figure 1.1 Mass chromatograms of ethyl glucuronide (EtG, 5 ng) (a) and ethyl sulphate (EtS, 5 ng) (b) without ACN post-column addition and of the same amounts of the two analytes (c and d, respectively) with 0.1 mL/min ACN post-column addition. Mobile phase: 0.1% v/v formic acid:ACN 99:1, 0.2 mL/min; column: C_{18} (100 × 3 mm i.d., 3 μm particle size); ESI MS-MS detection in negative-ion MRM mode; transitions monitored: m/z 221 → 75 for EtG; m/z 125 → 97 for EtS.

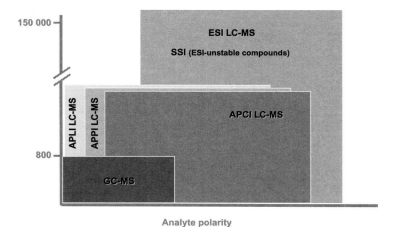

Figure 1.2 Range of application of different hyphenated chromatographic and MS techniques.

applicability of LC-MS to compounds exhibiting instability under ESI conditions (Figure 1.2). However, it has been observed that, as a consequence of the lower energy input, cluster ions may be formed more readily. This can be reduced by applying an external electrical field, but this essentially transforms this ion source back into an ESI source [23].

Atmospheric pressure chemical ionisation

As its name implies, APCI is, like ESI, based on the API strategy, which makes it possible to configure both these ion sources on the same instrument (the latest generation instruments even allow rapid switching between the two ionisation modes) [26]. However, APCI produces ions through a completely different process. Here, the LC eluent is sprayed into a heated chamber (around 400–500°C) by means of a capillary and a coaxial flow of nitrogen (nebuliser gas). The high temperature causes the immediate evaporation of the solvent and the analyte. This means that ionisation occurs in the gas phase, which differs from ESI. In addition to volatility at the applied temperature, thermal stability of the analyte is also a prerequisite for the successful application of APCI (e.g. glucuronides may break down and appear in the mass spectrometer in the form of protonated aglycone) [27].

A corona discharge electrode (2–5 kV) placed close to the tip of the capillary acts as a source of electrons that causes the ionisation of the atmosphere surrounding the tip which is made up of solvent vapours, nitrogen and oxygen. At atmospheric pressure, extensive interactions occur between the excess reagent ions and the analyte, which becomes ionised following a process analogous to classic chemical ionisation (CI). In the positive-ion mode, ionisation occurs when the proton affinity of the analyte is higher than that of the reagent ion(s). In the negative-ion mode, the analyte donates a proton to the reagent ion(s) if the gas phase acidity of the latter is higher. Ionisation by charge transfer, producing M^+ and M^- ions, may also take place.

As ionisation occurs in the gas phase, the pH of the mobile phase should be selected in order to allow the analyte to exist in the neutral form in the liquid phase, as this obviously favours its volatilisation. This makes APCI more suitable than ESI for low-polarity compounds, but less suitable for highly polar ones (see Figure 1.2).

If the analyte is volatile enough and its proton affinity relative to the mobile phase components favours ionisation in the gas phase, ionisation efficiency can be very high, approaching 100%. Also, APCI appears to be less susceptible to chemical interference than ESI, conferring more ruggedness to the analytical procedure with reference to the ME [11, 28–30]. It is for these reasons that authors with sufficient experience with APCI tend to prefer this ion source to ESI for the sensitive and robust analysis of small, thermally stable molecules.

As in ESI, volatile buffers are required in the mobile phase in order to get the right pH (although the purposes are different, as stated earlier). In addition, the mobile phase must be suitable for the production of reagent ions. APCI tolerates higher flow rates much more easily than ESI because of the immediate evaporation of the mobile phase.

Atmospheric pressure photo-ionisation and atmospheric pressure laser ionisation

Developed by Bruins *et al.* in 2000 [31], APPI is basically a modified APCI source where the corona-discharge electrode is replaced by a gas-discharge lamp generating UV photons of a specific energy, depending on the gas used (argon, 11.2 eV; krypton, 10.0 eV; xenon, 8.4 eV) [23]. Direct ionisation of the analyte occurs when its first ionisation potential (typically in the range 7–10 eV for most organic compounds) is lower than the photon energy. The interaction with photons causes the extraction of an electron and the formation of a radical molecular ion $[M]^{+\bullet}$.

However, most of the published work with APPI involves the use of a directly ionisable compound, a dopant (e.g. acetone, toluene, anisole), in order to promote the ionisation process by exploiting more efficiently the beam of photons into the source [32]. In the positive-ion mode, the dopant radical ion may react directly with the analyte through charge exchange or indirectly through the intermediate action of protonated solvent molecules by proton transfer. Therefore, depending on the analyte's proton affinity relative to the composition of the mobile phase, either a radical molecular ion or a protonated molecular ion is obtained. Typically, non-polar compounds appear as $[M]^{+\bullet}$, whereas polar compounds are observed as $[M + H]^+$ ions. In the negative-ion mode, ionisation may occur by proton transfer, charge exchange or electron capture. Dopant-assisted APPI can improve ionisation efficiency, but it may also result in a more difficult interpretation of mass spectra, e.g. because of adduct formation.

Although the potential of APPI has not yet been fully explored, some advantages over APCI appear of interest:

- the higher ability of APPI to promote the formation of radical cations further extends the range of molecules amenable to LC-MS in the low-polarity region (*see* Figure 1.2)
- the energy of photons is usually lower than the ionisation potential of the typical mobile phase components (methanol, 10.8 eV; ACN, 12.2 eV; water, 12.6 eV), resulting in a considerable reduction of the chemical noise
- it is possible to apply normal-phase chromatography using solvents with dopant properties.

Recently, an improvement in the detection sensitivity of apolar analytes (e.g. polycyclic aromatic hydrocarbons) has been achieved by promoting ionisation using a fixed-frequency excimer laser operating at 248 nm within an API source (so-called APLI), thus obtaining a significantly increased photon flux during the ionisation process [33]. Detection limits in the femtomole range have been achieved. Moreover, mechanisms and processes described for dopant-assisted APPI also apply to APLI. Although still at an experimental level, this new entry in the family of API sources will probably further extend the ionisation potential of API sources.

Ion separation

The mass analyser represents the heart of a mass spectrometer, i.e. the device able to measure the m/z ratios of gas-phase ions. In order to allow a free path of the ions through the analyser towards the detector, the analyser must be operated under vacuum. The lower the pressure (typically in the range 10^{-4} to 10^{-7} torr), the longer the mean free path of the gas-phase ions. This has an obvious positive effect on sensitivity and mass resolution, although these parameters ultimately also depend on the type of mass analyser and on the instrumental design adopted by the manufacturer.

Together with sensitivity, other parameters are important in order to decide whether a particular mass analyser fulfils the requirements of a laboratory. These are mass range, scan speed, mass resolution and mass accuracy.

Mass range is defined as the interval of m/z ratios delimited by the lower and upper m/z ratios that the analyser is able to transmit to the detector. As stated earlier, compounds with a molecular weight over the upper limit of the mass range can still be handled if they are multiply charged in the ion source so that the m/z ratio of the resulting gas-phase ions falls within the mass range.

A mass analyser is also able to scan ions within a defined mass range, so that a 'full-scan' spectrum of the compound can be obtained. The higher the *scan speed*, the higher the number of data points per unit time in the chromatogram, meaning that the actual LC signal will be more accurately monitored, thus taking full advantage of chromatographic separation.

Similar to chromatographic resolution, *mass resolution* defines the ability of the analyser to separate adjacent ions. In other words, mass resolution reflects the sharpness of mass peaks, which can be expressed as the full width of the mass peak at half of its maximum height (FWHM or $\Delta m_{50\%}$) or, sometimes, as the width at 5% of the maximum height. Alternatively, mass resolution may be expressed as the distance between the apices of two adjacent mass peaks separated so that the valley between them at its highest point is 10% of the maximum height of either peak. The resolving power of a mass analyser is the value of an observed mass divided by mass resolution ($m/\Delta m$). Therefore, a resolving power of 500 means that m/z 50.0 can be separated from m/z 50.1 (50/0.1 = 500) or that m/z 500 can be separated by mass 501 (500/1 = 500). As the actual shape of a mass peak is Gaussian (due to the small energy differences of isobaric ions), if mass resolution is increased (i.e. by narrowing the mass window around the mass peak apex) an improvement in selectivity is obtained (interfering ions will have less probability to be detected), but sensitivity is reduced.

Mass accuracy, which is strictly linked to mass resolution, expresses the accuracy of a mass measurement. This is best defined as the difference between measured mass and theoretical mass calculated from the elemental composition of the ion. It is expressed in atomic mass units (amu) or, more often, in parts per million (ppm). For example, if the measured mass is 500.0 and the theoretical mass is 500.1, the achieved mass accuracy is (500.1 − 500.0) = 0.1 amu or $(500.1 − 500.0)/500.0 \times 10^6 = 200$ ppm.

As illustrated in Chapter 6, a mass analyser providing high mass accuracy (better than 2–5 ppm) enables the identification of the elemental composition of small ions (m/z 200–500), or at least allows the reduction of the number of matching molecular formulae to a few, by simply measuring the mass of the molecular or pseudomolecular ion [34, 35]. For example, if an unknown compound appears at mass 180.094 amu in the positive-ion mode as the [M + H]$^+$ ion and all possible formulae (e.g. containing C, H, O, N, P, S, F, Cl, Br atoms) matching a molecular weight of 179.094 (neutral species) with a 5 ppm tolerance are searched by means of formula-finding software, only the following formulae will be selected:

(1) $C_4H_{14}N_5OP$ $M_r = 179.0935924$
     $\Delta m = -2.3$ ppm
(2) $C_5H_{12}FN_4O_2$ $M_r = 179.0944244$
     $\Delta m = 2.4$ ppm
(3) $C_5H_{14}ClN_5$ $M_r = 179.0937674$
     $\Delta m = -1.3$ ppm
(4) $C_8H_{11}N_4O$ $M_r = 179.0932816$
     $\Delta m = -4.0$ ppm
(5) $C_{10}H_{13}NO_2$ $M_r = 179.0946238$
     $\Delta m = 3.5$ ppm

In order to reduce the list of possible candidate formulae further, a search in a formula database of drugs/toxic organic compounds (e.g. the Merck Index [36]) can be performed. It is also possible to extend the formula database by adding formulae of possible metabolites, using a strategy similar to that used for metabolite identification (described later in this chapter). In the example above, only formula (5) matches possible candidates in the Merck Index. These are dihydrotuberin, homarylamine, 3,4-methylenedioxyamphetamine (MDA), phenacetin, phenprobamate, propham, risocaine, *N*-isopropyl salicilamide and iproniazid. Therefore, with such an automated system, it is possible to reduce the list of candidates to a few just by measuring the accurate mass of the unknown analyte. It will not be difficult, at this point, to discriminate between the different candidate structures by looking at the in-source CID or product-ion spectrum of the analyte in question. In addition, chromatographic information (retention behaviour), too often neglected when MS information is available, can be effectively used for this purpose. Clinical data/post mortem findings are also paramount in this process as they often provide the analytical toxicologist with helpful information in order to reduce the range of possible candidates. A similar result can be also achieved in terms of identification power by applying a medium/high mass accuracy, i.e. 50–100 ppm, together with the measure of the [M + 1] and [M + 2] isotope ratios (within a 10–15% tolerance). The combination of these two pieces of information (measured mass and isotope ratios) enables the identification of a number of candidate molecular formulae of the same order as that obtained by high mass accuracy measurement.

Quadrupole

At the time of writing, the quadrupole is the most common mass analyser for bench-top MS instruments. The most likely reason for this resides in the fact that quadrupoles offer a good compromise of mass range covered, reproducibility of mass spectra, mass resolution and precision for quantification purposes at reasonable prices.

A quadrupole consists of four parallel rods or poles equally spaced around a central axis. An electrical potential is applied to each rod so that each two adjacent poles have opposite polarities. By applying a precisely controlled combination of two electrostatic fields – one direct current (DC) and one at varying radiofrequency (RF) – a resonance frequency for a specific m/z ratio is obtained so that ions at that m/z ratio can reach the detector, whereas ions with lower and higher m/z ratios are discarded.

A quadrupole acts as a continuous mass filter, which means that most of the ions that are continuously transferred to the quadrupole from the ion source are lost on their way to the detector and only a few ions (the ions at resonant frequency) are translated into measurable electric signal. By varying the resonant frequency of the quadrupole, a complete set of masses can be scanned at speeds of up to 4000–5000 amu/s, thus obtaining a 'full-scan' mass spectrum.

In addition to the above scan mode (typically adopted for the detection and identification of unknown compounds, see Chapter 6), a quadrupole can also be operated in selected-ion monitoring (SIM) mode. In this mode, the quadrupole is set at one or a few resonant frequencies (depending on the number of ions to be monitored). As a result, both sensitivity and precision in quantification are improved. In fact, since in SIM mode the analyser spends more time on a specific ion, the amount of signal belonging to this ion reaching the detector is larger than in scan mode. The increase in the S/N ratio is therefore due to a true increase in signal and should not be confused with the increase in the S/N ratio due to the elimination of the huge amount of chemical noise resulting from not monitoring the full mass range. This second component of the improvement in the S/N ratio is actually the same as what can be obtained by so-called selected-ion retrieval (SIR), i.e. by extracting *a posteriori* (after the run) a mass chromatogram from the TIC. In addition, the overall scan cycle is much faster in SIM mode than in scan mode and, as illustrated earlier, this provides a more precise measurement of the chromatographic peak area/height. Together with better sensitivity, this is the reason why quantification with quadrupoles is typically carried out in SIM mode.

A typical quadrupole covers a mass range of up to 1000–4000 amu at low resolution (0.7 amu FWHM) and with mass accuracy of 0.1 amu.

Ion trap

Designed as a variation of the quadrupole, the ion trap (IT) is also sometimes referred to as 'quadrupole IT'. An IT consists of (i) a 'doughnut' electrode to which a RF voltage is applied, and (ii) two hemispherical electrodes (end caps) placed above and below the ring electrode and held near ground potential. Ions are trapped in the centre of the ring electrode by the applied RF potentials and by means of a 'bath gas' (typically helium at about 1 mmHg) that dumps down and stabilises the oscillations of ions by colliding with them. The more tightly focused the ions are, the higher the mass resolution and sensitivity.

An IT holds ions in a static position (hence the name trap), in contrast to quadrupoles where ions follow a trajectory towards the detector. In order to scan the mass range, trapped ions are ejected towards the detector selectively by mass by the ramping of the amplitude of the RF voltage on the ring electrode. After the scan cycle is completed other ions are admitted to the trap and the process is repeated. Due to this operating mode, IT is defined as a pulsed mass analyser (in contrast to continuous ones, as quadrupoles) and the amount of signal lost during a scan cycle is much less than in a quadrupole. If a specific ion is trapped by ejecting all the other ions during the admission period in the trap, SIM mode can be implemented in an IT. However, as the trapping mechanism is basically the same, there is not much improvement in sensitivity between scan and SIM mode.

Space-charge effects (i.e. ion–ion repulsions) may severely limit the dynamic range of an IT. In order to overcome this problem an auto-ranging procedure is adopted consisting of:

- performing a pre-scan (to determine the admission time into the trap, which is inversely proportional to the amount of signal detected by the pre-scan)
- allowing ions to enter the IT for the defined admission time
- performing the scan.

The obvious consequence of this procedure is that the duration of the scan cycle is not constant, but depends on the amount of signal measured during the pre-scan. When a highly abundant signal is detected (i.e. around a peak apex), the scan cycle is shorter and the number of data points per unit time in the chromatogram is higher, and vice versa, when a low abundance signal is measured (i.e. before and after the elution of a chromatographic peak), the number of data points per unit time is lower. Thus, IT may provide worse performance in quantification than quadrupoles.

The 'self-CI' effect that may be observed due to concentration-dependent reactions of ions with neutral analyte molecules within the IT and leading to concentration-dependent changes in the mass spectrum is much less of a problem with LC-MS as in this case ions are formed in an external source (the API source) and then transferred to the trap.

ITs are compact mass analysers, usually cheaper than quadrupoles, providing similar scan speeds, mass ranges and mass resolution (resolving power = 1000–2000). In addition, as discussed later in this chapter, IT mass analysers allow MS-MS analysis simply by holding specific ions in the trap and fragmenting them by collisional dissociation.

Linear ion trap

Many of the limitations of the IT mass analyser have been ironed out with the introduction of the linear IT (LIT). It consists of a linear quadrupole mass filter operated in RF-only mode [37, 38]. Ions are trapped in the radial (x, y) direction by the RF quadrupole field, whereas the trapping in the axial direction (z) is accomplished by applying blocking DC potentials at the exit and entrance of the quadrupole. Since there is no quadrupolar electric field in the z direction, the efficiency of the injection and extraction of ions is much higher than in an IT, accounting for better absolute sensitivity. Furthermore, because of the greatly enhanced trapping volume, a LIT provides greater ion capacity, accounting for a more extended linear dynamic range. A LIT can be operated in all of the MS and MS-MS modes of an IT.

LIT mass analysers have been adopted so far in LIT-only instruments (with radial ion ejection) or in hybrid configurations (e.g. LIT-Fourier-transform ion cyclotron resonance [FTICR]). A particularly interesting configuration is a quadrupole MS-MS where the second mass analyser can be operated either as a traditional quadrupole or as a LIT with axial ion ejection (QqQ_{LIT}), thus making available in one instrument the features of the two types of mass analysers, in addition to those resulting from their combination (see the MS-MS section of this chapter) [37].

Ion cyclotron resonance

This is another trap-type mass analyser consisting of a cell kept within a very high magnetic field, giving rise to its nickname of a 'magnetic IT'.

The magnetic field constrains ions to move in a circular path in a plane that is perpendicular to the magnetic field (cyclotron motion). During their motion, ions pass by two opposite electrode plates (receiver plates), thus inducing a small alternating electric current (called 'image' current) in a circuit connecting the two plates. The amplitude of this current is related to the number of ions within the cell, whereas its frequency is the same as the frequency of rotation of ions around the axis of the magnetic field (cyclotron frequency). Therefore, detection does not require collision of ions with a detector as per previously illustrated mass analysers. This non-destructive detection allows a single ion to be repeatedly measured with a consequent significant sensitivity enhancement.

By applying RF electric potentials, trapped ions are excited into larger circular orbits and their cyclotron frequency is inversely related to their m/z ratio. As different ion masses are simultaneously present in the cell, the recorded frequency will be the sum of all cyclotron frequencies. Fourier transformation, analogous to Fourier-transform nuclear magnetic resonance, allows one to mathematically extract frequencies and relative abundances of the different masses, and to obtain a mass spectrum. Once detection is accomplished, ions are ejected from the cell and another bundle of ions is admitted (another pulsed mass analyser).

FTICR mass analysers have very high vacuum requirements (10^{-8} to 10^{-9} torr) in order to avoid collisions between ions and the background gas which causes the scatter of frequencies of isobaric ions (poor resolution). In addition, higher resolving power can be obtained by increasing the strength of the magnetic field (resulting in an increase both in cyclotron frequencies and in the frequency difference between ions). FTICR instruments using superconducting magnets provide the highest mass resolution ever recorded (resolving power of up to 1 000 000 at mass 1000) and accurate mass measurement with internal mass calibration. Similar to ITs, FTICR can perform MS-MS analysis using a single mass analyser.

Time of flight

This is in theory a very simple mass analyser, as the principle on which different ions are separated is that, the higher the m/z ratio, the longer the ion will take to cover a defined distance (i.e. the path from the ion source to the detector, or the flight tube). Typical flight times are 5–100 µs. In order to obtain an accurate measure of the flight time, the starting time of ions (i.e. the time at which ions enter the flight tube) needs to be precisely defined. Therefore, coupling with LC requires rapid electric field switching ('gating') so that ions produced in the ion source enter the flight tube in a very short and defined time. Due to this, time-of-flight (TOF) analysers are, like ITs and FTICR, pulsed mass analysers with a very high efficiency of ion transmission to the detector.

A simple linear TOF, with a 1-m long flight path, may provide a resolving power of around 1000. Better resolutions require rather long flight tubes, which are not practical in a routine laboratory. In fact, isobaric ions may differ not only in the time of formation, but also in location in space and in initial kinetic energy, and these differences in initial conditions will cause these ions to reach the detector at different times (i.e. poor mass resolution). Uncertainties in temporal, spatial and kinetic energy distribution can be reduced by applying a so-called delayed extraction (i.e. ions are held for a certain time before entering the analyser). Delaying may allow ions to undergo collisions before extraction, which may be a disadvantage for some applications and an advantage for others as it allows fragmentation and structural information to be obtained.

Another very effective way to enhance mass resolution of a TOF analyser is to use one or more ion mirrors (reflectrons). A reflectron is an ion optical device able to reflect (reverse) the flight of an ion. Two important advantages are obtained:

- a long flight path can be attained within a small bench-top instrument

- the difference in kinetic energies of the ions can be compensated.

In fact, an ion with a higher kinetic energy will penetrate into the reflectron deeper that an ion having the same m/z ratio, but lower kinetic energy. As a consequence, the first ion will cover a longer distance and will reach the detector at the same time as the second ion.

The reflectron-type TOF, currently the most common TOF analyser in analytical toxicology applications, can reach very high mass resolutions (resolving power up to 10 000) and mass accuracies better than 2–5 ppm (with reference mass) [34, 35].

Another very interesting feature of TOF analysers is that, due to the simple operating mode, there is theoretically no limitation in the mass of ions reaching the detector. However, as the m/z ratio increases, it becomes increasingly difficult to discriminate between different flight times – a practical limit of 500 000 amu can therefore be defined. The reason for using an ESI source in combination with TOF for the LC-MS analysis of large molecules is therefore not due to a limitation in mass range of the analyser (as stated earlier, molecules with multiple charges appear at a lower molecular weight as their mass is divided by the number of charges), but to the difficulty in ionising these non-volatile compounds with other ionisation mechanisms.

Another important difference with other mass analysers is that TOF are scan-only devices and SIM would not bring any advantage in terms of sensitivity. The upside is that TOF mass analysers can perform extremely high-speed scans (higher than 10^6 amu/s).

Ion detection

The detector is the device where ions separated by the mass analyser are converted into a measurable electric signal (current). The analogue current is then further converted to a digital signal (counts) stored by the data system. Apart from FTICR, which is by definition a mass analyser and a detector at the same time, all the other mass analysers require an ion-detection device. The most important characteristics for an ion-detection device are speed, dynamic range and sensitivity (gain).

The most common detection device is the electron multiplier. In an electron multiplier, ions are converted into electrons by means of a dynode electrode – when an ion strikes the dynode surface (made of secondary emitting materials, e.g. CsSb, GaP or BeO), electrons are emitted and the induced current is recorded. Signal amplification can be obtained either by means of a series of dynodes maintained at increasing potentials, where the secondary electrons produced by each dynode are multiplied in the following one, or by means of a horn-shaped continuous dynode (a so-called channel electron multiplier [CEM]) where amplification occurs by repeated collisions of electrons with the internal surface of the detector. In both cases a cascade of electrons is produced with a typical signal gain of 10^6–10^7.

A less common ion-detection system is the photo-multiplier. In a photo-multiplier, ions are initially converted by a dynode into electrons. These electrons are then converted into photons by means of a phosphorous screen. A photo-multiplier, operating in a cascading mode similar to that of the CEM, provides signal amplification. The photo-multiplier detector is sealed and kept under vacuum (photons pass through glass), and therefore may have a longer lifespan; this is different to the electron multiplier, which is exposed to the internal environment of the mass spectrometer.

TOF mass analysers have a very high scan speed, and require fast detectors with a large and plane detection area so that many different ions can be detected at the same time. Multi-channel plate detectors providing time responses lower than 1 ns and high gain (more than 50 mV per single ion) are typically used for this purpose. Each channel works similarly to a small CEM; because of the easier saturation and also the rapidity of the time-to-digital conversion required, the dynamic range of this type of detector is typically lower than that of the electron and photo-multiplier. Nevertheless, saturation can be observed with any type of ion detector and should be taken into account when a lower than linear response is observed. Apart from affecting quantification, detector saturation may also influence the appearance of the full-scan mass spectrum, thus impairing the performance of library search-based identification.

Tandem mass spectrometry

The ability to perform tandem MS analysis (MS-MS) is of crucial importance when making the decision to purchase a LC-MS apparatus. In fact, given the lower chromatographic resolution of LC compared with capillary GC, on the one hand, and the poor structural information provided by the 'soft' API techniques, on the other hand, the overall selectivity of LC-MS is lower compared with GC-MS. As illustrated earlier, in-source CID can be used to obtain structural information through fragmentation. However, an in-source CID mass spectrum is typically affected by high chemical noise and by the possible presence of interfering mass fragments (particularly when biological extracts are analysed) that may complicate its interpretation/identification. On the other hand, the selectivity enhancement offered by MS-MS allows full exploitation of the potential of LC-MS.

Traditionally, MS-MS has been implemented in quadrupole instruments where two different quadrupoles, one performing mass selection of precursor (parent) ion(s) and the other mass selection of product (daughter) ion(s), are separated by a collision cell. Here, the precursor ion is accelerated and fragmented into product ions by collision with an inert gas. As the collision cell is also a quadrupole, although simply acting as an ion guide, the two quadrupole mass analysers are usually referred to as Q1 and Q3, Q2 being the collision cell, and the instrument is referred to as a 'triple quadrupole' (QqQ). Fragmentation can be favoured by increasing either the kinetic energy (collision energy) of the precursor or the gas pressure in the collision cell.

This MS-MS configuration, where precursor and product ions are produced and analysed at different places (ion source/Q1 and collision cell/Q3, respectively) takes the name of 'in-space' MS-MS.

A triple-quadrupole instrument can be operated in a number of different ways, the simpler one being to resemble a single-quadrupole instrument by using either Q1 or Q3 as a pass-all filter (Figure 1.3). One of the most typical MS-MS scanning modes is the so-called product-ion scan, where one precursor ion is selected by Q1 and the

Figure 1.3 Different operating modes of a tandem mass spectrometer. Injection of a mixture of amphetamine (AMP) and MDA, 5 ng each. Mobile phase: 0.1% v/v formic acid:ACN 90:10, 0.2 mL/min; column: C_{18} (100 × 3 mm i.d., 3 μm particle size); ESI-MS-MS positive-ion mode. Chromatographic peaks: 1, AMP; 2, MDA. (a) *Full-scan mode*: Q1 (or Q3) is scanning and the other quadrupole is operated as a pass-all filter (declustering potential, 30 V). In this operating mode a tandem mass spectrometer works just like a single-stage MS. Note the noisy chromatogram (left) and the interferent mass peaks (one is indicated by the arrow) in the full-scan spectrum of AMP (right). More fragmentation could be obtained by increasing the declustering potential. (b) *Product-ion scan*: Q1 is selecting one precursor ion (m/z 136, $[M + H]^+$ of AMP) and Q3 is scanning (declustering potential, 30 V; collision energy, 15 eV). Note the absence of the MDA peak in the chromatogram (left), and the absence of interferent mass peaks and of the isotopic peaks of $[M + H]^+$ in the product-ion spectrum of AMP. (c) *Precursor-ion scan*: Q1 is scanning and Q3 is selecting one product ion (m/z 77) (declustering potential, 30 V; collision energy, 50 eV). Note that both AMP and MDA are present in the chromatogram (left) as both fragment to m/z 77 in the collision cell. The precursor-ion spectrum of AMP (right) contains only the precursor ions fragmenting to m/z 77 in the collision cell (i.e. m/z 136 and 119). (d) *Neutral-loss scan*: both Q1 and Q3 are scanning with a fixed mass difference of 45 (declustering potential, 30 V; collision energy, 25 eV). Note the presence of both AMP and MDA in the chromatogram (left) as both lose a neutral fragment of 45 amu in the collision cell. The neutral-loss scan spectrum of AMP contains only the precursor ions losing a neutral fragment of 45 amu (i.e. m/z 136 and 102). (e) *SRM*: Q1 is selecting one precursor ion and Q3 is selecting one product ion. Other transitions of the same and of other compounds can be monitored simultaneously (transitions monitored in the example: m/z 136 → 119 and 136 → 91 of AMP; declustering potential, 30 V; collision energy, 15 eV). The chromatogram of the transition m/z 136 → 91 of AMP (left) and the corresponding MS 'spectrum' with mass peaks at the two product ions monitored only (right) are shown. This is the most selective and sensitive operating mode of a MS-MS instrument (compare the S/N ratio with the chromatograms above).

Tandem mass spectrometry

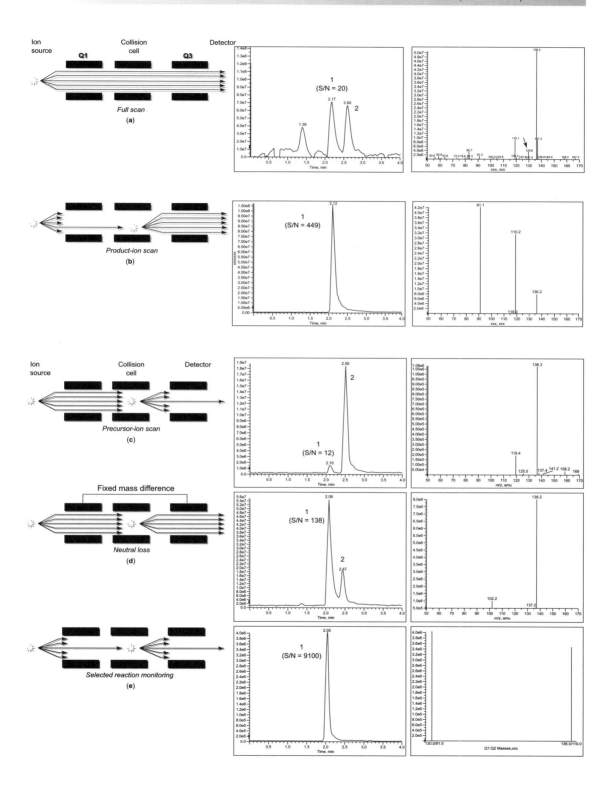

fragment (product) ions obtained after collision are scanned by Q3, thus obtaining a product-ion spectrum. This differs from an in-source CID spectrum in that all fragments detected necessarily originate from the precursor ion.

Other scanning modes, which are very helpful in metabolite profiling studies, are precursor-ion scan and neutral-loss scan [39–41]. Both these scan modes are unique to mass spectrometers that provide MS-MS in space. Precursor-ion scan allows one to screen for homologous or similar compounds having a common fragment ion (e.g. different metabolites of a drug): Q1 scans the mass range, whereas Q3 filters a unique product ion and associates it back to the precursor ion that it originated from (Figure 1.3). The latter is useful for screening compounds with a common neutral fragment loss (e.g. loss of glucuronic acid). In this operating mode, both Q1 and Q3 are synchronously scanned with a fixed mass difference (e.g. 176), and a mass is assigned only to those precursor ions that have lost the predetermined neutral mass (Figure 1.3). Figure 1.4 shows an example of the identification of metabolites of a compound by neutral-loss scan.

MS-MS quantification is typically carried out in non-scanning mode by selecting one precursor ion in Q1 and one specific product ion in Q3. This operating mode resembles the SIM mode of single-quadrupole instruments, although, in this case, a reaction (precursor ion → product ion) instead of an ion is monitored (so-called selected-reaction monitoring [SRM] or, if different reactions are monitored, multiple-reaction monitoring [MRM]). In this configuration, an MS-MS instrument associates extreme selectivity of detection (the probability that two compounds will share the same precursor AND the same product ion being quite low) with the highest possible sensitivity (Figure 1.3). The increase in sensitivity with respect to the SIM mode is actually a relative increase due to the dramatic reduction of the chemical noise. In fact, the addition of a further MS stage implies that a lower number of analyte ions will reach the ion detector. However, as the noise is largely reduced, the resulting S/N ratio will be enhanced.

When working in SRM mode, the best results in terms of sensitivity are typically obtained by reducing in-source fragmentation as much as possible, so that most of the analyte's signal before CID is kept by one ion (usually the quasi-molecular ion) or a few ions. This is usually not difficult to achieve with API sources, where the amount of energy transferred to the analyte during the ionisation process is relatively low. For the same reason, CID conditions should be optimised, whenever possible, in order to obtain a few abundant product ions. This may be more difficult to achieve, depending on the chemical stability of the analyte. Some analytes (polycyclic structures are a good example) are rather difficult to fragment efficiently – no fragmentation is observed at low collision energy until a threshold collision energy value is reached, where the precursor ion 'explodes' into many low-abundance fragments. As described in Chapter 8, this effect may be more or less pronounced depending on the instrumentation used. However, in such situations, a possible alternative strategy is so-called 'surviving-ion' detection. A full explanation of that approach as well as examples of applications are given in Chapters 3 and 10. Briefly, it consists of optimising the collision energy at the highest value at which no significant fragmentation of the precursor ion is observed in the collision cell. The 'surviving' ion is then selected by Q3 and reaches the detector, whereas possible isobaric, less-stable interferences are fragmented in the collision cell. By doing so, an improvement in sensitivity can be obtained, but obviously at the expense of selectivity [42].

One important practical advantage of LC-MS over GC-MS is that the optimisation of ionisation and MS-MS parameters (e.g. orifice/fragmentor voltage, nebuliser gas flow, source temperature, collision energy, collision gas pressure, etc.) can be carried out without chromatography, i.e. by continuous infusion of a diluted solution of the analyte into the ion source or by flow injection analysis (FIA) where the LC injector is connected directly to the ion source. This enables the optimisation of every single MS parameter within a few minutes and, most importantly, allows the design of a set of experiments in order to evaluate the combined effect of the different parameters. However, infusion and FIA experiments should be carried out using a mobile-phase composition as similar as possible to that of the defined LC conditions as this parameter may affect ionisa-

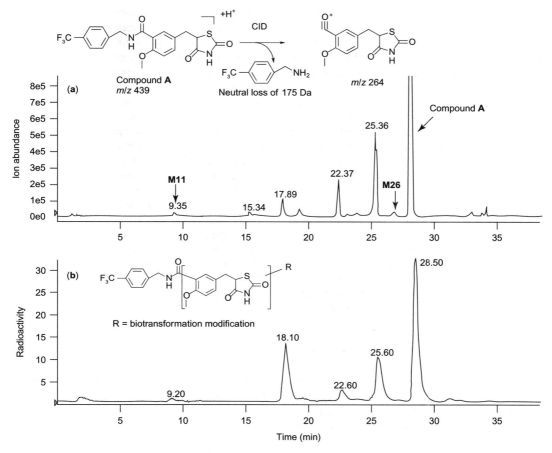

Figure 1.4 Example of metabolite identification using neutral-loss scan. (a) LC-MS TIC of neutral loss of 175 and (b) LC-radioactivity profile of a rat liver microsomal incubation of radioactive compound **A**. The metabolites M11 and M26 were detected in the TIC, but not the radioactivity trace due to the loss of the ^{14}C radiolabel. (Reprinted from [3] with permission from Elsevier.)

tion. Similarly, separation conditions can be optimised using a traditional and less-expensive detector (e.g. a fixed-wavelength spectroscope), thus saving MS operating time, and accounting for a cost-effective and fast overall method optimisation.

When different reactions are monitored consecutively, as in MRM, care should be taken in order to prevent unwanted contributions to a monitored signal (i.e. the product ion of one reaction) from other reactions monitored in the same acquisition cycle. This phenomenon, called cross-talk, occurs if a precursor ion enters the collision cell when product ions from the previous reaction monitored are still present [16]. If the two precursor ions share common fragments, the signal pertaining to one precursor may be overestimated due to the contribution from the other. Different strategies allow one to control this problem, such as:

- adding a pause time between two consecutive transitions (though this will obviously increase the acquisition cycle and will reduce the number of data points per unit time in the chromatogram);

- avoiding the consecutive monitoring of two compounds sharing a common product ion (i.e. by monitoring a third reaction in between); or, even better
- optimising LC separation in order to avoid chromatographic overlapping between the cross-talking compounds [43].

Recently designed instruments take care of this problem by speeding up the exit of fragments from the collision cell. Cross-talk should always be considered when developing an analytical method as it is obviously a source of error in quantification.

MS-MS can be also implemented in IT instruments, with the considerable advantage that in this case MS-MS analysis can be carried out in a single mass analyser (with quadrupoles, two mass analysers in tandem are required). IT mass analysers carry out MS-MS analysis sequentially in time ('MS-MS in time'):

(i) the precursor ion is kept in the trap while all other ions are ejected;
(ii) a resonating frequency is applied to the end caps of the IT so that isolated ions start oscillating and collide with helium bath gas molecules;
(iii) the internal energy acquired by the precursor ion by repeated collisions is dissipated by fragmentation;
(iv) since fragment-resonating frequencies are different from those of the precursor, fragment ions are kept within the trap during collisional dissociation;
(v) after a given time, a scan is performed by ramping the amplitude of the RF voltage on the ring electrode so that a product ion spectrum of the precursor is obtained.

Theoretically, the described process can be repeated indefinitely by trapping a different product ion each time, so that an MS^n analysis can be carried out. However, the loss of signal occurring at each step and practical reasons limit the number of steps to $n = 3$ and applications of MS^4 are very rare in the literature [44–46].

FTICR mass analysers can also perform MS^n analysis using a sequential trap-and-scan process similar to that described for ITs. In this case, however, the collision gas needs to be introduced into the cell after the precursor has been isolated. Alternatively, the internal energy of isolated ions can be increased by means of a laser (photo-ionisation) without the need for collision gas.

MS-MS analysis can also be carried out using hybrid configurations, such as LIT-FTICR, QLIT and QTOF. Among these, the QqQ_{LIT} hybrid configuration is very interesting as it has been implemented so far in triple-quadrupole instruments where Q3 can work either as a traditional quadrupole or as an LIT, thus expanding the number and type of MS-MS experiments that can be performed. An interesting feature of this type of instrument is the possibility of performing a product-ion scan experiment (or other MS-MS scanning modes) using the LIT as the second mass analyser, thus achieving higher sensitivity compared with a QqQ instrument (due to the different operating modes of quadrupoles and ITs, and in particular to the higher efficiency of ion transmission to the detector of the latter) and also with a greater ion capacity when compared with a traditional IT.

QTOF, although rather expensive, is also an interesting configuration for MS-MS analysis as it combines the high efficiency of selected ion monitoring in the first mass analyser (Q) with the high scan rate and high resolving power of the second (TOF).

Acquiring and processing LC-MS data

Software application is among the most active areas of LC-MS technical development. Together with traditional software tools, such as background subtraction, peak/spectral deconvolution and library searches, new tools have been recently developed that have greatly expanded the analytical potential of LC-MS. Among these, one of the most interesting is the so-called information (or data)-dependent acquisition software (IDA or DDA), allowing one to automatically switch between different MS and MS^n modes within a single chromatographic run based on data acquired.

LC-MS is probably the most powerful technique for metabolite identification and profiling, and most of the manufacturers offer additional software tools for this purpose. Finally, protein sequencing and identification require specific software applications able to handle the large amount of complex data obtained in LC-MS protein analysis.

Library search

Library search software is required to automatically identify compounds by comparing the unknown spectrum with spectra contained in a database, generally going through the following steps:

(i) detection of the peak and generation of a 'clean' mass spectrum (by baseline noise subtraction or spectral deconvolution)
(ii) search of the 'clean' spectrum against a spectral library (or vice versa) and list of the most feasible candidates; and, if possible
(iii) comparison of non-spectral information of unknown and MS retrieved compounds, such as chromatographic or UV spectroscopy/diode array detector parameters.

A match factor is typically returned as a measure of the certainty of the library search result.

Deconvolution is the isolation of the components of a signal so that each of them can be examined individually. This is a very useful tool in LC-MS analysis of biological extracts where the huge amount of interferents present and low selectivity of LC make peak overlapping a rather frequent occurrence. Deconvolution algorithms allow the location and detection of peaks, and the extraction of clean spectra by subtraction of background noise (which is frequently high in in-source CID LC-MS spectra) and of the signal of co-eluting compounds. The software is expected efficiently to resolve compounds with close retention times, although, when the overlapping compounds share common ions and spectra are acquired in low-resolution mode, the deconvoluted spectrum may appear distorted. Deconvoluted or 'clean' spectra are automatically compared with commercial (at the time of writing, not available for LC-MS, at least in the field of toxicology) or built-in-house libraries, using either forward fit (experimental data searched against the library entries) or reverse fit (library entries searched against experimental spectra). Reverse search, ignoring all the peaks in the spectrum that are not contained in the library entry, is usually preferred when high interfering m/z signals are present in the acquired spectrum which can lead to false identification. Among the reverse search algorithms, the probability-based matching (PBM) algorithm is one of the most powerful [47]. The use of library search algorithms requires careful optimisation of the search parameters. Using library search as a black box may in fact lead to false-positive or -negative results (see Chapter 5). Among the search parameters, those permitting one to attribute a low weight to relative intensities of the mass peaks are most important as they allow the reduction of the inter-instrument and inter-laboratory variability of LC-MS spectra (see Chapter 6).

Library entries are usually stored in the software supplied with the instrument, using the same type of interface and analyser, and using standardised conditions (determined on the basis of instrument performance, e.g. comparable relative intensities of a calibrating compound, rather than instrument settings) [48–50]. Some authors have created libraries of either in-source CID or product ion spectra by summing the low- and high-fragmentation voltage (or collision energy) spectra of each compound in either positive- or negative-ionisation mode (see Chapter 6).

As illustrated earlier, accurate mass measurements may enable the identification of low-molecular-weight compounds. Gergov et al. [34] developed a post-run library screening software using Microsoft Visual Basic, with each record containing the compound name and elemental formula. Library search was performed by searching the accurate mass measured by LC-TOFMS analysis against the masses of the formula entries in the database calculated on the basis of the elemental composition. Recently, Thurman et al. [51] proposed to use a digital database containing several thousand relevant molecules to widen the possibilities of an unknown screening of environmental contaminants. In fact, this could be the way to release identification from the *a priori* selection of substances. The authors performed an LC-TOFMS screening on the basis of the accurate mass measurement and the database search of the empirical formula on the Merck Index and ChemINDEX databases [36, 52]. The authors did not specify if the search was automated or performed manually.

Information-dependent acquisition

IDA (or DDA) is one of the most promising features of LC-MS concerning the rapid identification of unknown compounds. Data-dependent experiments involve (Figure 1.5):

(i) acquisition of survey data
(ii) handling of data in order to identify the ions of interest
(iii) acquisition of one or more dependent scans on the ions of interest during the same chromatographic run.

Ions are selected on the basis of specific criteria such as intensity, charge state and redundancy of that ion throughout the chromatogram. The dependent scan(s) frequently involve the modification of the collision cell settings to enhance fragmentation and to maximise mass spectral information during a single run. Due to the flexibility of the acquisition mode, IDA experiments have been used in toxicological screening methods (*see* Chapter 6) and in metabolite identification studies, where both full-scan and neutral-loss experiments can be used as a survey to trigger MS^n-dependent experiments. Xia *et al.* [53] used constant neutral-loss scans of *m/z* 176 and 128 as survey scans for detection of phase II glucuronide conjugates and phase I oxidative metabolites, respectively. These experiments allowed the identification of five different gemfibrozil metabolites in a single run.

Metabolite identification

LC-MS-MS is probably the best-suited technique for the analysis, detection and characterisation of metabolites due to its high potential in the analysis of polar drugs. Nevertheless, the identification of unknown drug metabolites is a challenging topic due to their frequently low and varying concentrations, and to the interference from endogenous compounds in biological fluids. However, the identification of unknown metabolites can explain the mechanism of action of a drug (in the case of active metabolites), or extend the time window of detectability of a drug [19, 39, 42].

Neutral-loss and precursor-ion scans can help to tackle the interpretation of unknown peaks in metabolite studies through the identification of characteristic neutral losses or characteristic fragment ions in the MS-MS spectrum. Metabolite interpretation software tools attempt to correlate mass variations of the pseudo-molecular ion to expected biotransformations of the parent molecule (e.g. hydroxylation, oxidation, conjugation, etc.), and to confirm the identity of the candidate metabolite and the site of modification by comparison with the fragmentation pattern of the parent drug. The identification of characteristic neutral losses or fragment ions in the MS-MS spectra can trigger new product-ion scan experiments to further elucidate the structure of the candidate metabolite. Moreover, high mass accuracy instruments may help in the interpretation of the metabolite structure through the discrimination of isobaric ions and the determination of elemental formulae.

Protein identification tools

Search engines able to identify proteins from primary sequence databases have recently lightened the load of raw data generated by LC-MS in protein studies. Usually, enzyme-digested proteins are injected into the LC-MS-MS system. Peptides are separated and fragmented under CID conditions to produce peptide-dependent MS-MS information, and the MS-MS data are searched against a primary sequence database. After signal de-noising and selection of the appropriate peaks, the protein identification software is required to determine the precursor ion of peptides and then to relate peptides to known proteins or to close homologous, often equivalent proteins from related species, if the protein is not contained in the database. Searches can be performed on the basis of mass spectra of different origin:

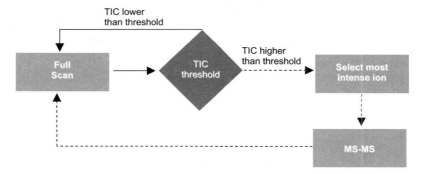

Figure 1.5 Example of IDA. The mass spectrometer is initially set to operate in full-scan mode (solid line). When the TIC reaches a predefined threshold level, the most intense ion in the full-scan spectrum is selected, and the acquisition is automatically switched to MS-MS mode in order to obtain the product-ion spectrum and then switched back to full-scan mode (dotted line).

(i) peptide mixtures resulting from the digestion of a protein (fingerprint)
(ii) peptide mass data together with amino acid sequence and composition information (generally from MS-MS spectrum interpretation)
(iii) uninterpreted MS-MS data from one or more peptides (identification from multiple matches to peptides from the same protein, without determining the amino acid sequence).

Conclusions

Many techniques are currently available in LC-MS enabling us to tackle almost any analytical problem. Small and large, thermally stable or labile, apolar or highly polar compounds can be ionised and fragmented using different types of ion sources. A range of mass analysers provides sensitive detection for target analysis and high selectivity for identification purposes (by fragmentation and/or accurate mass measurement). MS-MS greatly enhances the potential of LC-MS in terms both of sensitivity and selectivity of detection, and, for tandem in-space instruments, of metabolite profiling and identification. Data acquisition and processing (not to mention operators) take great advantage of the newest software tools available. In addition, rapid and cost-effective method development and optimisation can be achieved by carrying out separate LC and MS experiments, thus adding further versatility to the technique.

This variety of solutions offered makes LC-MS an extremely powerful analytical tool, and counterbalances the relatively expensive purchasing and maintenances cost of the equipment, which still remains higher than GC-MS despite the trend of decreasing prices.

References

1. Willoughby R, Sheehan E, Mitrovich S. *A Global View of LC/MS*, 2nd edn. Pittsburgh, PA: Global View, 2002.

2. Whitehouse CM, Dreyer RN, Yamashita M, *et al.* Electrospray interface for liquid chromatographs and mass spectrometers. *Anal Chem* 1985; 57: 675–679.

3. Yamashita M, Fenn JB. Electrospray ion source. Another variation of the free-jet theme. *Phys Chem* 1984; 88: 4451–4459.

4. Iribarne IV, Thomson BA. On the evaporation of small ions from charged droplets. *J Chem Phys* 1976; 64: 2287–2294.

5. Kebarle P. A brief overview of the present status of the mechanisms involved in electrospray mass spectrometry. *J Mass Spectrom* 2000; 35: 804–817.

6. Mortier KA, Zhang GF van Peteghem CH, et al. Adduct formation in quantitative bioanalysis: effects of ionization conditions of paclitaxel. *J Am Soc Mass Spectrom* 2004; 15: 585–592.

7. Stefansson M, Sjoberg PJR, Markides KE. Regulation of multimer formation in electrospray mass spectrometry. *Anal Chem* 1996; 68: 1792–1797.

8. Jemal M, Almond RB, Teitz DS. Quantitative bioanalysis utilizing high-performance liquid chromatography electrospray mass spectrometry via selected ion monitoring of the sodium ion adduct [M + Na]+. *Rapid Commun Mass Spectrom* 1997; 11: 1083–1088.

9. Weinmann W, Stoertzel M, Vogt S, et al. Tune compounds for electrospray ionisation/in-source collision-induced dissociation with mass spectral library searching. *J Chromatogr A* 2001; 926: 199–209.

10. Hopfgartner G, Bean K, Henoin J, et al. Ion spray mass spectrometric detection for liquid chromatography: a concentration- or a mass-flow-sensitive device? *J Chromatogr* 1993; 647: 51–61.

11. Dams R, Benijts T, Gunther W, et al. Influent of the eluent composition no the ionization efficiency for morphine of pneumatically assisted electrospray, atmospheric-pressure chemical ionization and sonic spray. *Rapid Commun Mass Spectrom* 2002; 16: 1072–1077.

12. Bruins AP. Liquid chromatography-mass spectrometry with ionspray and electrospray interfaces in pharmaceutical and biomedical research. *J Chromatogr* 1991; 554: 39–46.

13. Zhou S, Cook KD. Probing solvent fractionation in electrospray droplets with laser-induced fluorescence of a solvatochromic dye. *Anal Chem* 2000; 72: 963–969.

14. Zhou S, Prebyl BS, Cook KD. Profiilng pH changes in the electrospray plume. *Anal Chem* 2002; 74: 4885–4888.

15. Mei H, Hsieh YS, Nardo C, et al. Investigation of matrix effects in bioanalytical high-performance liquid chromatography-tandem mass spectrometric assays. Application to drug discovery. *Rapid Commun Mass Spectrom* 2003; 17: 97–103.

16. Matuszewski BK, Constanzer ML, Chavez-Eng CM. Matrix effect in quantitative LC/MS/MS analyses of biological fluids: a method for determination of finasteride in human plasma at picogram per milliliter concentrations. *Anal Chem* 1998; 70: 882–889.

17. Bonfiglio R, King RC, Olah TV, et al. The effects of sample preparation methods on the variability of the electrospray ionization response for model drug compounds. *Rapid Commun Mass Spectrom* 1999; 13: 1175–1185.

18. Liang HR, Foltz RL, Meng M, et al. Ionization enhancement in atmospheric pressure chemical ionization and suppression in electrospray ionization between target drugs and stable-isotope-labeled internal standards in quantitative liquid chromatography/tandem mass spectrometry. *Rapid Commun Mass Spectrom* 2003; 17: 2815–2821.

19. Politi L, Morini L, Groppi A, et al. Direct determination of the ethanol metabolites ethyl glucuronide and ethyl sulfate in urine by liquid chromatography/electrospray tandem mass spectrometry. *Rapid Commun Mass Spectrom* 2005; 19: 1321–1331.

20. Naidong N. Bioanalytical liquid chromatography tandem mass spectrometry methods on underivatized silica columns with aqueous/organic mobile phases. *J Chromatogr B* 2003; 796: 209–224.

21. Hirabayashi A, Sakairi M, Koizumi H. Sonic spray ionization method for atmospheric pressure ionization mass spectrometry. *Anal Chem* 1994; 66: 4557–4559.

22. Hirabayashi A, Sakairi M, Koizumi H. Sonic spray mass spectrometry. *Anal Chem* 1995; 67: 2878–2882.

23. Van Berkel GJ. An overview of some recent development in ionization methods for mass spectrometry. *Eur J Mass Spectrom* 2003; 9: 539–562.

24. Benijts T, Gunther W, Lambert W, et al. Sonic spray ionization applied to liquid chromatography/mass spectrometry analysis of endocrine-disrupting chemicals in environmental water samples. *Rapid Commun Mass Spectrom* 2003; 17: 1866–1872.

25. Dams R, Benijts T, Gunther W, et al. Sonic spray ionization technology: performance study and application to a LC/MS analysis on a monolithic silica column for heroin impurity profiling. *Anal Chem* 2002; 74: 3206–3212.

26. Gallagher RT, Balough MP, Davey P, et al. Combined electrospray ionization-atmospheric pressure chemical ionization for use in high throughput LC-MS applications. *Anal Chem* 2003; 75: 973–977.

27. Keski-Hynnila H, Kurkela M, Elovaara E, et al. Comparison of electrospray, atmospheric pressure chemical ionization, and atmospheric pressure photoionization in the identification of

apomorphine, dobutamine and entacapone phase II metabolites in biological samples. *Anal Chem* 2002; 74: 3449–3457.

28. Hsieh Y, Chintala M, Mei H, *et al*. Quantitative screening and matrix effect studies of drug discovery compounds in monkey plasma using fast-gradient liquid chromatography/tandem mass spectrometry. *Rapid Commun Mass Spectrom* 2001; 15: 2481–2487.

29. Tiller PR, Romanyshyn LA. Implications of matrix effects in ultra-fast gradient or fast isocratic liquid chromatography with mass spectrometry in drug discovery. *Rapid Commun Mass Spectrom* 2002; 16: 92–98.

30. Henion J, Brewer E, Rule G. Sample preparation for LC/MS/MS: analyzing biological and environmental samples. *Anal Chem* 1998; 70: 650A–656A.

31. Robb DB, Covey TR, Bruins AP. Atmospheric pressure photoionization: an ionization method for liquid chromatography mass spectrometry. *Anal Chem* 2000; 72: 3653–3659.

32. Raffaelli A, Saba A. Atmospheric pressure photoionization mass spectrometry. *Mass Spectrom Rev* 2003; 22: 318–331.

33. Constapel M, Schellentrager M, Schmitz OJ, *et al*. Atmospheric-pressure laser ionization: a novel ionization method for liquid chromatography/mass spectrometry. *Rapid Commun Mass Spectrom* 2005; 19: 326–336.

34. Gergov M, Boucher B, Ojanpera I, *et al*. Toxicological screening of urine for drugs by liquid chromatography/time-of-flight mass spectrometry with automated target library search based on elemental formulas. *Rapid Commun Mass Spectrom* 2001; 15: 521–526.

35. Ojanpera L, Pelander A, Laks S, *et al*. Application of accurate mass measurement to urine drug screening. *J Anal Toxicol* 2005; 29: 34–40.

36. The Merck Index. http://chemfinder.cambridgesoft.com/reference/TheMerckIndex.asp (accessed 27 June 2005).

37. Hager JW, Le Blanc JCY. Product ions scanning using a Q-q-Q linear ion trap (Q TRAPTM) mass spectrometer. *Rapid Commun Mass Spectrom* 2003; 17: 1056–1064.

38. Douglas DJ, Frank AJ, Mao D. Linear ion traps in mass spectrometry. *Mass Spectrom Rev* 2005; 24: 1–29.

39. Canezin J, Cailleux A, Turcant A, *et al*. Determination of LSD and its metabolites in human biological fluids by high-performance liquid chromatography with electrospray tandem mass spectrometry. *J Chromatogr B* 2001; 765: 15–27.

40. Liu DQ, Hop CECA. Strategies for characterization of drug metabolites using liquid chromatography-tandem mass spectrometry in conjunction with chemical derivatization and on-line H/D exchange approaches. *J Pharm Biomed Anal* 2005; 37: 1–18.

41. Kostiainen R, Kotiaho T, Kuuranne T, *et al*. Liquid chromatography/atmospheric pressure ionization-mass spectrometry in drug metabolism studies. *J Mass Spectrom* 2003; 38: 357–372.

42. Polettini A, Huestis MA. Simultaneous determination of buprenorphine, norbuprenorphine, and buprenorphine-glucuronide in plasma by liquid chromatography-tandem mass spectrometry. *J Chromatogr B* 2001; 754: 447–459.

43. Gergov M, Ojanpera I, Vuori E. Simultaneous screening for 238 drugs in blood by liquid chromatography-ion spray tandem mass spectrometry with multiple reaction monitoring. *J Chromatogr B* 2003; 795: 41–53.

44. Baseski HM, Watson CJ, Cellar NA, *et al*. Capillary liquid chromatography with MS3 for the determination of enkephalins in microdialysis samples from the striatum of anesthetized and freely-moving rat. *J Mass Spectrom* 2005; 40: 146–153.

45. Hsu FF, Turk J, Rhoades ER, *et al*. Structural characterization of cardiolipin by tandem quadrupole and multiple-stage quadrupole ion-trap mass spectrometry with electrospray ionization. *J Am Soc Mass Spectrom* 2005; 16: 491–504.

46. Collin F, Khoury H, Bonnefont-Rousselot D, *et al*. Liquid chromatographic/electrospray ionization mass spectrometric identification of the oxidation end-products of metformin in aqueous solutions. *J Mass Spectrom* 2004; 39: 890–902.

47. Stauffer DB, McLafferty FW, Ellis ED, *et al*. Probability-based-matching algorithm with forward searching capabilities for matching unknown mass spectra of mixtures. *Anal Chem* 1985; 57: 1056–1060.

48. Hough JM, Haney CA, Voyksner RD, *et al*. Evaluation of electrospray transport CID for the generation of searchable libraries. *Anal Chem* 2000; 72: 2265–2270.

49. Marquet P, Venisse N, Lacassie E, *et al*. In-source

CID mass spectral libraries for the 'general unknown' screening of drugs and toxicants. *Analusis* 2000; 28: 41–50.

50. Weinmann W, Lehmann N, Renz M, *et al.* Screening for drugs in serum and urine by LC/ESI/CID-MS and MS/MS with library searching. *Prob Forensic Sci* 2000; 22: 202–208.

51. Thurman EM, Ferrer I, Fernandez-Alba AR. Matching unknown empirical formulas to chemical structure using LC/MS TOF accurate mass and database searching: example of unknown pesticides on tomato skins. *J Chromatogr A* 2005; 1067: 127–134.

52. ChemINDEX database. http://chemfinder.cambridgesoft.com/reference/chemindex.asp (accessed 27 June 2005).

53. Xia YQ, Miller JD, Bakhtiar R, *et al.* Use of a quadrupole linear ion trap mass spectrometer in metabolite identification and bioanalysis. *Rapid Commun Mass Spectrom* 2003; 17: 1137–1145.

2

Method development and optimisation of LC-MS

Paul J. Taylor

Introduction

The development of a quantitative liquid chromatography–mass spectrometry (LC-MS) assay, as with all other chromatographic methods, requires the optimisation of various parameters. In general there are five steps in the quantification of drugs: sampling, sample preparation, separation, detection and data analysis. The focus of this chapter will be on sample preparation, separation and detection. The number of parameters to be investigated and optimised in each of these three steps is extensive, with unique aspects that must be understood.

In quantitative LC-MS, the most commonly used ion source is atmospheric pressure ionisation (API) and the most commonly used mass analyser is the triple quadrupole instrument. This combination may be substituted by other components, such as matrix-assisted laser desorption ionisation (MALDI), or time-of-flight (TOF) or linear ion trap (LIT) mass analysers. The combination of instrumentation is potentially large and each of these components has specific needs for optimisation. As the most frequently used combination of instrumentation for quantitative bioanalysis is a LC-API-quadrupole tandem mass analyser (LC-MS-MS) with selected-reaction monitoring (SRM) for detection, the focus of this chapter will be on the development and optimisation of this equipment (Figure 2.1).

The development of API techniques (electrospray [ESI], atmospheric pressure chemical [APCI] and atmospheric pressure photo-ionisation [APPI]) has enabled the robust coupling of LC with MS. The use of hydrophobic separation combined with selective MS detection makes this a versatile analytical tool. LC-MS is now considered the benchmark for measurement of drugs and their metabolites in biological matrices [1–3]. The high selectivity of tandem mass spectrometry (MS-MS), with successive mass filtrations, leads to little or no observed interference even though there may be relatively high concentrations of co-extracted and co-eluted matrix components present. These characteristics have led to a growing trend of high-throughput analysis that incorporates little or no sample preparation and minimal chromatographic retention [4–6]. While this approach is applicable in some situations, caution is warranted as unseen co-eluting molecules can have deleterious effects on results [7, 8]. These important phenomena, known as matrix effects (MEs), will be discussed in detail later.

Method development is costly and time-consuming, with every analytical problem having a unique set of problems. Applying a systematic approach to methods development compared with an *ad hoc* approach can save time. In this chapter, a step-by-step approach to developing a successful quantitative LC-MS method will be outlined (Table 2.1). The first, 'general' step in establishing a method is all about understanding your analytical problem and gathering resources together to find a solution. Mass spectrometry is next – without detection of the analyte(s), the process of methods development cannot proceed. The establishment of chromatography will change the conditions under which the mass spectrometer was optimised initially (flow rates, mobile phase composition, etc.). Thus, the mass spectrometer may require re-optimisation. Similarly, when sample

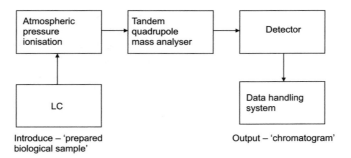

Figure 2.1 A schematic of the LC-MS system.

Table 2.1 The stepwise process for LC-MS method development

Step	Parameter	
1	General	define the problem: which analytes, analytical range, lower LOQ, matrix
		physicochemical properties of the analytes
		literature search
		IS selection
2	Mass spectrometry	select ionisation mode
		optimisation of ionisation parameters (ESI voltage, gas flows)
		pH effects on ionisation
		choose a fragment ion and optimise collision energy
3	Chromatography	select column
		optimise mobile-phase composition
		optimise flow rate
		check stability of system with multiple injections of a pure mixture of analyte(s) and IS; for a mid-range concentration, the coefficient of variation should be less than 2% under these conditions return to step 2 and re-optimise MS conditions
4	Sample preparation	select a suitable sample preparation method
		prepare and extract samples in the matrix of interest
		optimise sample preparation particularly recovery of analytes
		investigate MEs using post-column infusion technique
		if MEs are an issue, alter chromatography (return to step 3 or possibly step 2) and move analyte away from affected area; alternatively, look at a more extensive sample preparation technique
5	Preliminary validation	a standard curve is assayed in quadruplicate, with the first set of standards used to establish the calibration curve and the others being quality controls; the accuracy and precision of the method, over the analytical range, can be determined ($n = 3$)
		investigate between subject accuracy and precision
6	Full validation	see Chapter 4

preparation or preliminary validation is performed, chromatography and MS may need re-optimisation. It can be seen that the whole process is iterative – as each step is finished, the previous may necessitate re-evaluation.

It is difficult to cover all aspects of methods development in this chapter as there are currently no definitive recommendations on the development of a robust LC-MS method. However, an overview of these important steps will be discussed along with practical examples. This chapter can be considered the basis of a general standard operating procedure for methods development. It is hoped that the reader will be able to apply this approach to their particular analytical problem.

General

The core to developing a LC-MS method is centred on the three processes of sample preparation, chromatographic separation and mass spectrometric detection. Before starting on this three-step process it is necessary to perform background work. Method development starts with a clear understanding of what is required of the analysis:

- Are you required to measure a parent drug and several of the metabolites?
- Are the concentrations of interest for the drug different to those of the metabolites?
- Will more than one matrix be investigated?

The following points must be addressed to obtain a clear understanding of the problem with defined goals:

(i) what to measure (drug ± metabolites?)
(ii) select matrix (whole blood, plasma, serum, tissue)
(iii) sample limitations (volume, sample size)
(iv) linear range
(v) lower limit of quantification (LOQ).

Once the goals of the method have been established, research on the analyte(s) is required as the nature of the analyte dictates the inlet and ionisation mode used for LC-MS. Of particular importance is having an understanding of the physicochemical properties of the compounds under study. The physical and chemical properties such as molecular weight, boiling point, functional groups, stability and fragmentation characteristics are major factors in determining the best analytical option.

A literature search on the analyte and any chemically related compound should be performed. Information should be assembled on mass spectrometric methods (including potential ionisation mode), LC conditions and sample preparation. If this process is done thoroughly and effectively, a tremendous amount of resources can be saved and one can avoid the time-consuming problem of 're-inventing the wheel'. Finally, this information can also be used for the selection of a suitable internal standard (IS).

Internal standards

Selection of an appropriate IS can be difficult, but it is vital to the success of a method. ISs are used in quantitative LC-MS methods to compensate for sample-to-sample differences not only in extraction efficiency during sample preparation, as for classic LC-UV, but also for ionisation efficiency during the transfer of the analyte from the liquid phase to the gas phase. There are three types of ISs that may be used in LC-MS methods: the analyte labelled with several stable isotopes (^{18}O, ^{15}N, ^{13}C or ^2H or D), a structural analogue of the analyte or any other chemical.

Labelled ISs are the preferred IS for LC-MS quantification. They are chemically identical to the analyte, and thus have the advantages of the same extraction efficiency, ionisation efficiency and co-elution with the analyte. As the labelled IS and the compound of interest co-elute, the mass difference between these two compounds must be at least 3 amu. If the mass difference is less than 3 amu, the isotope peaks of the analyte may interfere with the IS signal. While labelled ISs are preferred, problems have been found with deuterium-labelled ISs [9].

Chavez-Eng et al. [9], in establishing a method for the cyclooxygenase-2 inhibitor, rofecoxib, reported that the isotope-labelled [^{13}C, D_3]rofecoxib was unstable in plasma and water-containing solvents (including their mobile phase). After a 6-h incubation of [^{13}C, D]rofecoxib in plasma at room temperature, the concentration of unlabelled rofecoxib had increased by 28%. The loss of deuterium from the IS was due to exchange with hydrogen atoms on water molecules. As an alternative IS, [$^{13}C_7$]rofecoxib was investigated. This IS showed no degradation into partially labelled or unlabelled drug. Thus, if using deuterated ISs it is important to establish their stability in solution and through all the processes of a method. Further, to remove potential instability problems, ^{13}C-labelled ISs are preferred over deuterated species.

Labelled isotope ISs are expensive and not always available. Thus, as an alternative, an analyst may have to utilise an analogue of the compound of interest. To be suitable, the ionisation of the analogue must be compared with the analyte. Stokvis et al. [10], in a case study of a matrix metalloprotease inhibitor, ABT-518 (Figure 2.2), and one metabolite (formed after reduction of the N-hydroxy moiety followed by hydrolysis of the amide), showed the importance of the analogue needing the same ionisation efficiency as the compounds of interest.

The analogue was used to quantify ABT-518 and its metabolite. Validation of a method to measure ABT-518 was successful, but the accuracy and precision for the metabolite was unacceptable (above 15%). Infusion of equimolar amounts of the three compounds revealed that the response for ABT-518 and the IS were similar, but the metabolite had approximately four times the response (Figure 2.3). This difference in response occurred due to the higher proton affinity of the amine group on the metabolite compared with the amide group on the parent drug.

Thus, when an analyte and analogue IS differ in functional groups, the IS is less likely to be appropriate than when the difference is in the carbon backbone of the molecule. Further, substitution or rearrangement of atoms such as oxygen, nitrogen, sulphur and halogens is more likely to result in altered charge distribution of the molecule than changes to carbon–hydrogen groups [10]. These factors must be taken into account when selecting and assessing an analogue as the IS.

An important factor that needs investigation when using an analogue is whether interference will occur from metabolites of the parent drug. Oxidative metabolism, such as demethylation or hydroxylation, may result in metabolites that are of similar molecular weight to the analogue. As an example, Streit et al. [11] has reported interferences from ciclosporin metabolites when using the analogue, ciclosporin D, as an IS to measure ciclosporin. Therefore, 'real' patient samples should be screened for potential interference to the IS during method development.

The third alternative as an IS is to select any structurally unrelated compound. Using this type of IS can be fraught with danger and is not recommended unless necessary. Volosov et al. [12] successfully used ritonavir as the IS to measure the immunosuppressant drugs ciclosporin, sirolimus and tacrolimus. Two reasons for this structurally unrelated compound being suitable as an IS may be due to the simple sample preparation (protein precipitation and injection of supernatant) and the use of APCI as the ion source. Extraction efficiency using protein precipitation may prove similar and consistent for the analytes and IS even though they have very different chemical structures. APCI is not influenced greatly by MEs (discussed later).

Further information on ISs for LC-MS analysis can be found in Chapter 3.

The background information has been gathered and a possible IS selected – it is now time to optimise the mass spectrometric conditions for the analyte.

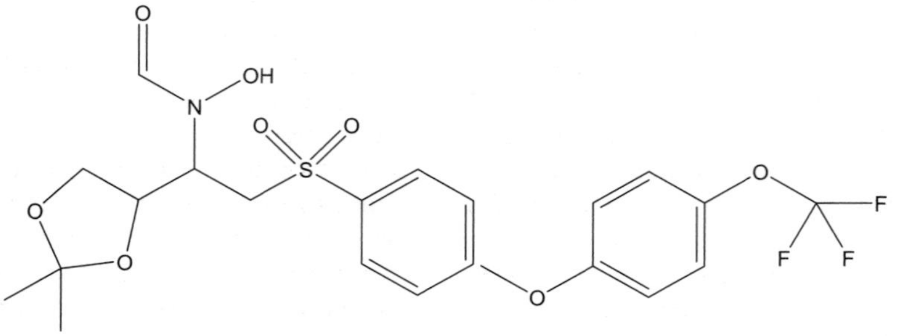

Figure 2.2 The chemical structure of ABT-518.

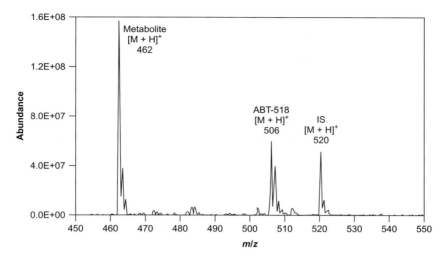

Figure 2.3 A mass spectrum of ABT-518, a metabolite of ABT-518 and a structural analogue IS. All three were infused at equimolar concentrations. (Reprinted from Stokvis et al. [10] with permission from John Wiley & Sons.)

Mass spectrometry

Mass spectrometers work by ionizing molecules and then sorting the results according to their respective mass-to-charge (m/z) ratios. The two components of this process, i.e. ionisation and ion transmission, are important to the success of a method and both require optimisation during method development. Optimisation of these two processes is facilitated by the infusion of the test article directly into the mass spectrometer.

Ionisation

A great amount of research has taken place over the last 20 years on the development of ion sources to improve the process of transferring the compound of interest from the mobile phase into the gas phase as a charged species. There have been rapid advances in source design, making them more rugged, efficient and sensitive.

Currently, two major ion sources are used – ESI and APCI. These techniques are complementary, with most compounds being sufficiently ionised by either or both of these two ion sources. ESI and APCI are very mild ionisation techniques which provide little or no fragmentation, making them both suitable for quantitative methods. The best choice of ionisation type will depend on the compound(s) of interest and the experimental conditions under consideration. There are no rules for ionisation selection apart from larger molecules such as peptides and proteins which work best under ESI conditions. A review of the literature on drug metabolism studies (drugs and xenobiotics) by Oliveira and Watson published in 2000, reported that ESI was used in 73% of methods and APCI was used in 23% of methods [13]. Using these data as a general guide, ESI can be used for small molecules in most cases (almost three-quarters) and is a suitable starting point for methods development unless other information suggests that APCI is a better choice.

Some highly non-polar compounds can be difficult to ionize using ESI or APCI (flavonoids, vitamins and steroids). A newly developed ion source, APPI, has shown the potential to ionise this type of compound [14, 15]. The introduction of this new ion source further adds to the analyst's options. All three ion sources require optimisation of source temperature and gas flows. These parameters are different for each manufacturer's instrument. However, it can be said that the optimisation of these parameters is not only to obtain maximum signal for the analyte, but also to attain stable signal.

Matrix effects

One of the major problems with generating ions using API techniques is MEs. MEs occur when molecules co-eluting with the compound(s) of interest alter the ionisation efficiency of ESI. This phenomenon was first described by Kebarle and Tang [16], who showed that ESI responses of organic bases decreased as the concentration of other organic bases was increased. The exact mechanism of MEs is unknown, but it probably originates from the competition between an analyte and the co-eluting, undetected matrix components in the ion source.

King et al. [17] have shown through a series of experiments that MEs are the result of competition between non-volatile matrix components and analyte ions for access to the droplet surface for transfer to the gas phase. Although they conclude that the exact mechanism of the alteration of analyte release into the gas phase by these non-volatile components is unclear, they postulate '. . . a likely list of effects relating to the attractive force holding the drop together and keeping smaller droplets from forming should account for a large proportion of the ionisation suppression observed with electrospray ionisation'. Depending on the environment in which the ionisation and ion evaporation processes take place, this competition may effectively decrease (commonly known as ion suppression) or increase (ion enhancement) the efficiency of formation of the desired analyte ions present at the same concentrations in the ion source. Thus, the efficiency of the analyte ions to form is very much dependent on the matrix entering the ion source.

MEs are also compound dependent. Bonfiglio et al. [18] reported that the chemical nature of a compound has a significant effect on the degree of MEs. In a study of four compounds of different polarities, under the same mass spectrometric conditions, the most polar was found to have the largest ion suppression and the least polar was affected less by ion suppression. These findings of differential MEs have important ramifications particularly when selecting a suitable IS for quantification purposes (see above). For example, if a drug and a glucuronide metabolite were quantified by internal standardisation against a close analogue of the parent drug and MEs were slightly different between samples, then the change in ionisation of the more polar glucuronide metabolite would probably not be compensated by the IS. Thus, if there are multiple analytes to be quantified, with varying degrees of polarity, there may be requirements for multiple ISs [10, 19].

The importance of MEs on reliability of LC-MS has been shown in terms of accuracy and precision [20]; when ion suppression occurs, the sensitivity and lower LOQ of a method may be adversely affected [21]. Thus, experiments should be performed to understand these MEs in order to develop a reliable LC-MS method.

It has been shown that APCI is less prone to MEs than ESI [22, 23]. The presence of excess reagent ions to produce charged species means that the ionisation process of APCI is less susceptible to MEs. Therefore, the problem of MEs may be removed by changing the ionisation source. However, some differences in ME for APCI have been found between manufacturer's ion sources [22].

The two main techniques used to determine the degree of MEs on an LC-MS method are (i) post-extraction addition and (ii) post-column infusion. The post-extraction addition technique requires sample extracts with the analyte of interest added post-extraction compared with pure solutions prepared in the mobile phase containing equivalent amounts of the analyte of interest [8, 20, 21, 24, 25]. The difference in response between the post-extraction sample and the pure solution divided by the pure solution response determines the degree of ME occurring to the analyte in question under chromatographic conditions.

An example of calculations from post-extraction addition experiments is shown in Table 2.2. These data are adapted from a study by Buhrman et al. [21] who showed, using this technique, the various amounts of MEs that may occur using different sample preparation methods of a platelet-activating factor receptor antagonist from human plasma. The example shown is for the liquid–liquid extraction (LLE) of this compound using hexane. The ME was determined as −26%. This value represents a loss of 26% of the analyte signal (ion suppression) due to alterations in ionisation efficiency. A calculated value of 0% would represent no MEs – the ideal scenario. Additional data from this experiment can be obtained from the comparison of the ion intensity obtained from extracts with analyte added pre-extraction with pure solutions to give an overall value termed 'process efficiency' (Table 2.2). Process efficiency represents the combination of MEs and recovery of the analyte from the matrix by the sample extraction process.

Low process efficiency can be deleterious to the accuracy and, in particular, the lower LOQ of a method [21]. Fierens *et al.* [26] reported preliminary data on the measurement of urinary C-peptide by LC-MS with an ESI ion source. Using aqueous standards it was determined that by direct injection of urine a limit of detection (LOD) of 0.2 ng could be achieved. Comparing these standards with urine, it was found that the signal was suppressed by 70–85%. Thus, without removing these MEs, via sample preparation, the desired lower limits of detection could not be attained.

The post-extraction addition technique can be considered a static technique that provides information about MEs at the point of elution of the analyte of interest. A more dynamic technique for determining MEs is post-column infusion [17, 18, 27–29]. An infusion pump is used to deliver a constant flow of analyte into the LC eluent at a point after the chromatographic column and before the mass spectrometer ionisation source (Figure 2.4). A sample extract (without added analyte) is injected under the desired chromatographic conditions and the response from the infused analyte recorded. The post-infusion technique enables the influence of the matrix on analyte response to be investigated over the entire chromatographic run.

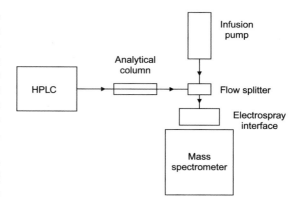

Figure 2.4 A schematic of the post-column infusion system.

Results of post-column infusion experiments enable the scientist to evaluate the influence of different sample extraction techniques on MEs, the appropriate analytical column, where MEs occur and are absent during a chromatographic run, and the influence of mobile additives on response. The use of this technique in methods development will be discussed later.

Electrospray

ESI is currently the most widely used ionisation source. ESI produces singly or multiply charged ions directly from an aqueous/organic solvent system by creating a fine spray of highly charged droplets in the presence of a strong electric field, with the assistance of heat and/or pneumatics [30–33].

ESI can ionise a wide variety of analytes from small drugs to macromolecules; however, it can be inefficient at ionising some non-polar compounds.

Table 2.2 Calculation of MEs, recovery and process efficiency using the post-extraction addition technique (data modified from [21])

Experiment	Ion intensity (cps)	Parameter	Calculation and result (%)
Post-extraction addition	6390	ME	(6390 – 8580)/8580 × 100 = –26
Pre-extraction addition	4050	process efficiency	4050/8580 × 100 = 47
Pure solution	8580		

ESI is widely used in bioanalytical studies because it is a water-based ionisation process. It is ideally suited to reversed-phase (RP) separations where formation of ions in solution is possible. One issue with ESI is the formation of adducts in solution. Typically, sodium and potassium cations or ammonium buffers in the mobile phase will form adducts with many compounds. Thus the precursor ion for a compound of interest may be in three forms: $[M + H]^+$, $[M + Na]^+$ and $[M + K]^+$ (in the positive-ionisation mode). It is highly advantageous to force the analyte into one form and this will give the maximum response. Larger molecules, such as peptides and proteins, generate multiply charged ions. Factors that influence ionisation efficiency for ESI are mobile-phase composition (volatility, surface tension, viscosity, pH and electrolyte concentration) and analyte properties (pK_a, hydrophobicity, surface activity, gas-phase basicity and ion-solvation energies) [34].

Atmospheric pressure chemical ionisation

A corona discharge needle is used as the primary ionisation source for the formation of reactant ions for most analytical applications of APCI. A voltage of 5–10 kV is applied for a corona discharge current of 1–5 µA. These values are dependent on the shape and position of the discharge needle. Two types of reactions occur under these conditions – charge and proton transfer.

For generation of positive reactant ions, the discharge needle is at a higher voltage with respect to the counter-electrode; alternatively for negative ionisation. APCI is well suited to both normal and RP-LC applications. The APCI process for LC-MS is dominated by reactant ions produced from the solvents used in the mobile phase. Usually, the analyte is in the lower mass range of under 1000 amu. Ions generated by APCI are singly charged and most commonly used for volatile, less-polar compounds, i.e. compounds that can be vaporised and contain few if any acidic or basic sites (alcohols, aldehydes, ketones and esters). The establishment of optimal conditions for APCI (gas flow rates, discharge needle current and source temperature) depend on the instrument and compound.

APCI is a complementary method to ESI, giving the analyst a wide coverage of potential analytes. With careful selection of chemical conditions robust quantitative methods can be developed using APCI.

Atmospheric pressure photo-ionisation

APPI is the newest of the soft ionisation techniques to be developed. This new ion source has the potential to facilitate the measurement of non-polar compounds that are inefficiently ionised by ESI or APCI. The process of photo-ionisation involves two main steps – vaporisation of the LC eluent and production of the photo-ions by interaction between a photon emitted by a UV source [15].

For this process to take place the ionisation energy of the molecule must be less than the energy of the lamp. Table 2.3 shows a list of compounds with their respective ionisation energies,

Table 2.3 Ionisation energies of some commonly used compounds and the energies of lamps (data modified from [15])

Lamp	Compound	Ionisation energy (eV)
	nitrogen	15.58
	water	12.62
	ACN	12.20
	oxygen	12.07
Argon 11.2 eV		
	methanol	10.84
	isopropanol	10.17
	hexane	10.13
Krypton 10.0 eV		
	heptane	9.93
	isooctane	9.80
	acetone	9.70
	pyridine	9.26
	benzene	9.24
	furan	8.88
	toluene	8.83
Xenon 8.4 eV		
	naphthalene	8.14
	triethylamine	7.53

along with the energies of some discharge lamps used in photo-ionisation [15]. As can be seen, the judicious selection of a lamp, in particular a krypton lamp, will avoid the ionisation of the commonly used solvents in LC-MS, i.e. methanol, acetonitrile (ACN) and water, thus minimising the chemical noise observed with mass spectrometric detection.

Further reactions occur and it is postulated that the formation of protonated species, [M + H]$^+$, occurs in the presence of protic solvents (S) by the mechanism [35]: M$^+$ + S → MH$^+$ + S(–H). The addition of a dopant to the LC eluent has been employed to act as an intermediate between photons and analyte, and thus increase the probability of the ionisation of the analyte [14, 36].

There are limited applications currently reported in the literature for APPI. It can be said that this ion source will have specific applications where ESI and APCI do not perform adequately. One example is where APPI has been compared with APCI for the measurement of idoxifene and two of its metabolites [37]. APPI was found to be six to eight times more sensitive for idoxifene and one of the metabolites. Interestingly, MEs were similar for both ion sources. The role of APPI in quantitative LC-MS is yet to be established, but it is another option in the analyst's armament of ion sources.

Ion transmission for selected-reaction monitoring

The focus of this chapter is optimisation of LC-MS using SRM for quantification. In the first mass filter the ions introduced into the vacuum travel through their respective regions, but only the ions with the selected m/z (precursor ion) are allowed into the collision cell. The collision cell is pressurised with either argon or nitrogen. The precursor ion collides with these gas molecules and a portion of the ion translational energy is converted to internal energy. The internal energy causes dissociation of the ion and produces an array of lower m/z ions (product ions) which travel through the second mass filter, with one or more of these product ions selected to be monitored. The relationship between precursor and product ion(s) is representative of the molecular characteristics, and is unique for the compound being studied. Thus, the combination of LC retention and characteristic mass transition(s) gives this technique high selectivity and, as consequence, the signal-to-noise (S/N) ratio is increased. This is the major reason why SRM is considered the acme of analytical techniques for the quantification of small molecules in biological samples.

The analyst is limited in the parameters that can be optimised for the efficient transfer of ions. The efficiency of transmission is mainly dependent on the instrumentation, with each manufacturer having instruments with different capabilities. One exception is the compound-specific parameter, i.e. the collision energy. Collision energy influences the translational energy that an ion has on entry to the collision cell. The higher the collision energy, the higher the translational energy and thus the more fragmentation is achieved. For SRM, a predominant fragment (product ion) is selected and the collision energy optimised to give the highest response for the chosen precursor-to-product ion transition. As a rule of thumb, the collision energy is optimised when the precursor ion is approximately 10% of the product ion.

Use of the immunosuppressant drug, tacrolimus, as an example of the effect of collision energy on fragmentation is shown in Figure 2.5. The precursor ion for tacrolimus is the ammoniated species [M + NH$_4$]$^+$ (m/z 821.4). Tacrolimus was infused directly into the mass spectrometer and the fragmentation studied under five collision energies (10, 20, 30, 40 and 50 eV). At a collision energy of 10 and 20 eV (Figure 2.5a and b), little or no fragmentation occurs. At 30 eV (Figure 2.5c), the precursor ion is approximately 10% of the major product ion (m/z 768.4). According to the rule of thumb this would be the optimum voltage. A considerable number of fragments are produced at 40 and 50 eV (Figure 2.5d and e). Using these higher collision energies gives informative structural data, but less sensitivity due to loss of signal via the multiple fragments. Of the five collision energies studied, it appears that 30 eV is optimum for the mass transition m/z 821.4 → 768.4.

Another experiment available to the analyst to

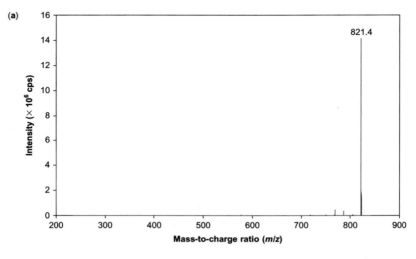
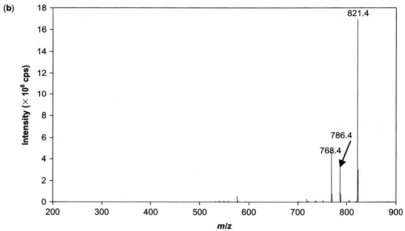
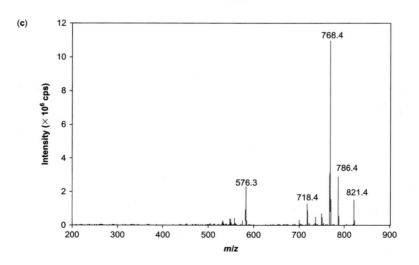

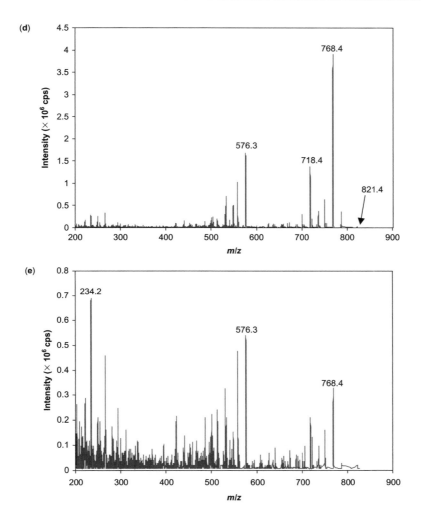

Figure 2.5 The influence of collision energy on the fragmentation of tacrolimus at (a) 10, (b) 20, (c) 30, (d) 40 and (e) 50 eV.

optimise collision energy is to infuse the compound of interest, monitor the mass transition of interest and ramp the collision energy in steps. This type of experiment was performed on tacrolimus over the range of 5–70 eV in 0.1-eV increments. The results are shown in Figure 2.6, with the maximum intensity at approximately 30 eV. Mass spectrometric conditions for ion production and detection have been selected and optimised. The next step is to investigate potential chromatography.

Chromatography

As discussed previously, the presence of co-eluting compounds causes MEs. Thus, to obtain a robust LC-MS method there is a need to remove or minimise their presence. The source of these interfering matrix components must also be considered. The interference may come from the current sample being injected, a previously injected sample (as a late-eluting interference) or build-up and overload of the analytical column [18, 19]. The two approaches to remove or

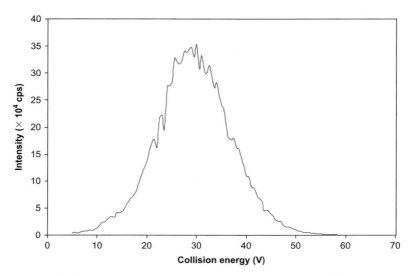

Figure 2.6 The ramping of collision energy from 5 to 70 eV in 0.1-eV increments while monitoring infused tacrolimus using the mass transition m/z 821.4 → 768.4. The maximum intensity for this transition was at approximately 30 eV.

minimise MEs are modification to the sample extraction methodology and/or improved chromatographic separation [28]. These two parameters are linked together in developing a quantitative LC-MS method.

The majority of MEs occur in the solvent front of a chromatographic run [27]. Thus, intuitively, if the analytes can be retained chromatographically to some degree then MEs can be minimised. Data obtained from post-column infusion experiments can provide data on where MEs are occurring. Thus, using the data to move the analytes away from these affected areas provides for a more stable method. These changes may limit the aim of high-throughput analysis. An alternative approach is to apply a rapid gradient or 'ballistic gradient' to separate the analytes from the solvent front while maintaining high throughput [6].

The degree of chromatographic retention is related to the cleanliness of the sample. In the following example, two sample preparation techniques, i.e. protein precipitation or 'protein crash' and C_{18} solid-phase extraction (SPE), are compared using the post-column infusion technique (Figure 2.7). For the sample prepared by protein precipitation (Figure 2.7a) there is a large amount of ion suppression occurring from the solvent front to 5 min into the chromatographic run. If the analyte(s) or IS had a retention time in this region then results would be adversely affected and chromatographic conditions would require adjustment. Alternatively, the sample prepared by SPE (Figure 2.7b) shows a small amount of ion suppression in the solvent front, but after a retention time of 3 min the response has returned to baseline. When performing such experiments it is advisable to inject the extracted sample solvent as a control (Figure 2.7c). Thus, it is clear that selection of the appropriate chromatographic conditions is intrinsically linked to the prepared sample.

The transfer of ions from the liquid phase to the gas phase via an API source does limit the analyst in what additives can be used in the mobile phase. Typically, methanol and ACN are used for the organic phase, and volatile additives such as acetic and formic acid and ammonium hydroxide are used in the aqueous phase along with buffers such as ammonium acetate and formate. The selection of which additive to use and at what pH the mobile phase should be buffered for a specific analyte is not known. A general rule is to have 0.1% or less of an additive [29]. The concentration range for buffers is 2–50 mmol/L. Some signal suppression may

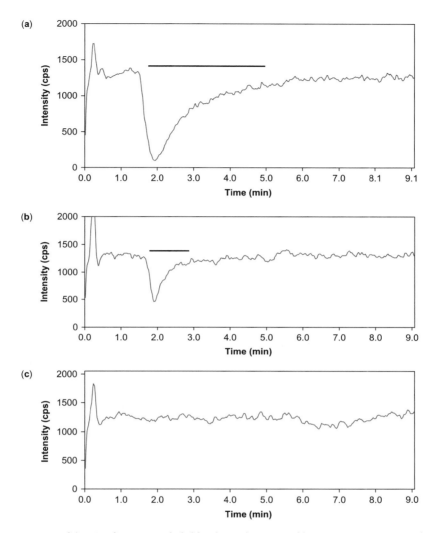

Figure 2.7 Comparison of the MEs from (a) a whole-blood sample prepared by protein precipitation, (b) a whole-blood sample prepared by SPE and (c) a mobile phase using the post-column infusion method. The solid line indicates the regions of altered ionization due to MEs.

occur if a higher buffer concentration is used. Common additives used in traditional LC, such as trifluoroacetic and phosphoric acids, are not recommended for they have been shown to suppress responses [19, 29]. The selection of the pH to be used should be assessed on a compound-to-compound basis. Mallet et al. [29] studied the effect of pH additives on the mass spectrometric response of some acidic and basic drugs. Interestingly, they found that the basic compound terfenadine gives the best response at a higher pH. This result is counter-intuitive, as terfenadine is being detected in its protonated form. The authors reported similar findings for other compounds studied. Thus, the effect of the whole pH range on the analyte being determined should be examined.

The number and variety of LC columns at the analyst's disposal are enormous. A consideration in selecting the appropriate column for an application is that the higher organic concentration in the mobile phase, the greater the ionisation efficiency. The use of a more retentive column will allow an increase in organic strength in the mobile phase without losing chromatographic retention. It has been shown that using the normal phase for highly ionic species enables increased organic composition with similar or better retention compared with the RP [38, 39]. Similarly, Needham et al. [40] compared the retention characteristics of cocaine on a C_{18} column with a pentafluorophenylpropyl column. Both columns had the same dimensions and the same flow rate was used. The mobile-phase composition was adjusted for each column so that cocaine had the same retention time; for the C_{18} and pentafluorophenylpropyl columns, the ACN content in the mobile phase was 12 and 90%, respectively. By having a higher organic content the response for cocaine was 12 times higher on the pentafluorophenylpropyl column. Thus, judicious selection of analytical column can greatly improve the sensitivity of an assay.

Sample preparation

The sample preparation process is a method to convert the biological matrix into a form that is suitable for LC-MS while retaining as much of the analyte as feasible. It is well known that sample preparation is required for classic LC with the saying 'garbage in, garbage out' [41]. It has been suggested that, due to the high selectivity of mass spectrometric detection, little or no sample preparation is required. The composition of biological samples is complex, containing numerous different compounds. These compounds range from large proteins to simple organic salts and are potential interferents to successful analysis. Thus, it is important that sample preparation is used to minimise the effects of the biological matrix on results. As an example, human plasma contains on average 3.2 g/L of sodium and 160 mg/L of potassium. It is known that salts will preferentially form alkali adducts in the mass spectrometer ion source. Such adduct formation results in loss of sensitivity and potentially reproducibility for a quantitative LC-MS method. It is now recognised that thorough sample preparation is often essential for reliable quantitative results and to obtain the lower limits of quantification often required [20, 21, 42]. In general, poor sample treatment or sample preparation will lead to poor results.

Keeping this point in mind, the goals of sample preparation for LC-MS are:

(i) sample loss to be minimal with a good yield of analytes of interest
(ii) efficient removal of other potential interfering components
(iii) concentration of the analyte(s)
(iv) results to be independent of matrix variability
(v) the final extract to be compatible with both the chromatographic and mass spectrometric systems
(vi) conversion of the analyte(s) to a suitable form for detection
(vii) to be a convenient and fast process.

The following are the most frequently used sample preparation techniques for LC-MS: dilution, protein precipitation, LLE, SPE and on-line two-dimensional chromatography (column switching). Several of these sample preparation options have been evaluated in terms of their degree of MEs on LC-MS [18, 21, 27, 28]. It is considered that protein precipitation using an organic solvent or dilute-and-shoot is the 'dirtiest' sample preparation technique and thus produces the most MEs compared with SPE or LLE (Figure 2.7) [18, 27]. More recently, on-line two-dimensional chromatography using a column-switching device has been shown to provide an alternate sample preparation that gives minimal MEs [43]. As previously discussed, the selection of which extraction technique to use for an application is not just limited to providing the cleanest sample, but also linked to the chromatographic separation when optimising a method.

Sample dilution and protein precipitation

These types of sample preparation are generally used only for cases where the matrix is relatively clean or the required measured concentrations are relatively high. MEs should be investigated thoroughly when developing methods using these sample preparation techniques.

As an example, Fierens et al. [26] developed a method for urinary C-peptide by LC-MS over the clinical range (20–120 µg/L). Under the study conditions the compound of interest was well retained chromatographically, with a $k' = 2.0$. Initial studies using pure solutions indicated that the method would have a LOD of 0.2 ng. Thus, no sample preparation and direct injection of urine was considered. The injection of urine resulted in up to seven-fold suppression of signal compared with water-based standards. Only a 10-fold dilution of the urine restored the full response. The authors developed the method further by using ultra-filtration of the urine followed by reconstitution of the ultra-filtrate in water. Interestingly, the authors advocate the use of a stable isotope-labelled IS, $[^2H_{14}]$C-peptide, for controlled quantification. Similarly, a simple 1:10 dilution of urine samples has been used as a sample preparation for the measurement of cocaine and benzoylecgonine [40]. Using this approach, the lower LOQ was 1.0 µg/L.

Protein precipitation is considered as fast, easy and applicable to a wide range of analytes. Its major disadvantage is its low selectivity. Protein precipitation of biological samples is typically performed using an organic solvent (methanol or ACN), a mixture of organic solvent combined with aqueous zinc sulphate, an acid (perchloric or trichloroacetic acid) or salts (ammonium sulphate). The resultant supernatant is suitable for injection on most analytical columns. A recent study by Souverain et al. [44] compared three protein precipitation procedures, i.e. ACN, perchloric acid and trichloroacetic acid, for the analysis of three pharmaceutical compounds and their primary metabolites in human plasma. All three sample preparation methods were effective in removing proteins and compatible with LC-MS, but only ACN gave adequate recovery of the drugs studied. As an example, Wood et al. [45] used simple precipitation as sample preparation for the measurement of amphetamines in plasma and oral fluid. This rapid sample preparation enabled a total analysis time of less than 20 min. It should be noted that a deuterated IS was used in this study and, while the absolute results may not be affected by MEs, the lower LOQ may be compromised given this type of sample preparation.

Another approach with protein precipitation is to use rapid gradient chromatography on a small dead volume column. In using a gradient on this type of column, the majority of interference occurs in the dead volume and the analyte can be separated from potential problems. To add stability to the ion source, a dump to waste of the initial part of the chromatographic run can be employed. Keevil et al. [46–48] have successfully applied this technique for the therapeutic drug monitoring of ciclosporin, tacrolimus and voriconazole. Similar results have been obtained for sirolimus [49].

Protein precipitation has the advantages of simplicity and is relatively cheap to perform; however, analysts must be aware that the dirty nature of the extract may require more or specialised chromatography.

Liquid–liquid extraction

LLE is a classic and mature technique that has been applied to all areas of analytical chemistry. The process of trapping the analyte in an immiscible solvent has greater selectivity than the previous methods. Further selectivity can be gained by adjusting the pH of the extraction process or by a further back-extraction of the analyte into an aqueous phase. The drawbacks of this methodology are that the process can be slow, with little or no ability to automate, emulsion formation may occur, and solvents are environmentally unfriendly and expensive to dispose of.

LLE has been reported extensively with LC-MS and discussing the merits of all techniques is beyond the scope of this chapter. However, it can be said that this technique has a place in LC-MS methods development. It has been shown that using the back-extraction technique, samples can be cleaner in terms of MEs than SPE [21]. To obtain the cleanest extracts with the highest analyte recovery, the correct selection of solvent and pH is required. The addition of an ionic neutral salt, such as sodium chloride, will increase the ionic strength of the aqueous layer and further improve the extraction of the analyte into the organic phase.

Solid-phase extraction

SPE is used extensively for the analysis of liquid samples such as xenobiotics in biological matrices. The wide range of phases available, RP and normal phase and ion exchange, provides a variety of isolation mechanisms by hydrophobicity, polarity or ionisation. Such is the selection of combinations that there are a large number of LC-MS applications in the current literature.

SPE can process large numbers of samples and be automated, particularly in the 96-well format [2, 50, 51]. These characteristics make this extraction technique desirable for high-throughput LC-MS methodologies. Peng *et al.* [51] reported a high-throughput LC-MS method (3 min/sample) that employed an Oasis HLB 96-well extraction plate for the measurement of matrix metalloprotease inhibitors. By the use of a 96-channel robotic liquid handling workstation they found that sample preparation was faster than a protein precipitation method and also with added selectivity.

Thus, SPE is well suited for sample preparation of xenobiotics from biological matrices. The additional selectivity provided by this technique results in reduced MEs (*see* Figure 2.7) and this may allow the analyst to reduce the chromatographic analysis time. In general, the more time spent on optimising this extraction technique, since there are many parameters to improve selectivity, the less time is required with chromatography.

Two-dimensional chromatography (column switching)

Using one column to perform sample preparation and submitting the appropriate fraction to a second column is a well-developed technique. Selectivity in this process can be obtained by using columns of different characteristics and mobile phases of different composition. A clear advantage of this approach is the ability to automate the analytical process. However, compared with SPE or LLE, the disadvantage is that two-dimensional chromatography requires potential LC-MS analysis time to prepare the sample. This may create a bottleneck in a busy laboratory with limited LC-MS instrumentation.

One strategy that is commonly employed is to trap the analyte on the first 'preparative' column, usually a short column with minimal dead volume, and to wash the column with high flow rates (up to 5 mL/min) while dumping the eluent to waste. This process effectively removes many potential matrix interferents, particularly salts, and is performed quickly due to the high flow rate and minimal dead volume. After sufficient rinsing, the flow is diverted to the analytical column, either back- or forward-flushing, and the analyte is eluted using an LC-MS-compatible mobile phase. This second mobile phase has typically high organic content, which gives better ionisation efficiency.

There are many applications reported in the literature with some examples of this approach being used successfully for the measurement of immunosuppressant drugs [12, 52], the endothelin antagonist bosentan [53] and complex mixtures such as testosterone metabolites [54].

Preliminary method validation

A preliminary validation is a limited set of experiments to confirm that the method has adequate accuracy and precision over the required analytical range. For a thorough discussion on method validation the reader is referred to Chapter 4.

The first step is to look at MEs. The post-column infusion experiments, described above, can be considered as semi-quantitative, and confirm the presence or absence of MEs and aid in minimising their influence on results. However, these data do not provide evidence that a validated analytical method is acceptable in terms of these effects. While the US Food and Drug Administration states that there is a need to investigate MEs for LC-MS methodologies [55], there is no clear guidance on how this should be performed during method validation.

Most, if not all, method validations are performed using standard and quality control samples prepared from a pooled source of matrix (whole blood, urine, etc.). In such a case, the matrix is homogeneous and thus it can be assumed that prepared samples will have the same MEs. Using these homogeneous samples for validation does not take into account the inter- and intra-patient matrix variability present in the clinical setting. It is likely that an extract of a plasma sample collected from a uraemic patient will be of a different composition to a healthy subject and thus may have variable MEs. This inter-patient variability is a typical problem faced, but seldom addressed, by the clinical scientist establishing an LC-MS method.

Matuszewski *et al.* [25] have shown the influence of different subjects on assay imprecision for a quantitative LC-MS method by comparing standard samples prepared from one pool of plasma with those from multiple sources. Standard curves were prepared in pooled plasma and also different individuals, and subjected to the same LLE and chromatographic conditions with multiple extractions performed for each concentration (n = 5). For the standards prepared from pooled plasma, across an analytical range of 0.5–100 µg/L, the variability at each concentration in terms of coefficient of variation was 4.6–11.3%. This variability would be considered acceptable for a method validation based on current opinion [56]. For the standard samples prepared from different lots of plasma, across this same concentration range, the variability was 11.6–23.8%. A significant increase in variability due to inter-subject differences in MEs was reported and with this decrease in performance the method would now fail validation. It can be speculated that many LC-MS quantification methods may fall into this same scenario if validation is solely based on one pooled matrix.

It is clear that at some stage during method validation the variability of a method must be assessed in samples taken from a variety of subjects in the patient population of interest. Furthermore, it has been shown that endogenous sources of MEs may come from the types of specimen containers and preservatives within these containers [22]. Thus, ME studies during validation must be carried out in the collection tube used for the method to be established. If different types of anti-coagulants are used, all of these should be evaluated.

While it is not practical to prepare all calibration standards and quality control samples from individual sources, some assessment of patient variability must be undertaken. One approach is to prepare and analyse three quality control concentrations, over the analytical range, in replicates of five prepared using different individuals for each aliquot (each individual is used only once). The individuals used should be representative subjects of the clinical/forensic investigation being undertaken (transplant recipients, HIV patients, uraemic patients, drug addicts, healthy subjects, etc.). The effects of inter-subject variability on accuracy and precision can then be assessed from these data. If the inter-subject variability is greater than 15%, changes to the extraction procedure and chromatography would have to be undertaken.

Conclusion

A systematic approach has been described for the development of LC-MS methods. The development is an iterative process – at the completion of each step it may be a requirement to re-assess the previous step or steps. Throughout this method development there is a need to understand the limitations of LC-MS in terms of MEs. Clearly, to obtain a robust quantitative LC-MS method there is a need to balance sample preparation and chromatography. By careful assessment of MEs and the judicial use of appropriate sample preparation coupled with adequate chromatography, LC-MS can provide a robust analytical platform. In developing methods analysts must acknowledge MEs and build assessment systems into their validation protocols. If these procedures are not performed there can be doubt in the validity of results produced.

References

1. Brewer E, Henion J. Atmospheric pressure ionization LC/MS/MS techniques for drug disposition studies. *J Pharm Sci* 1998; 87: 395–402.

2. Jemal M. High-throughput quantitative bioanalysis by LC/MS/MS. *Biomed Chromatogr* 2000; 14: 422–429.

3. Plumb RS, Dear GJ, *et al.* Quantitative analysis of pharmaceuticals in biological fluids using

high-performance liquid chromatography coupled to mass spectrometry: a review. *Xenobiotica* 2001; 31: 599–617.

4. Bayliss MK, Little D, *et al*. Parallel ultra-high flow rate liquid chromatography with mass spectrometric detection using a multiplex electrospray source for direct, sensitive determination of pharmaceuticals in plasma at extremely high throughput. *Rapid Commun Mass Spectrom* 2000; 14: 2039–2045.

5. Grant RP, Cameron C, *et al*. Generic serial and parallel on-line direct-injection using turbulent flow liquid chromatography/tandem mass spectrometry. *Rapid Commun Mass Spectrom* 2002; 16: 1785–1792.

6. Hopfgartner G, Bourgogne E. Quantitative high-throughput analysis of drugs in biological matrices by mass spectrometry. *Mass Spectrom Rev* 2003; 22: 195–214.

7. Kebarle P. A brief overview of the present status of the mechanisms involved in electrospray mass spectrometry. *J Mass Spectrom* 2000; 35: 804–817.

8. Annesley TM. Ion suppression in mass spectrometry. *Clin Chem* 2003; 49: 1041–1044.

9. Chavez-Eng CM, Constanzer ML, *et al*. High-performance liquid chromatographic-tandem mass spectrometric evaluation and determination of stable isotope labeled analogs of rofecoxib in human plasma samples from oral bioavailability studies. *J Chromatogr B* 2002; 767: 117–129.

10. Stokvis E, Rosing H, *et al*. Stable isotopically labeled internal standards in quantitative bioanalysis using liquid chromatography/mass spectrometry: necessity or not? *Rapid Commun Mass Spectrom* 2005; 19: 401–407.

11. Streit F, Armstrong VW, *et al*. Mass interferences in quantification of cyclosporine using tandem mass spectrometry without chromatography [abstract]. *Ther Drug Monit* 2003; 25: 506.

12. Volosov A, Napoli KL, *et al*. Simultaneous simple and fast quantification of three major immunosuppressants by liquid chromatography-tandem mass-spectrometry. *Clin Biochem* 2001; 34: 285–290.

13. Oliveira EJ, Watson DG. Liquid chromatography-mass spectrometry in the study of the metabolism of drugs and other xenobiotics. *Biomed Chromatogr* 2000; 14: 351–372.

14. Robb DB, Covey TR, *et al*. Atmospheric pressure photoionization: an ionization method for liquid chromatography-mass spectrometry. *Anal Chem* 2000; 72: 3653–3659.

15. Raffaelli, A, Saba S. Atmospheric pressure photoionization mass spectrometry. *Mass Spectrom Rev* 2003; 22: 318–331.

16. Tang, L, Kebarle P. Dependence of ion intensity in electrospray mass spectrometry on the concentration of the analytes in the electrosprayed solution. *Anal Chem* 1993; 65: 3654–3668.

17. King R, Bonfiglio R, *et al*. Mechanistic investigation of ionization suppression in electrospray ionization. *J Am Soc Mass Spectrom* 2000; 11: 942–950.

18. Bonfiglio R, King RC, *et al*. The effects of sample preparation methods on the variability of the electrospray ionization response for model drug compounds. *Rapid Commun Mass Spectrom* 1999; 13: 1175–1185.

19. Lagerwerf FM, van Dongen WD, *et al*. Exploring the boundaries of bioanalytical quantitative LC-MS-MS. *Trends Anal Chem* 2000; 19: 418–427.

20. Matuszewski BK, Constanzer ML, *et al*. Matrix effect in quantitative LC/MS/MS analyses of biological fluids: a method for determination of finasteride in human plasma at picogram per milliliter concentrations. *Anal Chem* 1998; 70: 882–889.

21. Buhrman DL, Price PI, *et al*. Quantification of SR27417 in human plasma using electrospray liquid chromatography-tandem mass spectrometry: a study of ion suppression. *J Am Soc Mass Spectrom* 1996; 7: 1099–1105.

22. Mei H, Hsieh Y, *et al*. Investigation of matrix effects in bioanalytical high-performance liquid chromatography/tandem mass spectrometric assays: application to drug discovery. *Rapid Commun Mass Spectrom* 2003; 17: 97–103.

23. Schuhmacher J, Zimmer D, *et al*. Matrix effects during analysis of plasma samples by electrospray and atmospheric pressure chemical ionization mass spectrometry: practical approaches to their elimination. *Rapid Commun Mass Spectrom* 2003; 17: 1950–1957.

24. Fu I, Woolf EJ, *et al*. Effect of the sample matrix on the determination of indinavir in human urine by HPLC with turbo ion spray tandem mass spectrometric detection. *J Pharm Biomed Anal* 1998; 18: 347–357.

25. Matuszewski BK, Constanzer ML, *et al*. Strategies for the assessment of matrix effect in quantitative bioanalytical methods based on HPLC-MS/MS. *Anal Chem* 2003; 75: 3019–3030.

26. Fierens C, Thienpont LM, et al. Matrix effect in the quantitative analysis of urinary C-peptide by liquid chromatography/mass spectrometry. *Rapid Commun Mass Spectrom* 2000; 14: 936–937.

27. Muller C, Schafer P, et al. Ion suppression effects in liquid chromatography-electrospray-ionisation transport-region collision induced dissociation mass spectrometry with different serum extraction methods for systematic toxicological analysis with mass spectra libraries. *J Chromatogr B* 2002; 773: 47–52.

28. Dams R, Huestis MA, et al. Matrix effect in bioanalysis of illicit drugs with LC-MS/MS: influence of ionization type, sample preparation, and biofluid. *J Am Soc Mass Spectrom* 2003; 14: 1290–1294.

29. Mallet CR, Lu Z, et al. A study of ion suppression effects in electrospray ionization from mobile phase additives and solid-phase extracts. *Rapid Commun Mass Spectrom* 2004; 18: 49–58.

30. Bruins AP, Covey TR, et al. Ion spray interface for combined liquid chromatography/atmospheric pressure ionization mass spectrometry. *Anal Chem* 1987; 59: 2642–2646.

31. Fenn JB, Mann M, et al. Electrospray ionization for mass spectrometry of large biomolecules. *Science* 1989; 246: 64–71.

32. Gaskell SJ. Electrospray: principles and practice. *J Mass Spectrom* 1997; 32: 677–688.

33. Cech NB, Enke CG. Practical implications of some recent studies in electrospray ionization fundamentals. *Mass Spectrom Rev* 2001; 20: 362–387.

34. Sjoberg PJR, Bokman CF, et al. Factors influencing the determination of analyte ion surface partitioning coefficients in electrosprayed droplets. *J Am Soc Mass Spectrom* 2001; 12: 1002–1010.

35. Syage JA. Mechanism of [M + H]$^+$ formation in photoionization mass spectrometry. *J Am Soc Mass Spectrom* 2004; 15: 1521–1533.

36. Hanold KA, Fischer SM, et al. Atmospheric pressure photoionization. 1. General properties for LC/MS. *Anal Chem* 2004; 76: 2842–2851.

37. Yang, C, Henion J. Atmospheric pressure photoionization liquid chromatographic-mass spectrometric determination of idoxifene and its metabolites in human plasma. *J Chromatogr A* 2002; 970: 155–165.

38. Naidong W, Shou W, et al. Novel liquid chromatographic-tandem mass spectrometric methods using silica columns and aqueous-organic mobile phases for quantitative analysis of polar ionic analytes in biological fluids. *J Chromatogr B* 2001; 754: 387–399.

39. Naidong W, Shou WZ, et al. Liquid chromatography/tandem mass spectrometric bioanalysis using normal-phase columns with aqueous/organic mobile phases – a novel approach of eliminating evaporation and reconstitution steps in 96-well SPE. *Rapid Commun Mass Spectrom* 2002; 16: 1965–1975.

40. Needham SR, Jeanville PM, et al. Performance of a pentafluorophenylpropyl stationary phase for the electrospray ionization high-performance liquid chromatography-mass spectrometry-mass spectrometry assay of cocaine and its metabolite ecgonine methyl ester in human urine. *J Chromatogr B* 2000; 748: 77–87.

41. Smith RM. Before the injection – modern methods of sample preparation for separation techniques. *J Chromatogr A* 2003; 1000: 3–27.

42. Matuszewski BK, Chavez-Eng CM, et al. Development of high-performance liquid chromatography-tandem mass spectrometric methods for the determination of a new oxytocin receptor antagonist (L-368,899) extracted from human plasma and urine: a case of lack of specificity due to the presence of metabolites. *J Chromatogr B* 1998; 716: 195–208.

43. Pascoe R, JP Foley JP, et al. Reduction in matrix-related signal suppression effects in electrospray ionization mass spectrometry using on-line two-dimensional liquid chromatography. *Anal Chem* 2001; 73: 6014–6023.

44. Souverain S, Rudaz S, et al. Protein precipitation for the analysis of a drug cocktail in plasma by LC-ESI-MS. *J Pharm Biomed Anal* 2004; 35: 913–920.

45. Wood M, De Boeck G, et al. Development of a rapid and sensitive method for the quantitation of amphetamines in human plasma and oral fluid by LC-MS-MS. *J Anal Toxicol* 2003; 27: 78–87.

46. Keevil BG, McCann SJ, et al. Evaluation of a rapid micro-scale assay for tacrolimus by liquid chromatography-tandem mass spectrometry. *Ann Clin Biochem* 2002; 39: 487–492.

47. Keevil BG, Newman S, et al. Validation of an assay for voriconazole in serum samples using liquid chromatography-tandem mass spectrometry. *Ther Drug Monit* 2004; 26: 650–657.

48. Keevil BG, Tierney DP, et al. Rapid liquid

chromatography-tandem mass spectrometry method for routine analysis of cyclosporin A over an extended concentration range. *Clin Chem* 2002; 48: 69–76.

49. Wallemacq PE, Vanbinst R, *et al.* High-throughput liquid chromatography-tandem mass spectrometric analysis of sirolimus in whole blood. *Clin Chem Lab Med* 2003; 41: 921–925.

50. Allanson JP, Biddlecombe RA, *et al.* The use of automated solid phase extraction in '96 well' format for high throughput bioanalysis using liquid chromatography coupled to tandem mass spectrometry. *Rapid Commun Mass Spectrom* 1996; 10: 811–816.

51. Peng SX, King SL, *et al.* Automated 96-well SPE and LC-MS-MS for determination of protease inhibitors in plasma and cartilage tissues. *Anal Chem* 2000; 72: 1913–1917.

52. Ceglarek U, Lembcke J, *et al.* Rapid simultaneous quantification of immunosuppressants in transplant patients by turbulent flow chromatography combined with tandem mass spectrometry. *Clin Chim Acta* 2004; 346: 181–190.

53. Lausecker B, Hess B, *et al.* Simultaneous determination of bosentan and its three major metabolites in various biological matrices and species using narrow bore liquid chromatography with ion spray tandem mass spectrometric detection. *J Chromatogr B* 2000; 749: 67–83.

54. Magnusson MO, Sandstrom R. Quantitative analysis of eight testosterone metabolites using column switching and liquid chromatography/tandem mass spectrometry. *Rapid Commun Mass Spectrom* 2004; 18: 1089–1094.

55. Anon. *Guidance for Industry. Bioanalytical Method Validation*, 2001. www.fda.gov/cder/guidance/guidance.htm.

56. Shah VP, Midha KK, *et al.* Bioanalytical method validation – a revisit with a decade of progress. *Pharm Res* 2000; 17: 1551–1557.

3

Quantification using LC-MS

Robert Kronstrand and Martin Josefsson

Introduction

Not only does the toxicologist have to determine what drugs are present in a sample, but he or she also has to quantify them in order to interpret their significance in the individual case. This is the goal of any quantitative analysis, and thus it is also important that the toxicologist has knowledge and understanding of method characteristics. As illustrated in Chapter 2, each step in the bioanalytical method should be investigated to determine the extent to which environmental, matrix, material or procedural variables, from the time of sample collection through to analysis, may affect the estimate of analyte in the matrix. In order to determine the concentration of an analyte, the response is compared with that of a well-defined reference substance, a calibrator, processed in the same way as the unknown sample. The concentration range over which the analyte will be determined must be defined based on statistical evaluation of calibrators over the total range. A method should be established and validated for the intended use and the intended matrix, not forgetting that variability of the matrix is likely to occur. In the case of liquid chromatography-mass spectrometry (LC-MS)-based procedures, it is essential that appropriate steps be taken to ensure the lack of matrix effects (MEs) throughout the application of the method. Some essential parameters to ensure the acceptance of the performance of a bioanalytical method are accuracy, precision, selectivity, sensitivity, recovery, stability and response function [1–3].

However, before setting up and validating a quantitative method, one has to define its purpose. Should it be a general or a specific method, will it be used routinely or only in certain cases? Should the method determine therapeutic or toxic concentrations, or both? The answers will have a great impact on how the extraction, chromatography and detection are designed. Also, the background information of a case or sample will define the strategy. In contrast to therapeutic drug monitoring (TDM) analysis, both clinical and forensic toxicological analysis may start with the question: 'What's in the sample?' rather than 'How much is in the sample?'.

A typical scenario observed in forensic post-mortem toxicology is when one indeed knows what to analyse, but there is no analytical method at hand. Here, LC-MS comes into its own with the possibility to directly inject reference compounds for optimisation of chromatography as well as the ease obtaining spectral information by direct infusion into the MS. Recently, several papers have shown that LC-MS is particularly suited to situations where a quantitative method has to be developed for casework and, hopefully, with a minimum of effort [4–11].

Selective chromatography in combination with mass-selective detection gives the flexibility needed for many different applications, even methods that screen for hundreds of substances [12], although they may not include quantification in the same injection. However, there are several applications described where a range of similar compounds are both screened for and then quantified using LC-MS. Kratzsch *et al.* [13, 14] have established validated methods for the screening and subsequent quantification of both neuroleptics and benzodiazepines in plasma. Wood *et al.* developed methods for a range of common amphetamines and benzodiazepines in different matrices. Six amphetamines were determined isocratically within 3 min with considerable co-elution [15], whereas 10 benzodiazepines were well separated within 12 min using gradient elution [16]. In summary, LC-MS methodology can easily be adapted to different demands depending on sample matrices and the aim of the investigator, with aims spanning from analysing hundreds of samples every day to one sample once and never again. No matter what the strategy, all of them require their own special considerations. In this chapter we will discuss how the choice of matrix, extraction, chromatography, ionisation and mass analysis may affect the quantitative results obtained.

Samples and matrices

The specimens available for analysis in toxicological cases can be anything from numerous in postmortem cases to only blood or urine from a living subject. Preferably, several body fluids or tissues are analysed to help the toxicologist interpret the results. Particularly for ante-mortem samples, plasma, serum or whole blood are the matrices of choice as there are large databases of information concerning the effects of drugs at different concentrations and combinations. For the clinical or forensic toxicologist, parent compounds and active metabolites in blood are of interest, since they may both contribute to impairment, adverse effects or death. LC-MS not only lends itself to the study of drug metabolites on a qualitative basis, as has been described [17–20], but it is also well suited for the quantification of both polar metabolites as well as less polar parent compounds. Several papers have been published where drugs were determined together with their glucuronides [21–28], sulphoxides [29], hydroxyls, de-alkylated and other polar metabolites [22, 30–32] in blood or urine.

Urine, however, is not suitable for establishing drug effects because the substances found are no longer in contact with the circulation. Rather, the concentrations of drugs in urine depend on the time when the urine was produced in relation to the intake of drug. However, urine may be useful to estimate the time when a drug was administered and also a suitable matrix for screening drugs before quantification in blood. As urine consists of 99% water, polar metabolites dominate. In urine drug testing, for example, the aim of the assay is to confirm a positive screening result (usually an immunochemical analysis), and it is usually sufficient to detect and quantify a limited number of well-defined analytes. LC-MS applications have lately been described for several drugs of abuse [28, 33–39]. Substances with extensive metabolism such as lysergic acid diethylamide (LSD) and Δ^9-tetrahydrocannabinol (THC) are excreted almost entirely as hydrophilic metabolites. The LSD metabolite, 2-oxo-3-hydroxy-LSD, is present in concentrations up to 40 times higher in urine than the parent LSD and thus constitutes a suitable analyte as the detection time can be increased [35]. One of the most frequently encountered drugs of abuse is THC, and its carboxylic acid metabolite 11-nor-9-carboxy-Δ^9-tetrahydrocannabinol (THC-COOH) requires both hydrolysis and derivatization before analysis with gas chromatography (GC)-MS. Weinmann *et al.* [33, 39] have shown that it can be readily quantified by LC-MS without any derivatization. The glucuronide was also included in their method, showing that it may be readily detected with LC-MS.

No matter what matrix is chosen, the integrity of the sample is as important as the choice of analytes and strategy of analysis.

Sample handling

In toxicological analysis the quality of the sample has great impact on analytical results. Important aspects for the evaluation of the reliability of quantitative results are:

(i) sample collection time
(ii) sample collection technique
(iii) quality and volume of the sample tubes
(iv) preservatives
(v) transportation
(vi) sample storage conditions.

The aim of quantitative toxicological analysis is often to reflect a condition in the body at the time of sampling or at the time of death. Thus, for accurate interpretation of the results, it is of great importance to have knowledge about how the samples were treated before analysis as well as the stability of the analytes in the particular matrix. If possible, sampling, transportation and storage should be performed according to guidelines issued by the laboratory for each assay. In reality, especially in forensic toxicology and in post-mortem work, this is seldom the case and stability studies under different conditions are very important to gain knowledge of the limitations that a result may have because of instability of analytes. Pre-analytical stability tests include short-term stability in the matrix, long-term stability and freeze/thaw stability.

Sample preparation

Reversed-phase (RP)-LC allows direct injections of any aqueous solution, including biological specimens. In such direct analysis no or little sample preparation is used before introduction to the LC-MS system for analysis. For biological samples, however, some pre-treatment is often needed. Urine is often filtrated or centrifuged and diluted only before analysis, whereas proteins in plasma or blood samples are removed by precipitation. This can be performed by addition of an organic solvent, e.g. acetonitrile (ACN) or ethanol, or by addition of a strong acid or a high-molarity salt solution, e.g. zinc sulphate. Accordingly to Blanchard, the volume needed to precipitate close to 100% of the proteins varies from only 0.2–0.4 parts of reagent for strong acids (e.g. trichloroacetic acid) to twice the volume of plasma for methanol [40]. Especially in urine analysis, cleavage of conjugates by hydrolysis is often performed before further preparation of the free analytes, even though conjugates are readily analysed with LC-MS. In tissue samples, the analytes must be liberated by hydrolysis, mechanical treatment and solvent extraction before analysis. Despite the fact that body fluids may be analysed directly, the advantages of sample preparation cannot be overlooked.

The main purposes of sample preparation in quantitative analysis are to:

- isolate the analyte from the matrix
- concentrate the analyte before analysis
- transfer the analyte to a solvent suitable for introduction into the chromatographic system.

The most common sample preparations used are isolation by liquid–liquid extraction (LLE) or solid-phase extraction (SPE).

In LLE, the analytes are isolated by their differences in affinity for an organic solvent. The affinity depends on the analyte's partition coefficient for a given solvent at a given pH. LLE can be performed in several steps at different pH values and with different solvents, resulting in very clean extracts. A drawback is that LLE is often time-consuming and tedious; however, it comes with very low reagent costs.

SPE is a flexible, rapid, simple method that can be applied to many isolation procedures [41]. The analytes are passed through a cartridge with a bed of packing material consisting of silica- or polymer-based structures with different functional groups or polarity. As the specimen passes through the sorbent bed, the analytes bind to the solid surface, while some of the sample components pass through. Interferences are subsequently removed from the cartridge by selective washings and, finally, the analytes are eluted into a volume of solvent for analysis. The stationary phases in most SPE cartridges resemble those used in LC. Historically, silica-based octadecasilane (C_{18}) columns have been widely used; however, because of its ability to retain many different compounds, it is one of the least selective phases. Thus, in toxicological analysis, mixed-mode cation-exchange SPE is often preferred instead. On a mixed-mode material the analytes are retained by a combination of non-selective hydrophobic interaction and more selective ion–ion interaction. More recently, polymeric SPE phases have been commercially available that have higher stability than silica phases, and thus can be used over the entire pH range and in combination with most solvents.

Sample preparation is one of the most important issues in quantitative chemical analysis of substances in body liquids and tissues. It makes the sample suitable for detection as well as concentrating the analytes that are often present in 'parts per billion' levels. Previously, when LC-MS was used, it was suggested that extraction was no longer needed and several applications were published where dilution of urine or protein precipitation of plasma was the only pre-treatment used [15, 28, 29, 42–46]. Indeed, with tandem LC-MS (LC-MS-MS), and the selection of a specific molecular ion and the subsequent isolated fragmentation to specific product ions that gave so much higher specificity compared with single LC-MS, it was possible to achieve high sensitivity allowing accurate quantification just by diluting the sample prior to injection. However, untreated samples will speed up processes that lead to poor chromatography and variations in ion formation, especially if large batches of samples are to be analysed on a routine basis. The risk for MEs (ion suppression or enhancement) by endogenous compounds and other drugs co-eluting with the analyte is also true for quantitative LC-MS-MS, so selective sample preparation is still important [47–52]. Recent studies show that MEs are dependent not only on ionisation mode (i.e. atmospheric pressure chemical ionisation [APCI] or electrospray ionisation [ESI]), but also on source design (i.e. Sciex, Finnigan, Micromass) [53]. Among the extraction methods used for preparation of urine, blood, tissues and oral fluid there is a fairly equal distribution between LLE and SPE (Table 3.1). A common belief is that SPE would solve the problems of ion suppression; however, Müller *et al.* [51] found that neither LLE nor SPE of plasma samples could remove ion suppression at the beginning of the chromatogram, but concluded that ion suppression is not generally present at retention times after the void when using RP chromatography.

Dams *et al.* [52], however, found that SPE could concentrate not only the analyte, but also more non-polar compounds responsible for ion suppression, and, in contrast to Müller *et al.* [51], that ion suppression was indeed present both at the beginning and throughout the chromatographic run depending on the procedure used. For direct injections of urine samples it is of certain importance to have a good separation of the analytes from the void in order to avoid ion suppression from early eluting polar compounds and salts. For blood samples, however, there might be a greater risk for ion suppression by more non-polar endogenous compounds eluting late in the chromatogram. Table 3.1 summarises the pre-treatment and extraction used for a variety of drugs and matrices.

Automation in SPE sample preparation has become very popular as unattended operation and minimal operator intervention allow for time-saving and safer handling of hazardous material [54]. In many automated systems individual samples are processed in series off-line from the analytical instrument. Although this is slower than manual operations in batches, time-savings are still achieved because they can operate continuously during the day, night or weekend. Recent automated systems can provide higher sample throughput by utilising the concept of automated parallel sample processing. A major drawback with automation is the risk of analyte contamination between samples due to carry-over. For this reason extensive validation for carry-over over a wide concentration range is needed for all quantitative assays.

Automated systems are available for the handling of traditionally sized SPE columns of 1 or 3 mL with various packing materials in amounts normally ranging from 30 to 150 mg. The 96-well format has been introduced for high-throughput analysis, allowing effective handling of large numbers of samples. The latter, however, has not been that widely applied in toxicological laboratories since normally fewer than 96 samples are handled on a daily basis (typically 10–50 samples). However, laboratories that need high-throughput systems have also exploited the sensitivity and selectivity of MS-MS, enabling them to analyse hundreds or even thousands of samples every day [38, 55–57]. These methods are usually run isocratically because no equilibration time is needed between sample injections and they have run times ranging from less than 30 s to about 3 min.

For a high-throughput system one not only needs fast chromatography, but also a rapid extraction procedure, and the use of the 96-well technique dominates.

Since LC-MS assays most often operate with aqueous solutions, the technique is suitable for combination with automated sample preparation on-line. In on-line applications the sample is injected directly on the chromatographic system or injected after a minor preparation such as protein precipitation or filtration.

Table 3.1 Processing, chromatographic and mass detection conditions from selected references

Analytes	MS Apparatus	Ionisation	Mode	Chromatography Column	Dimension (mm)	Particle size (µm)	Mode	Runtime (min)	Matrix	Extraction	Reference
Amphetamine and others	Q	ESI+	SIM	Supelcosil CN	4.6 × 33	3.0	isocratic	7	urine	on-line SPME	77
Amphetamine and others	IT	SSI	MRM	Hypersil BDS Ph	3.0 × 100	3.0	isocratic/gradient	35	blood/urine/tissue	LLE	72
Amphetamine and others	Q	ESI+	SIM	Xterra RP C$_{18}$	2.1 × 150	5.0	gradient	30	meconium	SPE Certify	78
Amphetamine	QqQ	ESI+	MRM	Hypersil BDS C$_{18}$	2.1 × 100	3.5	isocratic	4	plasma/oral fluids	PP	15
Methamphetamine and metabolite(s)	Q	ESI+	SIM	Capcell Pak SCX	1.5 × 150	5.0	isocratic	30	hair	on-line SPE	79
THC-COOH	Q	ESI+	SIM	Zorbax XDB C$_8$	3.0 × 150	5.0	gradient	15	urine	SPE, C$_{18}$	34
THC-COOH	QqQ	APCI−	MRM	Xterra-MS C$_{18}$	3.9 × 20	3.5	gradient	6.5	urine	SPE, C$_{18}$	33
THC-COOH	Q	ESI−	SIM	Zorbax SB Ph	2.1 × 50	3.5	isocratic	3	urine	SPE, C$_8$	
THC-COOH and glucuronides	QqQ	ESI+	MRM	RP-C$_8$-select B	2.0 × 125	5.0	gradient	37	urine	LLE	39
THC, THC-COOH, 11-OH-THC	QqQ	ESI+	MRM	Luna PhenylHexyl	2.0 × 50	3.0	gradient	11	plasma	SPE, C$_{18}$	36
Morphine, M3G, M6G	Q	ES	SIM	YMC ODS	4.0 × 100	5.0	isocratic/split	6	serum	SPE C$_{18}$	26
Morphine, codeine, 6-AM and glucuronides	Q	APCI+	SIM	Superspher RP$_{18}$	3.0 × 125	4.0	flow gradient	17	blood/serum/urine/vitreous humor	SPE, C$_{18}$	80
Morphine, M3G, M6G	QqQ	ESI+	MRM	Inertsil Si	3.0 × 50	5.0	isocratic	3	plasma	SPE, C$_{18}$	23
Morphine, M3G, M6G	QqQ	ESI+	MRM	Betasil Si	3.0 × 50	5.0	isocratic	3.5	plasma	SPE, C$_{18}$ 96-well	81
BP, NBP	Q	ESI+	SIM	Nucleosil C$_{18}$	1.0 × 150	–	isocratic	9	blood	LLE, Extrelut	11
BP, NBP	QqQ	APCI+	MRM	Purospher C$_{18}$	4.0 × 55	–	isocratic	15	plasma	SPE, various	27
BP, NBP, BPG	QqQ	ESI+	MRM	Inertsil ODS-3	3.0 × 100	3.0	gradient	12.7	plasma	SPE, C$_{18}$	25
BP, NBP, BPG, NBPG	QqQ	ESI+	MRM	Zorbax SB-Ph	2.1 × 50	3.5	gradient	6	urine	SPE, Certify	28
Methadone	Q	ES+	SIM	Chiralcel OJ-R	2.1 × 150	5.0	chiral	10	plasma	on-line SPE, HLB	42
Methadone	QqQ	ESI+	MRM	Chiral-AGP	2.0 × 50	5.0	chiral	6	plasma	LLE, 96-well	82
Fentanyl/norfentanyl	IT	ESI	MRM	Zirchrom-PBD	2.1 × 50	3.0	isocratic	5	plasma	LLE	83
Opioids	IT	APCI	MRM	Synergi Polar RP	2.0 × 150	4.0	gradient	35	urine	direct	84

Table 3.1 continued

Analytes	MS			Chromatography				Matrix	Extraction	Reference
	Apparatus	Ionisation	Mode	Column	Dimension (mm)	Particle size (µm)	Mode	Runtime (min)			

Analytes	Apparatus	Ionisation	Mode	Column	Dimension (mm)	Particle size (µm)	Mode	Runtime (min)	Matrix	Extraction	Reference
Cocaine and metabolite(s)	QqQ	ESI+	MRM	Zorbax XDB C_8	2.1 × 150	5.0	gradient	25	meconium	SPE, Certify	85
Cocaine and metabolite(s)	QqQ	ESI+	MRM	Advantage Basic	2.0 × 50	5.0	gradient/split	2	urine	direct 96-well	86
Cocaine and metabolite(s)	QqQ	ESI+	MRM	Zorbax XDB C_8	2.1 × 150	5.0	gradient/split	15	plasma	SPE, Certify/SCX	87
Cocaine and metabolite(s)	QTOF	ESI+	PIS	BDS C_{18}	2.1 × 100	3.0	gradient	18	oral fluid	SPE, HCX	88
Ketamine, norketamine	Q	ESI+	SIM	Zorbax XDB C_8	4.6 × 50	3.5	isocratic	3	urine	LLE	37
2-Oxo-3-hydroxy-LSD	Q	APCI+	SIM	Zorbax XBD C_{18}	4.6 × 150	3.5	gradient	21	urine	LLE/SPE	35
Illicit drugs	Q	TS	SIM	ODS	4.6 × 150	–	gradient	25	urine	SPE, C_{18}	89
Drugs	QqQ	ESI+	MRM	Zorbax SB-Ph	2.1 × 50	3.5	gradient	9	hair	LLE	90
Nicotine, cotinine	QqQ	APCI+	MRM	BDS Hypersil C_{18}	3.0 × 100	3.0	isocratic	1	plasma	LLE	91
Cotinine	QqQ	APCI+	MRM	PE BDS C_{18}	4.6 × 30	3.0	isocratic	1	plasma	LLE	92
L-Hyoscyamine	QqQ	APCI+	MRM	BDS C_{18}	3.0 × 50	3.0	isocratic	2	plasma	LLE	93
Benzodiazepines	QqQ	ESI+	MRM	Zorbax SB-C_{18}	2.1 × 15	3.0	isocratic	0.5	urine	LLE, 96-well	55
Benzodiazepines	QqQ	ESI+	MRM	Zorbax SB-Ph	2.1 × 50	3.5	gradient	9	hair	SPE, Certify	94
Benzodiazepines	QqQ	ESI+	MRM	Zorbax SB-Ph	2.1 × 150	5.0	gradient	15	larvae	PP, LLE, SPE	16
Benzodiazepines	IT	APCI	SIM	Superspher RP Select B	2.0 × 125	–	gradient	10	plasma	LLE	14
Benzodiazepines	IT	ESI+	MRM	Xterra C_{18}	2.1 × 150	3.5	gradient	45	blood	SPE, ChemElut	95
Benzodiazepines and metabolite(s)	QqQ	ESI+	MRM	Zorbax SB-Ph	2.1 × 5.50	3.5	gradient	10	urine	SPE, Certify	96
Alprazolam and metabolite(s)	Q	ESI+	SIM	Zorbax Rx-C_{18}	? × 150	–	isocratic	6	plasma	LLE	96
Bromazepam and others	Q	MB	SIM	Nucleosil C_{18}	4.0 × 300	–	isocratic	6	serum	not described	97
Bromazepam, clonazepam and metabolite(s)	QqQ	ESI+	MRM	Uptisphere ODB C_{18}	2.0 × 150	5.0	gradient	15	urine/hair	SPE, Toxitube A/LLE	98
Flunitrazepam and metabolite(s)	IT	ESI+	MRM	Alitma C_{18}	2.1 × 150	5.0	isocratic	10	blood/urine	SPE, Clean Screen	99
Triazolam and metabolite(s)	IT	ESI+	MRM	Mightysil RP-C_{18}	2.0 × 100	3.0	gradient	45	hair	LLE	100
Neuroleptics	IT	ESI+	MRM	Luna C_{18}	4.6 × 150	–	isocratic	15	hair	LLE	101

Table 3.1 continued

Analytes	MS			Chromatography				Runtime (min)	Matrix	Extraction	Reference
	Apparatus	Ionisation	Mode	Column	Dimension (mm)	Particle size (µm)	Mode				
Neuroleptics	QqQ	ESI+	MRM	RP-C_8-select B	2.0 × 125	5.0	gradient	45	hair	SPE, Chromabond	102
Neuroleptics and metabolite(s)	QqQ	ESI+	MRM	Zorbax SB-CN	2.1 × 50	3.5	gradient	10	blood/urine/hair	SPE, Certify	29
Neuroleptics, methadone	IT	APCI+	MRM	Superspher RP Select B	2.0 × 125	–	gradient	10	plasma	LLE	13
Clozapine and metabolite(s)	QqQ	ESI+	MRM	Phenomenex C_{18}	4.6 × 50	5.0	isocratic	4	plasma	LLE	31
Haloperidol	QqQ	ESI+	MRM	Symmetry C_{18}	4.6 × 100	3.5	isocratic/split	3	plasma	SPE, C_{18}, 96-well	57
Olanzapine	QqQ	APCI+	MRM	Metachem monochrom	4.6 × 150	5.0	gradient	9	plasma/serum	SPE, C_2, 96-well	103
Risperidone	QqQ	ESI+	MRM	Phenomenex Ph-Hexyl	4.6 × 50	5.0	isocratic	5	plasma	LLE	30
Risperidone and metabolite(s)	QqQ	ESI+	MRM	Zorbax SB-C_{18}	2.1 × 30	3.5	isocratic	4	plasma/serum	PP, On-line SPE	44
Zuclopenthixol	IT	ESI+	MRM	Symmetry	2.1 × 150	5.0	gradient	15	blood	LLE, Extrelut	104
Beta-blockers	IT	APCI+	MRM	Superspher RP Select B	2.0 × 125	–	gradient	10	plasma	SPE, HCX	105
Albuterol and other compounds	QqQ	ESI+	MRM	Inertsil ODS3/Beta Silica	3.0 × 50	5.0	gradient	4	plasma	SPE, HLB/C_{18}	106
Albuterol	IT	APCI+	MRM	Chirobiotic T	4.6 × 250	5.0	chiral	7	plasma	SPE, HLB	107
Sotalol	Q	ESI+	SIM	Teicoplanin	4.5 × 250	5.0	chiral	20	plasma	SPE, HLB	108
Antihistamines	QqQ	ESI+	MRM	Genesis C_{18}	2.1 × 100	4.0	gradient	10	blood	LLE	109
Methotrexate and metabolite(s)	QqQ	ESI+	MRM	Kromasil C_{18}	1.0 × 50	5.0	isocratic	2	urine/plasma	SPE, 384-well	38

11-OH-THC = 11-hydroxy-Δ^9-tetrahydrocannabinol; 6-AM = 6-acetylmorphine; BP = buprenorphine; BPG = BP glucuronide; IT = ion trap mass analyser; M3G, M6G = morphine-3 (or 6)-glucuronide; NBP = norbuprenorphine; NBPG = NBP glucuronide; QTOF = quadrupole time-of-flight tandem mass spectrometer; SPME: solid-phase microextraction; Q = quadrupole; QqQ = triple quadrupole; THC = Δ^9-tetrahydrocannabinol; THC-COOH = 11-nor-9-carboxy-THC.

The chromatography is performed in two dimensions where the analytes after injection are first retained on a cartridge column in a cartridge handling system or on a smaller pre-column (or even an analytical column) in a column-switching system. In the first case, a new cartridge is used for each injection; in the latter case, the same column is used for all samples in a series. Flarakos *et al.* [44] described a method for risperidone and its hydroxy-metabolite using only 25 μL of plasma. After protein precipitation with 75 μL ACN containing the internal standard (IS), an aliquot of the sample was introduced in a system containing a column-switching valve and a loading column used to clean the sample, guard the analytical column and thus prevent ion suppression. In addition to method controls, a system suitability check was run in each series to monitor MS performance.

The analytes in all on-line applications must be trapped on the pre-column or cartridge column. The analytes must then be eluted from the pre-column with a mobile phase compatible with the analytical column. Thus, the hydrophobicities of the pre-column and analytical column must be balanced in terms of retention and migration. The trade-off to be made is that the downstream separation and detection need to be adapted to the on-line extraction in terms of timing and solvent selection, and thus could be suboptimal. An on-line switching system with a single reused trapping column can easily be adapted to existing instrumental equipment, hardware and software. Often only an automatically actuated switching valve and an additional LC pump are needed.

When a pre-concentration is needed or there is a certain risk of cross-contamination between samples, off-line application is preferable. However, when large amounts of samples are handled, automation is indeed needed. Off-line automation can involve a number of steps such as filtration, precipitation, LLE or SPE, but the most time-consuming step is often solvent evaporation after extraction or elution. This step is commonly performed with a stream of nitrogen gas or vacuum centrifugation off-line. Sometimes special precautions must be taken during solvent evaporation. This is true when there is risk for co-evaporation or risk of extensive adsorption on surfaces of the analytes of interest. Acidification before evaporation and/or the use of silanised glassware could affect both.

In forensic analysis where the expected amount of drug is unknown and there is a wide range in concentrations between samples there is a certain risk of cross-contamination between samples. For instance, THC-COOH concentrations in urine may span over a 10 000-fold difference (from low micrograms per litre to above 10 mg/L). Several problems were identified when a column-switching LC-MS assay for direct analysis of THC-COOH in urine was evaluated at the author's laboratory. A phenyl column was used as the analytical column; and the adsorbents phenyl, C_8 and C_4 were tested in trapping columns. Even if extensive washing was performed between samples, a carry-over of up to 1% was seen. Thus, when a sample with a very high concentration was analysed prior to a negative sample, carry-over could reach levels of THC-COOH above the 5 μg/L cut-off. The source of contamination was found to be the valves and linings in the system. Both steel and polyetheretherketone (PEEK™) linings were tested with similar results. Moreover, high-density samples tended to clog filters and linings. We found instead that an automated off-line sample preparation was favoured over an on-line application. Using an automated parallel processing in serial with a four-needle preparation robot, an effective SPE was achieved without any carry-over. The serial processing, where four samples at a time were taken through all steps from activation of SPE columns to elution of analyte, was essential to exclude carry-over.

Sample introduction

Often less attention is focused on sample introduction in LC than in GC, probably because there are fewer instrumental parameters that can be optimised. However, the constitution of the sample solution injected will greatly affect the chromatographic performance in LC, for a single run or for the long-term stability of the assay. In RP chromatography, where aqueous mobile phases are used, the type and percentage of organic solvent and the pH of the solution that the sample is reconstituted in will influence the peak performance of the analytes. A high percentage of organic solvent, e.g. methanol or ACN, may cause peak broadening, resulting in reduced resolution of peaks and troublesome integration. The pH used will influence the degree of ionisation of the analytes, which will affect the degree of retention as well as the peak shapes. In the worst case, an unfavourable pH can cause twin peaks of a specific analyte. A good rule of thumb in LC is to reconstitute the sample in a solution similar to the mobile phase used. This may not always be possible in practice as the analytes may be unstable in mobile phase and degrade during storage in the auto-sampler prior to injection. These effects should be studied during method validation of a quantitative method. In some cases the percentage organic solvent in the mobile phase is unfavourably low for dissolution of the analytes and a higher amount of solvent has to be added. To avoid peak broadening when injecting solutions with high percentage organic solvent, one may compensate by using a lower injection volume.

In LC-ESI-MS, where narrow bore columns and low mobile phase flows are favourable, the injection volume will have even greater impact on the chromatographic performance. When direct injections are performed after a simple protein precipitation with an organic solvent, e.g. ACN, an injection volume of 5 μL or less (depending on mobile-phase composition) has to be used in order to avoid analyte peak broadening.

In quantitative analysis we recommend to have a strategy for the overall concept of sample preparation, solvent reconstitution and sample introduction. If a concentrated sample is reconstituted in 100 μL solvent and 10 μL is used for injection, 1:10 of the amount analyte will be injected on the column, resulting in a given analyte signal. If only a volume of 5 μL is used for injection, then 1:20 of the amount is injected, subsequently causing a decrease in analyte signal, but also fewer matrix components are introduced that may extend column life and enhance the overall chromatographic performance. In summary, the limit of quantification (LOQ), chromatographic performance as well as long-term stability of the assay will depend on the composition of the sample solution and the volume injected.

Chromatography

LC is a powerful separation technique that can be operated in normal-phase as well as RP mode. Normal-phase chromatography is often used for the analysis of relatively non-polar compounds, as the polarity of the stationary phase is higher than that of the mobile phase. Silica is used as adsorbent or stationary phase in combination with non-polar solvents like pentane, hexane and dichloromethane as mobile phase. Since these solvents are hazardous to health and environment, normal-phase application is unpopular and avoided in most laboratories. Nevertheless, normal-phase chromatography with an aqueous/organic mobile phase and a high percentage of organic modifier enables an improvement in detection sensitivity of polar analytes (e.g. glucuronides) with respect to RP chromatography [23].

In modern LC applications, however, RP chromatography with aqueous mobile phases is by far the most commonly used. RP chromatography is suitable for polar analytes such as biomolecules and drugs, and the solvents used, typically ACN and methanol, are more polar, less hazardous and easier to handle. Aqueous mixtures of methanol and ACN are conductive and volatile, and thus constitute mobile phases suitable for LC-MS. There is a diversity of non-polar silica-bonded stationary phases available and, even if C_{18} is the most popular, materials like C_8, cyano and phenyl are increasingly used in drug analysis. Moreover, a wide range of chemistries for end-capping, covering and linking is used in order to improve the performance and increase column stability.

Mobile-phase buffers are used in RP chromatography in order to control the ionisation of the analytes and improve chromatographic performance. Typically, in traditional LC, phosphate buffers have been used due to their high buffer capacity over a wide pH range.

However, when LC is combined with MS detection some new considerations must be taken into account. In LC-MS, non-volatile buffers will precipitate in the ion source causing clogging and impaired performance, e.g. reduced signal intensity. Thus, volatile buffers or additives such as ammonium acetate, ammonium formate, acetic acid and ammonia have to be used. Column dimensions, flow rate and solvent/buffer composition compatibility also need to be adapted to the interface used. There are various solutions if the original column flow rate (e.g. from a LC-UV application) is too high for MS analysis, such as more effective solvent evaporation and pumping capacity in the ion source housing, solvent splitting prior to sample introduction or miniaturisation of LC columns used. Narrow-bore columns are favourable for ESI. Changing the column bore size from 4 to 2 mm i.d. will result in a decrease of the column volume by a factor of 4. Since ESI is a concentration-dependent ionisation process, the analyte signal will increase due to an increase of the amount of analyte per volume solvent (i.e. mobile phase). With APCI-MS, though, which is a mass-dependent ionisation process, the total amount of sample introduced determines the final analyte signal intensity. Here, applications with traditional wide-bore columns using relatively high mobile-phase flow rates are favourable and method transfer from existing LC-UV assays might be more straightforward.

Chromatography set-up plays an important role especially for the selectivity and long-term robustness of the assay. An even peak distribution is a key factor for good performance, and can be optimised by selecting the right combination of mobile and stationary phases. The choice of column dimensions and particle size determines the maximum load on the column and the flow rate that can be used. Small particles in combination with narrow-bore columns improve selectivity and shorten the time for analysis. However, the consequent increase in backpressure over the column must be compensated by decreased flow rate.

Typically, peak fronting is seen when a chromatographic column is saturated; however, due to secondary interactions (e.g. residual silanol groups), peak tailing is often seen as well. Since column saturation not only depends on analyte concentration, but also on matrix components, the degree of sample preparation determines the chromatographic performance. This is true for the individual samples analysed as well as for the performance over time. Tightly retained matrix components may accumulate on the column and change performance over time. Thus, gradient elution with a wash-out step is highly recommended when sample preparation is simple.

The main purpose for selective chromatography in LC-MS is to separate the analytes from metabolites, analogues and interfering matrix components that could interfere with MS detection. If a highly efficient column is used and relatively selective chromatography is achieved, the retention time can be used as an additional criterion for identification, improving the overall selectivity and specificity of the LC-MS assay. It is important that the analytes of interest are sufficiently retained and are well separated from the major matrix components, mainly in the front of the chromatogram. A rule of thumb to use is that the retention should be at least two times the column dead volume.

Ionisation

Before introduction into the mass analyser the analytes must be ionised and transferred from solution to the gas phase. There are a number of different ion sources at atmospheric pressure (atmospheric pressure ionisation [API]) available on the market. ESI and APCI are widely used, and more recently atmospheric pressure photoionisation (APPI) and sonic spray ionisation (SSI) have been introduced. Since all these ion sources operate with aqueous solutions at relatively high flow rates, they can easily be combined on-line with LC to enable MS analysis without prior derivatisation of the analytes, making LC-MS a powerful alternative to GC-MS. They are all soft ionisation techniques by which stable quasi-molecular ions (e.g. $[M + H]^+$ or $[M - H]^-$) are readily formed, generally with only slight to moderate fragmentation. APCI, which is less soft than ESI, sometimes results in thermal artefacts. However, APCI is able to ionise molecules that are less basic or acidic and normally not amenable to ESI [58].

Ionisation can be performed in positive as well

as negative mode. Most often the best choice for ionisation is determined empirically. Since the majority of drugs are weak bases, applications in positive mode are often seen, whereas negative ionisation is common for acids and applications in both modes have been proposed for some analytes (e.g. THC-COOH) [33, 39]. Proton transfer is one of the commonly observed mechanisms for positive ion formation. This process is essentially a gas-phase acid–base reaction and therefore dependent on the gas-phase basicity and acidity of the species involved. The mobile-phase buffer constitution and concentration are thus of importance for the ionisation process. A drawback is that non-volatile buffers have to be avoided as they decrease sensitivity and can damage the orifice to the MS. Instead, volatile buffers and additives (e.g. ammonia or formic acid) have to be used which may cause, among other factors, changes in pH of the eluent during the ionisation process [59]. For robust quantitative assays the composition of the mobile phase must be optimised for ionisation performance as well as chromatography, which sometimes is contradictory. Post-column correction of pH, increasing ion strength and/or use of an organic modifier could be ways to overcome this problem.

In APCI, solvent evaporates when passing through a heated tube and a needle producing a corona discharge transfers charges. Dissolved solutes are ionised by gas-phase ion–molecule reactions at atmospheric pressure. APCI can be used with a wide range of solvents and solutes, and is compatible with chromatography in both normal-phase and RP modes. This is of special interest in chiral applications as more non-polar mobile phases are commonly used [60]. The APCI ion source is mass dependent and operates at relatively high flow rates (0.2–2 mL/min). Thus, for sensitive quantitative analyses the traditional 4- to 5-mm bore columns and higher flow rates can be favourable.

The ESI ion source, however, is concentration dependent and operates preferably at lower flow rates (0.05–0.5 mL/min), and thus narrow-bore columns are preferred (i.e. less than 3 mm i.d.). In ESI the solvent is pneumatically nebulised through a stainless steel needle maintained at a high potential, resulting in a spray of highly charged fine droplets. A restriction with ESI is that all analytes must dissolve in a solvent exhibiting moderate conductivity – typically water mixtures with methanol or ACN. Low flow rates can increase the overall ionisation efficiency and thereby improve detection sensitivity.

Even though API-based ion sources use a simple chemical process to ionise molecules, even a slight variation in the mobile-phase composition may greatly influence the response of the analytes. The ideal conditions for sample ionisation required by ESI or APCI can be in conflict with the composition of the mobile phase. Moreover, the ionisation efficiencies can vary over several orders of magnitude, even for compounds that are structurally very similar [29]. Therefore, the ion current cannot be used for quantification of analogues, even if the structure is very similar.

In addition, all of the species involved in the ionisation process compete for the charges available, sometimes unexpectedly suppressing or enhancing the signal response for another species [47, 61]. Suppression is seen more often in ESI than in APCI, where enhancement of the signal seems to be more common. This situation is particular true for samples of unknown composition, where ionisation cannot be optimised for all components. Therefore, in order to achieve robust quantitative LC-MS assays it is important to extensively evaluate chromatography and MEs on ionisation.

The formation of ions also depends on the mobile-phase composition, organic modifier, buffer strength and type. Tyrefors et al. [26] showed that formic acid gave a 10 times higher response than trifluoroacetic acid and that optimum ionisation was achieved at such a low buffer strength as 2–3 mmol/L. The separation power, however, decreases as the buffer strength decreases. Ion formation is also favoured by a high organic percentage of the mobile phase, whereas the separation of analytes will decrease as they are eluted more rapidly. Therefore, a compromise between chromatographic separation and ion intensity has to be made.

In some cases more or less stable adducts or clusters are formed, most commonly sodium adducts. Since sodium ions are present in the laboratory environment (e.g. glassware) they are difficult to avoid. Due to differences in interface design, adduct formation may be more often encountered on some instruments than on others. However, by addition of a salt (e.g. sodium acetate) or dodecylamine/acetic acid to the mobile phase, rugged bioanalytical methods have been developed for analysis with ESI [61–63]. Since interfaces are quite expensive and method development is time-consuming, more extensive evaluation of interfaces is still limited in the literature. However, some published evaluations, e.g. for anabolic steroids [64] and haloperidol [65], give important information on ionisation processes.

Fragmentation and mass analysis

MS is a sensitive method of molecular analysis that has the potential to yield information on the molecular mass as well as the structure of the analyte. In quantitative analysis MS detection ensures a high specificity and sensitivity of the assay. Introduced ions are separated according to their mass-to-charge (m/z) ratio and detected by means of an electron multiplier or another detecting device. Since API is a soft ionisation technique, the molecular ion (also called the parent ion) is commonly present and often used for quantification.

Additional information about the structure of a given compound can be generated by fragmentation, which can be achieved in the ion source by collision-induced dissociation (CID) controlled by voltage settings (so called in-source CID or up-front CID). Moreover, with tandem mass spectrometers, fragmentation can be achieved under more controlled conditions in a collision cell. Here, dissociation is achieved by collisions with gas molecules, and fragmentation is controlled by a combination of gas pressure and voltage settings. The degree of fragmentation can be enhanced by alteration of ion voltages, either in the instrument inlet or in the collision cell, to increase the kinetic energy. Another way to achieve fragmentation is to induce collisions by increasing gas pressure in the collision cell by the introduction of nitrogen or helium. Moreover, increased collision energy can be enhanced by the use of a heavier collision gas, e.g. argon or xenon. However, depending on the design of the instrument, exchange of gases may not be possible.

The molecular ion and fragments generated can be measured by recording a scan for masses over a given mass range or by a selective measurement of one or more selected masses, i.e. selected-ion monitoring (SIM). Since sensitivity is crucial, SIM LC-MS applications dominate in quantitative assays. Due to longer detector dwell times, SIM measurements are generally more sensitive than scans, at least with quadrupole mass analysers. Most often the quasi-molecular ions (e.g. $[M + H]^+$ or $[M - H]^-$) and selected specific ions are used in one-dimensional MS. Since fingerprint information of the analyte identity (full scan mass spectrum) is lost in SIM, co-determination of several specific ions (i.e. qualifiers) will increase the specificity of the assay. Due to markedly improved specificity and signal-to-noise (S/N) ratios, tandem MS has become more popular for quantification. In two-dimensional MS, a transition (SIM/SIM) from the quasi-molecular ion to an intense fragment ion is monitored.

For quantification, quadrupole or ion trap (IT) mass spectrometers in combination with various interfaces can be used. The main difference between the two MS techniques is that there is a continuous flow of ions through the MS in quadrupoles, whereas ions are pulsed through an ion gate in ITs. There are advantages and disadvantages with both techniques, e.g. the nominal MS range is wide on ITs, up to m/z 70 000 (although commercial equipment typically provides similar mass ranges as quadrupoles), whereas quadrupole instruments have a considerably lower m/z range, but unit mass resolution throughout the entire mass range. Generally the dynamic range is wider on quadrupole instruments [66], which also explains why quadrupoles may perform better in quantification (*see* Chapter 1). Narrow dynamic range has traditionally been a severe limitation on the usability of time-of-flight (TOF) MS for quantifications. New instruments are now being introduced with up to three decades or more of dynamic range. Certainly we will see TOF applications for drug screening in the future, but

Fragmentation and mass analysis 55

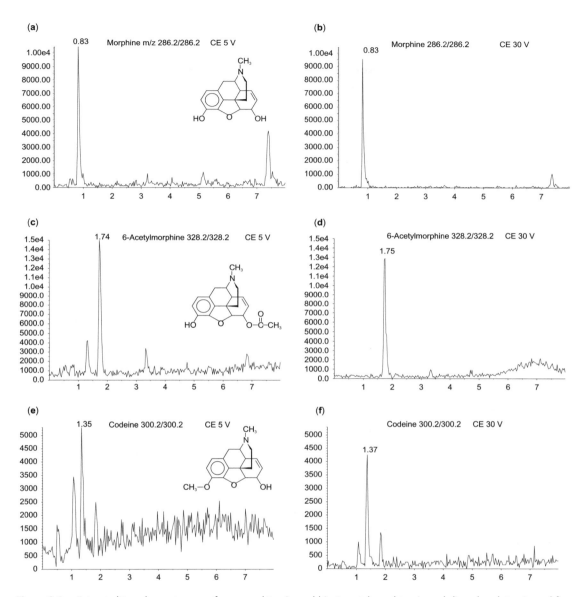

Figure 3.1 Extracted ion chromatograms from morphine (a and b), 6-acetylmorphine (c and d) and codeine (e and f) at collision energies of 5 and 30 eV. The S/N ratio is enhanced, whereas interfering peaks are reduced or absent in the 30-eV chromatograms.

whether there will be numerous applications for quantification purposes is yet to be seen.

A great number of analytes can be quantified in a single run with quadrupole instruments. Moreover, even if the chromatographic resolution is poor and some analytes co-elute, the selectivity of tandem MS (triple quadrupole MS) may compensate for incomplete chromatographic resolution of individual components, allowing a reduction in analytical run time and thus providing higher sample throughput. The discontinuous measurement in ITs enables full-scan spectra to be generated with high sensitivity. This is the biggest strength of the IT technique, together with the ability to perform multiple stages of MS (MS^n). Multi-dimensional MS can be performed as scan or SIM measurements in various combinations.

In quantification, two-dimensional SIM analysis is of particular interest due to the high specificity and high sensitivity achieved. Since the majority of the signal noise is filtered in the first dimension, a high S/N level is often achieved. SIM/SIM analysis is termed selected-reaction monitoring (SRM) or, when more than one reaction (precursor ion/product ion) is monitored, multiple-reaction monitoring (MRM).

For both IT and quadrupole instruments the total duty cycle and scan speed are important factors determining the total number of analytes that can be measured simultaneously or included in the assay, and also the LOQ achieved. Thus, the discontinued measurement of ions in IT instruments can be a drawback for quantifications if the method is complex, i.e. containing a great number of analytes, and/or the chromatography is rapid. However, this could partly be overcome by using several time windows for MS measurement during the chromatography [67].

In all multiple dimensional MS, ions are formed by molecular collisions in the IT or the collision cell. In these applications the speed of collision cell emptying is critical. Leakage of ions between measurements may cause contamination of ions, so called cross-talk, resulting in inaccurate determinations. Cross-talk is when the time between analytes present in the collision cell is too short and some old fragments are still in the cell when new ones enter the cell. Depending on instrumental design, special precautions may have to be taken in method design in order to avoid cross-talk [12]. This is especially true for assays including structurally related compounds. More recently, high-performance hybrid triple–quadrupole/linear IT instruments have been introduced on the market. These instruments combine the features of the two MS techniques and are promising tools for toxicological investigation.

Many of the steps performed to optimise and validate a quantitative method are aimed to secure an accurate and precise measurement of the concentration to be reported. Ion fragmentation is an essential step for any analyte characterisation. Analytes that have few or no intense fragments pose a special problem. Opiates and opioid derivatives have proved hard to fragment in LC-MS. Some of the published methods for buprenorphine deal with the difficulties in obtaining adequate fragmentation for MS-MS [25, 27, 28]. Papers describing methods for single MS detection of buprenorphine [68, 69] have used only the pseudo-molecular ion for quantification and confirmation of identification. The molecular ions of both buprenorphine and norbuprenorphine are readily formed, but seem very stable under varying conditions in the collision cell. This is probably due to the complex ring structure which is able to accommodate the excitation energy to a point where the molecule literally shatters with no significant high mass fragment ions. This problem has been overcome by using the parent ion for quantification, i.e. by letting the molecular ion pass through the mass filters to the detector as in single MS mode, although the instrument is operated under MS-MS conditions (the so-called 'surviving'-ion technique). To accomplish this, collision energy has to be selected low enough to avoid fragmentation of the molecular ion, but high enough to fragment interferent co-eluting compounds. Ceccato *et al.* demonstrated very clearly the difference between MS and MS-MS when choosing to monitor a $M + H^+/M + H^+$ transition. By ranging the collision energy from 0 to 60 eV the S/N could be enhanced six times (from S/N 50 to 300) for buprenorphine [27]. Figure 3.1 shows the chromatograms from morphine, codeine and 6-acetylmorphine as $M + H^+/M + H^+$ transitions at either 5 or 30 eV. Baseline noise, as well as adjacent peaks, are less pronounced at 30 eV and the intensity of the analyte peaks is of the same order as at 5 eV.

Figure 3.2 shows the S/N ratios for the same analytes in a hair extract at collision energies from 5 to 50 eV. The S/N ratio is more than doubled when increasing the collision energy. This approach can be used to lower the detection limits or improve precision for analytes that have stable molecular ions and few or no pronounced daughter ions, although it should be kept in mind that sensitivity is increased at the expense

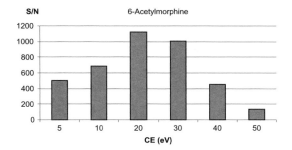
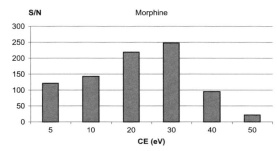
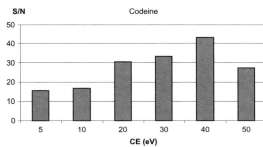

Figure 3.2 S/N ratios for 6-acetylmorphine, morphine and codeine at increasing collision energies (CEs). The high stability of the [M + H]$^+$ ions for opiates can be used to enhance the S/N ratio in tandem MS in a way not possible in single MS.

of selectivity of detection. When choosing the ion formation and separation parameters it is thus not only the intensity of the peak that is important, but also the overall signal compared with the noise.

Determining the concentration

In order to determine the concentration the detector response must be related to the actual concentration of analyte in the sample and this response–concentration relationship has to be characterised before quantification of unknown samples. This is usually referred to as method calibration and involves analysis of a well-defined reference substance added to negative matrix and processed in the same way as the unknown sample. These samples are known as calibrators. Calibrators should be spread out over the whole calibration range. The responses of the calibrators are then plotted against the added concentration of analyte and a function is fitted to the data as shown in Figure 3.3.

By tradition, a linear calibration function has been the first choice, but other, non-linear functions are also commonly used. When changing from linear to other functions one should consider adding new calibration levels as a non-linear function requires more data points to be accurately described (*see* Chapter 4).

In chromatographic methods, the response is normally based on the calculation of either peak areas or peak heights. Peak height is preferable in applications where baseline separation is not achieved and when the concentration range is limited. Peak area measurements have the advantage of being less susceptible to peak broadening than peak height measurements.

External standard calibration curves are performed by injection of processed calibrators prepared with different analyte concentrations in the target matrix and fitting the data to a response function. The analyte concentration in the unknown sample is then estimated by extrapolation of the peak area or height to the corresponding concentration as defined by the response function. When using external standard calibration, it is necessary to be very precise with volumes throughout the whole process in order to sustain precision and accuracy of the quantification. The tedious work of measuring volumes at every transfer of liquids and reconstitution can be avoided by the use of ISs. The concentration in an unknown sample is then estimated in the same way as for external standard calibration with the exception that one compares the response ratio of the analyte and IS between the sample and calibrators. The more extensive the sample preparation is, the higher the need for the addition of an IS to obtain accurate

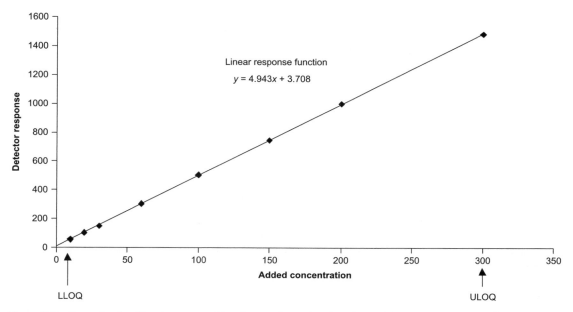

Figure 3.3 Example of calibration curve with a linear relationship ($y = kx + m$) between response and concentration. The lowest and highest calibrators (levels) define the lower LOQ (LLOQ) and upper LOQ (ULOQ), and together define the calibration range. Samples above the ULOQ should be re-analysed using less sample volume to be accurately quantified.

quantification. The IS should be added as soon as possible in the process to compensate for any losses or variations in the pre-treatment, extraction, evaporation step, reconstitution and injection of the sample. Moreover, the IS is also important in LC-MS in order to compensate for the differences in ionisation efficiency. The IS and the analyte should have similar structures and properties. Stokvis et al. [70] suggested that, for MS, ISs that differ in the functional groups are less appropriate than those that differ only in the carbon backbone. Ideally, in MS analysis, the corresponding deuterated (or other stable isotope) analogue is used. A wide range of deuterated analogues is available for drugs of abuse and certain metabolites, whereas there are only few medications and metabolites that have been commercially deuterated. The mobile-phase composition and matrix components are known to affect ionisation, and thus especially in gradient chromatography it is important to have the IS as close to the analyte as possible [15, 25, 71]. However, the deuterated analogues may suffer from ion suppression or enhancement by the analyte at high concentrations and vice versa. Figure 3.4 shows the effect on increasing haloperidol concentration on the area of $[D_4]$haloperidol and other ISs. The decrease in $[D_4]$haloperidol area may be the result of competition between the two co-eluting substances for charged sites on the droplet surface in the ESI ion source. This may have a great impact on the accuracy of other measurements in the same chromatographic method since it is not uncommon that one IS is used for several analytes.

Moreover, other exogenous or endogenous matrix components may also cause ion suppression on the IS. Selective sample preparation together with a selective gradient chromatography may partly overcome this, but a minimum acceptable IS signal should be determined and closely followed. One must know that there is a natural fluctuation in signal intensity in LC-MS since ionisation is highly dependent on interface conditions. By running a system suitability test on a daily basis or doing comparisons with prepared standards or controls in the actual batch, the expected IS signal intensity can be determined [44].

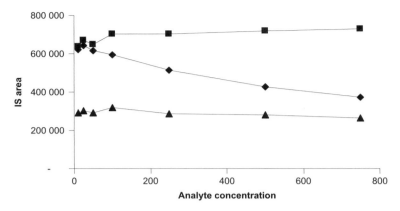

Figure 3.4 IS peak area depending on analyte (haloperidol) concentration: (♦) [D$_4$]haloperidol, (▲) [D$_3$]mianserine or (■) [D$_3$]chlorpromazine. The co-eluting [D$_4$]haloperidol shows a decrease in peak area at increasing haloperidol concentration (micrograms per litre), possibly due to competition between the co-eluting compounds for charged sites on the droplet surface.

Another aspect is that most deuterated analogues have been developed for the GC-EI-MS technique. This may pose some problems in single LC-MS because of the different fragmentation mechanisms in LC-MS than in GC-MS. An example is [D$_5$]fentanyl (M + H$^+$ = 342) for which the most abundant fragment in API-MS is m/z 188, which is also a fragment of fentanyl (M + H$^+$ = 337). Thus, one cannot use m/z 188 either as a quantification ion or as a qualifier in single MS. Because of the difference in molecular ion (M + H$^+$) there is no problem when performing the analysis in MS-MS mode, provided that no cross-talk is present in the collision cell.

Further information on how the selection of the IS impacts on the performance of a quantification method is available in Chapters 2 and 4.

When a number of analytes are included in the assay it may not be possible and certainly not cost-effective to use a deuterated standard for each analyte. However, it may be used in methods that will be used fairly seldom, as Wood et al. [16] described in their application for 10 benzodiazepines in larvae, where each analyte had its own deuterated IS providing very low variations as well as linear calibration curves. In general, if several analytes are included in a method, we recommend use of a limited number of ISs with a different polarity spread over the entire chromatographic run. If an IS is co-eluting with any analyte, dual quantification based on different ISs can be useful and may reveal ion suppression.

In an LC-MS-MS assay developed at the laboratory of the authors for the determination of 11 benzodiazepines and selected metabolites, a limited number of ISs are used. The five deuterated analogues used represent benzodiazepines of different polarity ([D$_5$]alprazolam, [D$_5$]nordiazepam and [D$_5$]oxazepam) and metabolites with different substituents ([D$_7$]7-aminoflunitrazepam and [D$_5$]α-hydroxy-alprazolam), and are suitable for quality control of the performance of sample preparation, chromatography, ionisation and mass detection. Linear calibration curves are achieved for all analytes and the performance of the method is followed by determination of control samples at two levels. The analytes that have their own corresponding deuterated IS generally have lower imprecision over time, illustrating the impact of deuterated ISs on method performance.

IS responses are also evaluated for each sample using a pre-defined threshold area as responses below the threshold may be caused by ion suppression. The most common reason for suppression is a high concentration of the co-eluting benzodiazepine that corresponds to the IS.

Dynamic range and response function

In toxicological analysis, quantification at toxic, therapeutic as well as subtherapeutic concentrations is often the objective. For this purpose assays covering concentrations of two to three magnitudes is desired. However, the dynamic range may be a limiting factor for quantification covering more than two magnitudes. It is often very difficult to extend the dynamic range; instead, one may consider re-running samples with very high concentrations using less sample for extraction to obtain a concentration within the range.

The dynamic range is the range where a correlation between concentration and signal intensity is achieved. The instrumentation used as well as the method set-up determine the dynamic range of a quantitative assay. In general, ITs have a more limited dynamic range compared with quadrupole instruments due to the discontinuous measurements of ions in the ITs. Moreover, the multiplier units have a limited dynamic range. Another important factor for the dynamic range is the ionisation process, where saturation is seen at higher concentrations due to competition of charges.

Figure 3.5 shows calibration curves for haloperidol with [D$_4$]haloperidol, [D$_3$]mianserine or [D$_3$]chlorpromazine as ISs. [D$_3$]Mianserine elutes 1 min before haloperidol and [D$_3$]chlorpromazine 1 min after, with complete baseline separation, and both give calibration curves with a significant curvature. The curvature is a consequence of the limited dynamic range of the assay, in this case mostly dependent on ion suppression effects in the ion source.

The differences in signal ratio (analyte/IS) observed between curves a–c depends on the difference in absolute signal intensity of the three different ISs where [D$_3$]chlorpromazine shows the most abundant signal (*see also* Figure 3.4). Interestingly, the calibration function is linear between 10 and 750 µg/L using the deuterated IS. This shows that some of the saturation effects normally seen can be compensated, but in a limited concentration range, by means of a deuterated IS. Since a deuterated analogue only differs slightly in its physical properties it co-elutes with its non-deuterated analogue. When the ionisation approaches saturation a proportional decrease in ionisation of the IS can compensate for the loss in ionisation of the analyte over a limited range. This is true for [D$_4$]haloperidol as shown in Figures 3.4 and 3.5. At higher concentration saturations these effects cannot be avoided and calibrations curves tend to be best fitted by non-linear equations (and also with deuterated ISs). Quadratic (or other non-linear) calibration functions will affect the precision in the upper range, as the change in response is small compared with the change in concentration. Secondly, one cannot extrapolate the calibration curve beyond the highest calibrator as one may (although not for accurate quantification) when the function is linear. Thirdly, the intercept, and the lower range, are affected by curvilinear functions. These effects can be reduced by weighting of the calibration curves and may still be better than when applying linear calibration curves [72]. Depending on the software available, the possibilities to choose curve-fit algorithms varies; linear, quadratic and power fits are usually sufficient. Changing the IS concentration will affect both the linear range and the detection limit as shown by Liang *et al.* [49] in their ion-suppression experiments. At low analyte concentrations (less than 50 times that of the IS) the signal was totally suppressed. A simple test is to perform calibrations with different amounts of IS and without IS. Comparing variations in accuracy between different calibration curves (and levels) will give information about the linear range and detection limit. Ion suppression is calculated by dividing the analyte area with IS present by that without IS. By adjusting the IS concentration to the lower part of the range one might improve method sensitivity as the ion suppression from the IS will be less pronounced.

Historically, it has been the custom in LC to perform daily calibrations due to detector instability and this good practice is still in use when the detectors have changed to MS [13, 23]. However, it is still possible to use historical calibrations in both single and tandem LC-MS. In the author's laboratory quantification of THC-COOH in urine is performed by single MS after SPE. Briefly, 1 mL of urine is hydrolysed with saturated sodium hydroxide in methanol at 60°C. The pH is adjusted to 3 with acetic acid, the samples are

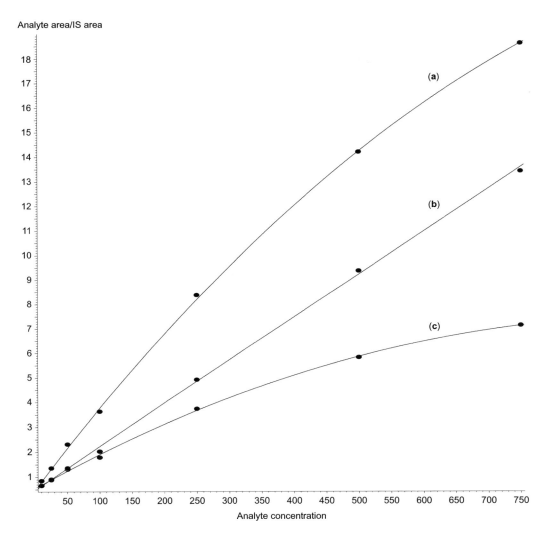

Figure 3.5 Calibration curves for haloperidol using [D$_4$]haloperidol, [D$_3$]mianserine or [D$_3$]chlorpromazine as ISs. Concentration range 10–750 µg/L. (a) [D$_3$]Mianserine: $y = -1.33\text{e-}005x^2 + 0.0342x + -0.0344$ ($r = 0.9996$). (b) [D$_4$]Haloperidol: $y = 0.0175x + -0.0337$ ($r = 0.9996$). (c) [D$_3$]Chlorpromazine: $y = -7.42\text{e-}006x^2 + 0.0144x + 0.0318$ ($r = 0.9997$).

extracted using an Aspec XL4 and the analyte is eluted in mobile phase with the addition of 30% diethyl ether. After evaporation of the ether, 10 µL is injected into the chromatographic system, which is run in isocratic mode with a 3-min run time. The method is normally calibrated every third month depending on the outcome of method controls run with each batch. Both calibrators and controls are made of certified reference material, enabling monitoring of accuracy through controls. Table 3.2 shows the precision and accuracy of controls at four levels during a 3-month period. During this period the controls showed coefficients of variation (CV) of less than 12% and accuracies of 89–98%.

In Figure 3.6, the 100 µg/L control plot depicts

Table 3.2 Control statistics for THC-COOH over a 3-month period

THC-COOH	n	Mean (µg/L)	Accuracy (%)	Precision (%)
5 µg/L	98	4.44	89	11.9
20 µg/L	100	18.33	91	8.5
100 µg/L	100	97.8	98	6.9
Hydrolysis control (approximately 330 µg/L)	100	328	–	6.2

The hydrolysis control is pooled from authentic samples and diluted to approximately 330 µg/L. During the period about 400 controls and 3000 authentic samples were analysed with the method, thus approximately 10% of the workload was quality controls.

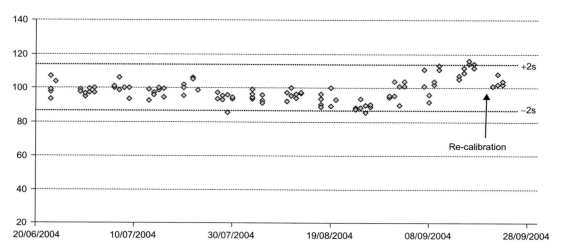

Figure 3.6 Between-day variation of a urine batch control at 100 µg/L THC-COOH over 3 months. The controls were quantified against a historical calibration curve. The dotted lines illustrate 2 standard deviations from the mean at a precision of 7% (see Table 3.2). An upward concentration trend is seen at the later part of the diagram and was the criterion for re-calibration as the controls cross the +2 standard deviations (+2s) line.

an upward trend in the later part indicating that it is time for re-calibration. This strategy demands the use of well-defined reference material not only for calibrators, but also for controls, otherwise the accuracy cannot be monitored over time. Also, the analyst has to continuously monitor the method performance through the control results. Historical calibration curves are often run in duplicates or triplicates at six to eight levels evenly spread out over the entire range, whereas daily calibrations can be less laboursome with few or even single-point calibrators if a linear relationship has been established during method validation.

Validation

This section briefly reviews the main aspects of method validation; the reader is referred to Chapter 4 for a thorough discussion on method validation.

The two most important criteria for evaluating a quantitative method are accuracy and precision, defining both the quality of the method and of the results. Sensitivity, selectivity, recovery and stability are terms that all have impact on accuracy and precision, and are also parameters necessary to evaluate during method validation. Accuracy and precision, however, should be evaluated continuously.

Accuracy

Accuracy is defined as the agreement between the measured value and the true value for a given analyte. Usually accuracy is reported as % Accuracy [(measured/true) × 100] or as % Bias [(measured − true/true) × 100], where bias is also known as the systematic error. As the accuracy of a method is affected by systematic errors it is best evaluated through inter-laboratory comparisons. For quantitative measurements of drugs in biological fluids there are numerous external quality assessment programmes available. However, if there are no programmes available, the laboratory may still test their method against a standard method or, as is common, by adding a certified reference standard to drug-free matrix and evaluating the closeness of the measured value to the theoretical (or true) value. Accuracy should be tested over the whole calibration range as the calibration function or detector response may vary non-linearly. A deviation of less than 15% from the true value is generally accepted, but at the lower LOQ the deviation should not exceed 20% [73].

Precision

Precision, as opposed to accuracy, does not relate to the true value, but rather to the scattering of repeated measurements around the mean value of those measurements. It is often expressed as the percentage coefficient of variation (%CV) calculated as [(standard deviation/mean value) × 100] and should be less than 15% [3, 73]. In addition to defining the precision during method validation, all quantitative methods should be continuously evaluated by analysing quality control samples. Controls are drug-free matrix with the addition of a known amount of drug, stored frozen in aliquots or prepared daily and analysed in each batch of authentic samples. Using aliquots of an authentic sample to determine the precision is also recommended, at least during the validation. In case a drug is conjugated in plasma or urine and the method includes a hydrolysis step, it is strongly recommended that a hydrolysis control be processed together with every batch of samples. In these cases, an authentic sample may be the only available source as those metabolites are seldom found as quantitative standards. It is generally agreed that about 10% of the batch should be control samples with varying concentrations spread out over the calibration range. The pre-treatment and extraction process usually accounts for the large part of the imprecision, and the use of an appropriate IS can reduce this significantly. However, one should not ignore the effect of chromatographic selectivity nor data acquisition and data manipulations such as bunching and smoothing as they have impact on peak shape and integration.

Selectivity and sensitivity

Sensitive quantitative assays for analytes present in a complex matrix are needed in toxicological analysis. In addition, the assay must be able to distinguish between a great number of structurally related analogues or endogenous compounds, and trends to more selective assays using multi-dimensional MS are seen.

Selectivity is the ability of a method to distinguish the analyte of interest from other components in the sample. When the selectivity reaches 100% the method is called *specific*, i.e. it produces a response for the analyte only [74].

Sensitivity is defined as the lowest concentration of analyte that is distinguished from the background noise. In method validation it is usually referred to by the 'limit of detection' (LOD). There are several ways to determine the LOD – one of the more widely used is to analyse lower and lower concentrations of analyte, and calculate the S/N ratio that should equal 3 at the LOD. Selectivity may also refer to a method's ability to produce a difference in response at different analyte concentrations, where a sensitive method produces a large change in response for a small change in concentration.

Selectivity and sensitivity can be achieved in several ways, e.g. during the extraction procedure, chromatography and detection. Selectivity is closely connected to sensitivity and one or other may be sacrificed depending on the particular demands, e.g. using a low-recovery extraction to gain selectivity or to use lower resolution at the detector to achieve higher sensitivity.

When evaluating selectivity one can

independently investigate possible interferences from the processing of samples and those originating from endogenous or other substances present in the samples. As a rule, one should use matrices from different sources to investigate inter-sample variations. At least 10 different sources should be used.

Recovery

Recovery is an important parameter in quantitative methods and high recovery is often desirable because of low concentrations of the analytes in the matrix. Causon [3] suggested that recoveries above 50% may be sufficient provided that good precision and sensitivity are still obtained with such a moderate recovery. Recovery can be estimated as absolute recovery by comparing the response of processed standards to that of unprocessed standards dissolved in the injection medium. Generally, recovery including the matrix components is also estimated by comparing processed standards with extracts of negative standards where analyte has been added prior to injection. If several steps are included in the processing of samples, one may estimate recovery over each of them to gain a better understanding of the process and to facilitate troubleshooting.

Stability

To be able to draw conclusions from the concentration of any analyte, one must ensure that analyte integrity has been maintained from sampling through analysis. Stability tests typically include short-term stability in the matrix, in-process stability, extract stability, long-term stability and freeze/thaw stability. There are several papers dealing with the aspects of testing stability that the reader can consult for further information [3, 75, 76].

Concluding remarks

A primary concern in toxicological analysis is that it should provide quantitative results for use in the evaluation of effects observed in the patient. Quantification of drugs in biological specimens requires well-defined methods. However, the quality of the results depends not only on analytical performance, but also on many other factors. As described in this chapter, all the steps from the sample integrity to mass detection are of paramount importance for the interpretation of results. Therefore, setting up a quantitative method is a matter of strategy. Before starting it is important to clearly define the purpose and scope of the method, defining sampling and storage conditions, what analytes to monitor, the number of samples to analyse, how often the method will be used, and what turn-around time is acceptable or desired.

Thereafter, evaluation of extraction, chromatography, ionisation and mass detection, individually as well as together, is needed for a successful result. Added to this is the proper choice of ISs, preparation of calibrators, calibration curve fit and periodicity together with an adequate quality control system.

References

1. Shah VP, Midha KK, Dighe S, et al. Analytical methods validation: bioavailability, bioequivalence and pharmacokinetic studies. Conference report. *Pharm Res* 1992; 9: 588–592.

2. Shah VP, Midha KK, Findlay JW, et al. Bioanalytical method validation – a revisit with a decade of progress. *Pharm Res* 2000; 17: 1551–1557.

3. Causon R. Validation of chromatographic methods in biomedical analysis. Viewpoint and discussion. *J Chromatogr B* 1997; 689: 175–180.

4. Beike J, Ortmann C, Meiners T, et al. LC-MS determination of oxydemeton-methyl and its main metabolite demeton-*S*-methylsulfon in biological specimens – application to a forensic case. *J Anal Toxicol* 2002; 26: 308–312.

5. Beike J, Karger B, Meiners T, et al. LC-MS determination of *Taxus* alkaloids in biological specimens. *Int J Legal Med* 2003; 117: 335–339.

6. Romhild W, Krause D, Bartels H, et al. LC-MS/MS analysis of pholedrine in a fatal intoxication case. *Forensic Sci Int* 2003; 133: 101–106.

7. Jones GR, Singer PP, Bannach B. Application of LC-MS analysis to a colchicine fatality. *J Anal Toxicol* 2002; 26: 365–369.

8. Arao T, Fuke C, Takaesu H, et al. Simultaneous determination of cardenolides by sonic spray ionization liquid chromatography-ion trap mass spectrometry – a fatal case of oleander poisoning. *J Anal Toxicol* 2002; 26: 222–227.

9. Dumestre-Toulet V, Cirimele V, Gromb S, et al. Last performance with Viagra: post-mortem identification of sildenafil and its metabolites in biological specimens including hair sample. *Forensic Sci Int* 2002; 126: 71–76.

10. Weinmann W, Bohnert M, Wiedemann A, et al. Post-mortem detection and identification of sildenafil (Viagra) and its metabolites by LC/MS and LC/MS/MS. *Int J Legal Med* 2001; 114: 252–258.

11. Hoja H, Marquet P, Verneuil B, et al. Determination of buprenorphine and norbuprenorphine in whole blood by liquid chromatography-mass spectrometry. *J Anal Toxicol* 1997; 21: 160–165.

12. Gergov M, Ojanpera I, Vuori E. Simultaneous screening for 238 drugs in blood by liquid chromatography-ion spray tandem mass spectrometry with multiple-reaction monitoring. *J Chromatogr B* 2003; 795: 41–53.

13. Kratzsch C, Peters FT, Kraemer T, et al. Screening, library-assisted identification and validated quantification of fifteen neuroleptics and three of their metabolites in plasma by liquid chromatography/mass spectrometry with atmospheric pressure chemical ionization. *J Mass Spectrom* 2003; 38: 283–295.

14. Kratzsch C, Tenberken O, Peters FT, et al. Screening, library-assisted identification and validated quantification of 23 benzodiazepines, flumazenil, zaleplone, zolpidem and zopiclone in plasma by liquid chromatography/mass spectrometry with atmospheric pressure chemical ionization. *J Mass Spectrom* 2004; 39: 856–872.

15. Wood M, De Boeck G, Samyn N, et al. Development of a rapid and sensitive method for the quantitation of amphetamines in human plasma and oral fluid by LC-MS-MS. *J Anal Toxicol* 2003; 27: 78–87.

16. Wood M, Laloup M, Pien K, et al. Development of a rapid and sensitive method for the quantitation of benzodiazepines in *Calliphora vicina* larvae and puparia by LC-MS-MS. *J Anal Toxicol* 2003; 27: 505–512.

17. Clarke NJ, Rindgen D, Korfmacher WA, et al. Systematic LC/MS metabolite identification in drug discovery. *Anal Chem* 2001; 73: 430A–439A.

18. Stanley SM. Equine metabolism of buspirone studied by high-performance liquid chromatography/mass spectrometry. *J Mass Spectrom* 2000; 35: 402–407.

19. Kerns EH, Rourick RA, Volk KJ, et al. Buspirone metabolite structure profile using a standard liquid chromatographic-mass spectrometric protocol. *J Chromatogr B* 1997; 698: 133–145.

20. Murphy AT, Lake BG, Bernstein JR, et al. Characterization of olanzapine (LY170053) in human liver slices by liquid chromatography/tandem mass spectrometry. *J Mass Spectrom* 1998; 33: 1237–1245.

21. von Euler M, Villen T, Svensson JO, et al. Interpretation of the presence of 6-monoacetylmorphine in the absence of morphine-3-glucuronide in urine samples: evidence of heroin abuse. *Ther Drug Monit* 2003; 25: 645–648.

22. Charles BK, Day JE, Rollins DE, et al. Opiate recidivism in a drug-treatment program: comparison of hair and urine data. *J Anal Toxicol* 2003; 27: 412–428.

23. Naidong W, Lee JW, Jiang X, et al. Simultaneous assay of morphine, morphine-3-glucuronide and morphine-6-glucuronide in human plasma using normal-phase liquid chromatography-tandem mass spectrometry with a silica column and an aqueous organic mobile phase. *J Chromatogr B* 1999; 735: 255–269.

24. Weinmann W, Schaefer P, Thierauf A, et al. Confirmatory analysis of ethylglucuronide in urine by liquid-chromatography/electrospray ionization/tandem mass spectrometry according to forensic guidelines. *J Am Soc Mass Spectrom* 2004; 15: 188–1893.

25. Polettini A, Huestis MA. Simultaneous determination of buprenorphine, norbuprenorphine, and buprenorphine-glucuronide in plasma by liquid chromatography-tandem mass spectrometry. *J Chromatogr B* 2001; 754: 447–459.

26. Tyrefors N, Hyllbrant B, Ekman L, et al. Determination of morphine, morphine-3-glucuronide and morphine-6-glucuronide in human serum by solid-phase extraction and liquid chromatography-mass spectrometry with electrospray ionisation. *J Chromatogr A* 1996; 729: 279–285.

27. Ceccato A, Klinkenberg R, Hubert P, et al. Sensitive determination of buprenorphine and its N-dealkylated metabolite norbuprenorphine in human plasma by liquid chromatography

coupled to tandem mass spectrometry. *J Pharm Biomed Anal* 2003; 32: 619–631.

28. Kronstrand R, Seldén TG, Josefsson M. Analysis of buprenorphine, norbuprenorphine, and their glucuronides in urine by liquid chromatography-mass spectrometry. *J Anal Toxicol* 2003; 27: 464–470.

29. Josefsson M, Kronstrand R, Andersson J, et al. Evaluation of electrospray ionisation liquid chromatography-tandem mass spectrometry for rational determination of a number of neuroleptics and their major metabolites in human body fluids and tissues. *J Chromatogr B* 2003; 789: 151–167.

30. Aravagiri M, Marder SR. Simultaneous determination of risperidone and 9-hydroxyrisperidone in plasma by liquid chromatography/electrospray tandem mass spectrometry. *J Mass Spectrom* 2000; 35: 718–724.

31. Aravagiri M, Marder SR. Simultaneous determination of clozapine and its N-desmethyl and N-oxide metabolites in plasma by liquid chromatography/electrospray tandem mass spectrometry and its application to plasma level monitoring in schizophrenic patients. *J Pharm Biomed Anal* 2001; 26: 301–311.

32. Maurer HH, Kratzsch C, Weber AA, et al. Validated assay for quantification of oxcarbazepine and its active dihydro metabolite 10-hydroxycarbazepine in plasma by atmospheric pressure chemical ionization liquid chromatography/mass spectrometry. *J Mass Spectrom* 2002; 37: 687–692.

33. Weinmann W, Goerner M, Vogt S, et al. Fast confirmation of 11-nor-9-carboxy-delta(9)-tetrahydrocannabinol (THC-COOH) in urine by LC/MS/MS using negative atmospheric-pressure chemical ionisation (APCI). *Forensic Sci Int* 2001; 121: 103–107.

34. Breindahl T, Andreasen K. Determination of 11-nor-delta9-tetrahydrocannabinol-9-carboxylic acid in urine using high-performance liquid chromatography and electrospray ionization mass spectrometry. *J Chromatogr B* 1999; 732: 155–164.

35. Klette KL, Horn CK, Stout PR, et al. LC-MS analysis of human urine specimens for 2-oxo-3-hydroxy LSD: method validation for potential interferants and stability study of 2-oxo-3-hydroxy LSD under various storage conditions. *J Anal Toxicol* 2002; 26: 193–200.

36. Maralikova B, Weinmann W. Simultaneous determination of delta9-tetrahydrocannabinol, 11-hydroxy-delta9-tetrahydrocannabinol and 11-nor-9-carboxy-delta9-tetrahydrocannabinol in human plasma by high-performance liquid chromatography/tandem mass spectrometry. *J Mass Spectrom* 2004; 39: 526–531.

37. Moore KA, Sklerov J, Levine B, et al. Urine concentrations of ketamine and norketamine following illegal consumption. *J Anal Toxicol* 2001; 25: 583–588.

38. Rule G, Chapple M, Henion J. A 384-well solid-phase extraction for LC/MS/MS determination of methotrexate and its 7-hydroxy metabolite in human urine and plasma. *Anal Chem* 2001; 73: 439–443.

39. Weinmann W, Vogt S, Goerke R, et al. Simultaneous determination of THC-COOH and THC-COOH-glucuronide in urine samples by LC/MS/MS. *Forensic Sci Int* 2000; 113: 381–387.

40. Blanchard J. Evaluation of the relative efficacy of various techniques for deproteinizing plasma samples prior to high-performance liquid chromatographic analysis. *J Chromatogr* 1981; 226: 455–460.

41. Moore CM. Principles of solid-phase extraction. *AACC TDM/Tox* 1994; 15: 205–214.

42. Souverain S, Eap C, Veuthey JL, et al. Automated LC-MS method for the fast stereoselective determination of methadone in plasma. *Clin Chem Lab Med* 2003; 41: 1615–1621.

43. Nordgren HK, Beck O. Multicomponent screening for drugs of abuse: direct analysis of urine by LC-MS-MS. *Ther Drug Monit* 2004; 26: 90–97.

44. Flarakos J, Luo W, Aman M, et al. Quantification of risperidone and 9-hydroxyrisperidone in plasma and saliva from adult and pediatric patients by liquid chromatography-mass spectrometry. *J Chromatogr A* 2004; 1026: 175–183.

45. Jemal M, Ouyang Z, Powell ML. Direct-injection LC-MS-MS method for high-throughput simultaneous quantitation of simvastatin and simvastatin acid in human plasma. *J Pharm Biomed Anal* 2000; 23: 323–240.

46. Sayer H, Quintela O, Marquet P, et al. Identification and quantitation of six non-depolarizing neuromuscular blocking agents by LC-MS in biological fluids. *J Anal Toxicol* 2004; 28: 105–110.

47. Annesley TM. Ion suppression in mass spectrometry. *Clin Chem* 2003; 49: 1041–1044.

48. Mallet CR, Lu Z, Mazzeo JR. A study of ion suppression effects in electrospray ionization from mobile phase additives and solid-phase extracts. *Rapid Commun Mass Spectrom* 2004; 18: 49–58.

49. Liang HR, Foltz RL, Meng M, et al. Ionization enhancement in atmospheric pressure chemical ionization and suppression in electrospray ionization between target drugs and stable-isotope-labeled internal standards in quantitative liquid chromatography/tandem mass spectrometry. *Rapid Commun Mass Spectrom* 2003; 17: 2815–2821.

50. Souverain S, Rudaz S, Veuthey JL. Protein precipitation for the analysis of a drug cocktail in plasma by LC-ESI-MS. *J Pharm Biomed Anal* 2004; 35: 913–920.

51. Müller C, Schafer P, Stortzel M, et al. Ion suppression effects in liquid chromatography-electrospray-ionisation transport-region collision induced dissociation mass spectrometry with different serum extraction methods for systematic toxicological analysis with mass spectra libraries. *J Chromatogr B* 2002; 773: 47–52.

52. Dams R, Huestis MA, Lambert WE, et al. Matrix effect in bioanalysis of illicit drugs with LC-MS/MS: influence of ionization type, sample preparation, and biofluid. *J Am Soc Mass Spectrom* 2003; 14: 1290–1294.

53. Mei H, Hsieh Y, Nardo C, et al. Investigation of matrix effects in bioanalytical high performance liquid chromatography/tandem mass spectrometric assays: application to drug discovery. *Rapid Commun Mass Spectrom* 2003; 17: 97–103.

54. Rossi DT, Zhang NJ. Automating solid-phase extraction: current aspects and future prospects. *J Chromatogr A* 2000; 885: 97–113.

55. Zweigenbaum J, Heinig K, Steinborner S, et al. High-throughput bioanalytical LC/MS/MS determination of benzodiazepines in human urine: 1000 samples per 12 hours. *Anal Chem* 1999; 71: 2294–2300.

56. Onorato JM, Henion JD, Lefebvre PM, et al. Selected reaction monitoring LC-MS determination of idoxifene and its pyrrolidinone metabolite in human plasma using robotic high-throughput, sequential sample injection. *Anal Chem* 2001; 73: 119–125.

57. Hempenius J, Steenvoorden RJ, Lagerwerf FM, et al. 'High throughput' solid-phase extraction technology and turbo ionspray LC-MS-MS applied to the determination of haloperidol in human plasma. *J Pharm Biomed Anal* 1999; 20: 889–898.

58. Tarr MA, Zhu J, Cole RB. Atmospheric pressure ionization mass spectrometry. In: Holyoake A, ed., *Encyclopedia of Analytical Chemistry*. New York: Wiley-Interscience, 2000: 11597–11623.

59. Wang G, Cole RB. Solution, gas-phase, and instrumental parameter influences on charge-state distributions in electrospray ionization mass spectrometry. In: Cole RB, ed., *Electrospray Ionization Mass Spectrometry*. New York: Wiley-Interscience, 1997: 137–174.

60. Bakhtiar R, Ramos L, Tse FL. Use of atmospheric pressure ionization mass spectrometry in enantioselective liquid chromatography. *Chirality* 2001; 13: 63–74.

61. Lambert W. Pitfalls in LC-MS-(MS) analysis. *Bull Int Ass Forensic Toxicol* 2004; 34: 59–61.

62. Jemal M, Almond RB, Teitz DS. Quantitative bioanalysis utilizing high-performance liquid chromatography/electrospray mass spectrometry via selected-ion monitoring of the sodium ion adduct [M + Na]+. *Rapid Commun Mass Spectrom* 1997; 11: 1083–1088.

63. Mortier KA, Zhang GF, van Peteghem CH, et al. Adduct formation in quantitative bioanalysis: effect of ionization conditions on paclitaxel. *J Am Soc Mass Spectrom* 2004; 15: 585–592.

64. Leinonen A, Kuuranne T, Kostiainen R. Liquid chromatography/mass spectrometry in anabolic steroid analysis – optimization and comparison of three ionization techniques: electrospray ionization, atmospheric pressure chemical ionization and atmospheric pressure photoionization. *J Mass Spectrom* 2002; 37: 693–698.

65. Arinobu T, Hattori H, Seno H, et al. Comparison of SSI with APCI as an interface of HPLC-mass spectrometry for analysis of a drug and its metabolites. *J Am Soc Mass Spectrom* 2002; 13: 204–208.

66. Jonscher KR, Yates JR, 3rd. The quadrupole ion trap mass spectrometer – a small solution to a big challenge. *Anal Biochem* 1997; 244: 1–15.

67. Dams R, Murphy CM, Choo RE, et al. LC-atmospheric pressure chemical ionization-MS/MS analysis of multiple illicit drugs, methadone, and their metabolites in oral fluid following protein precipitation. *Anal Chem* 2003; 75: 798–804.

68. Tracqui A, Kintz P, Mangin P. HPLC/MS determination of buprenorphine and norbuprenorphine

in biological fluids and hair samples. *J Forensic Sci* 1997; 42: 111–114.

69. Tracqui A, Kintz P, Ludes B. Buprenorphine-related deaths among drug addicts in France: a report on 20 fatalities. *J Anal Toxicol* 1998; 22: 430–434.

70. Stokvis E, Rosing H, Beijnen JH. Stable isotopically labeled internal standards in quantitative bioanalysis using liquid chromatography/mass spectrometry: necessity or not? *Rapid Commun Mass Spectrom* 2005; 19: 401–407.

71. Mortier KA, Clauwaert KM, Lambert WE, *et al.* Pitfalls associated with liquid chromatography/ electrospray tandem mass spectrometry in quantitative bioanalysis of drugs of abuse in saliva. *Rapid Commun Mass Spectrom* 2001; 15: 1773–1775.

72. Mortier KA, Dams R, Lambert WE, *et al.* Determination of paramethoxyamphetamine and other amphetamine-related designer drugs by liquid chromatography/sonic spray ionization mass spectrometry. *Rapid Commun Mass Spectrom* 2002; 16: 865–870.

73. Bressolle F, Bromet PM, Audran M. Validation of liquid chromatographic and gas chromatographic methods. Applications to pharmacokinetics. *J Chromatogr B* 1996; 686: 3–10.

74. Wessman J. Selectivity or sensitivity? Validation of analytical methods from the perspective of an analytical chemist in the pharmaceutical industry. *J Pharm Biomed Appl* 1996; 14: 867–869.

75. Dadgar D, Burnett PE, Choc MG, *et al.* Application issues in bioanalytical method validation, sample analysis and data reporting. *J Pharm Biomed Anal* 1995; 13: 89–97.

76. Dadgar D, Burnett PE. Issues in evaluation of bioanalytical method selectivity and drug stability. *J Pharm Biomed Anal* 1995; 14: 23–31.

77. Kataoka H, Lord HL, Pawliszyn J. Simple and rapid determination of amphetamine, methamphetamine, and their methylenedioxy derivatives in urine by automated in-tube solid-phase microextraction coupled with liquid chromatography-electrospray ionization mass spectrometry. *J Anal Toxicol* 2000; 24: 257–265.

78. Pichini S, Pacifici R, Pellegrini M, *et al.* Development and validation of a high-performance liquid chromatography-mass spectrometry assay for determination of amphetamine, methamphetamine, and methylenedioxy derivatives in meconium. *Anal Chem* 2004; 76: 2124–2132.

79. Miki A, Katagi M, Tsuchihashi H. Determination of methamphetamine and its metabolites incorporated in hair by column-switching liquid chromatography-mass spectrometry. *J Anal Toxicol* 2003; 27: 95–102.

80. Bogusz MJ, Maier RD, Erkens M, *et al.* Determination of morphine and its 3- and 6-glucuronides, codeine, codeine-glucuronide and 6-monoacetyl-morphine in body fluids by liquid chromatography atmospheric pressure chemical ionization mass spectrometry. *J Chromatogr B* 1997; 703: 115–127.

81. Shou WZ, Pelzer M, Addison T, *et al.* An automatic 96-well solid phase extraction and liquid chromatography-tandem mass spectrometry method for the analysis of morphine, morphine-3-glucuronide and morphine-6-glucuronide in human plasma. *J Pharm Biomed Anal* 2002; 27: 143–152.

82. Liang HR, Foltz RL, Meng M, *et al.* Method development and validation for quantitative determination of methadone enantiomers in human plasma by liquid chromatography/ tandem mass spectrometry. *J Chromatogr B* 2004; 806: 191–198.

83. Koch DE, Isaza R, Carpenter JW, *et al.* Simultaneous extraction and quantitation of fentanyl and norfentanyl from primate plasma with LC/MS detection. *J Pharm Biomed Anal* 2004; 34: 577–584.

84. Dams R, Murphy CM, Lambert WE, *et al.* Urine drug testing for opioids, cocaine, and metabolites by direct injection liquid chromatography/ tandem mass spectrometry. *Rapid Commun Mass Spectrom* 2003; 17: 1665–1670.

85. Xia Y, Wang P, Bartlett MG, *et al.* An LC-MS-MS method for the comprehensive analysis of cocaine and cocaine metabolites in meconium. *Anal Chem* 2000; 72: 764–771.

86. Jeanville PM, Estape ES, Needham SR, *et al.* Rapid confirmation/quantitation of cocaine and benzoylecgonine in urine utilizing high performance liquid chromatography and tandem mass spectrometry. *J Am Soc Mass Spectrom* 2000; 11: 257–263.

87. Klingmann A, Skopp G, Aderjan R. Analysis of cocaine, benzoylecgonine, ecogonine methyl ester, and ecgonine by high-pressure liquid

chromatography-API mass spectrometry and application to a short-term degradation study of cocaine in plasma. *J Anal Toxicol* 2001; 25: 425–430.

88. Clauwaert K, Decaestecker T, Mortier K, et al. The determination of cocaine, benzoylecgonine, and cocaethylene in small-volume oral fluid samples by liquid chromatography-quadrupole-time-of-flight mass spectrometry. *J Anal Toxicol* 2004; 28: 655–659.

89. Tatsuno M, Nishikawa M, Katagi M, et al. Simultaneous determination of illicit drugs in human urine by liquid chromatography-mass spectrometry. *J Anal Toxicol* 1996; 20: 281–286.

90. Kronstrand R, Nyström I, Strandberg J, et al. Screening for drugs of abuse in hair with ion spray LC-MS-MS. *Forensic Sci Int* 2004; 145: 183–190.

91. Xu AS, Peng LL, Havel JA, et al. Determination of nicotine and cotinine in human plasma by liquid chromatography-tandem mass spectrometry with atmospheric-pressure chemical ionization interface. *J Chromatogr B* 1996; 682: 249–257.

92. Bernert JT, Turner WE, Pirkle JL, et al. Development and validation of sensitive method for determination of serum cotinine in smokers and nonsmokers by liquid chromatography/atmospheric pressure ionization tandem mass spectrometry. *Clin Chem* 1997; 43: 2281–2291.

93. Xu A, Havel J, Linderholm K, et al. Development and validation of an LC/MS/MS method for the determination of L-hyoscyamine in human plasma. *J Pharm Biomed Anal* 1995; 14: 33–42.

94. Kronstrand R, Nyström I, Josefsson M, et al. Segmental ion spray LC-MS-MS analysis of benzodiazepines in hair of psychiatric patients. *J Anal Toxicol* 2002; 26: 479–484.

95. Smink BE, Brandsma JE, Dijkhuizen A, et al. Quantitative analysis of 33 benzodiazepines, metabolites and benzodiazepine-like substances in whole blood by liquid chromatography-(tandem) mass spectrometry. *J Chromatogr B* 2004; 811: 13–20.

96. Crouch DJ, Rollins DE, Canfield DV, et al. Quantitation of alprazolam and alpha-hydroxyalprazolam in human plasma using liquid chromatography electrospray ionization MS-MS. *J Anal Toxicol* 1999; 23: 479–485.

97. Tas AC, van der Greef J, ten Noever de Brauw MC, et al. LC/MS determination of bromazepam, clopenthixol, and reserpine in serum of a non-fatal case of intoxication. *J Anal Toxicol* 1986; 10: 46–48.

98. Cheze M, Villain M, Pepin G. Determination of bromazepam, clonazepam and metabolites after a single intake in urine and hair by LC-MS/MS: application to forensic cases of drug facilitated crimes. *Forensic Sci Int* 2004; 145: 123–130.

99. LeBeau MA, Montgomery MA, Wagner JR, et al. Analysis of biofluids for flunitrazepam and metabolites by electrospray liquid chromatography/mass spectrometry. *J Forensic Sci* 2000; 45: 1133–1141.

100. Toyo'oka T, Kanbori M, Kumaki Y, et al. Determination of triazolam involving its hydroxy metabolites in hair shaft and hair root by reversed-phase liquid chromatography with electrospray ionization mass spectrometry and application to human hair analysis. *Anal Biochem* 2001; 295: 172–179.

101. McClean S, O'Kane EJ, Smyth WF. Electrospray ionisation-mass spectrometric characterisation of selected anti-psychotic drugs and their detection and determination in human hair samples by liquid chromatography-tandem mass spectrometry. *J Chromatogr B* 2000; 740: 141–157.

102. Weinmann W, Muller C, Vogt S, et al. LC-MS-MS analysis of the neuroleptics clozapine, flupentixol, haloperidol, penfluridol, thioridazine, and zuclopenthixol in hair obtained from psychiatric patients. *J Anal Toxicol* 2002; 26: 303–307.

103. Berna M, Shugert R, Mullen J. Determination of olanzapine in human plasma and serum by liquid chromatography/tandem mass spectrometry. *J Mass Spectrom* 1998; 33: 1003–1008.

104. Kollroser M, Henning G, Gatternig R, et al. HPLC-ESI-MS/MS determination of zuclopenthixol in a fatal intoxication during psychiatric therapy. *Forensic Sci Int* 2001; 123: 243–247.

105. Maurer HH, Tenberken O, Kratzsch C, et al. Screening for library-assisted identification and fully validated quantification of 22 beta-blockers in blood plasma by liquid chromatography-mass spectrometry with atmospheric pressure chemical ionization. *J Chromatogr A* 2004; 1058: 169–181.

106. Naidong W, Bu H, Chen YL, et al. Simultaneous development of six LC-MS-MS methods for the determination of multiple analytes in human plasma. *J Pharm Biomed Anal* 2002; 28: 1115–1126.

107. Jacobson GA, Chong FV, Davies NW. LC-MS method for the determination of albuterol enantiomers in human plasma using manual solid-phase extraction and a non-deuterated internal standard. *J Pharm Biomed Anal* 2003; 31: 1237–1243.

108. Badaloni E, D'Acquarica I, Gasparrini F, *et al*. Enantioselective liquid chromatographic-electrospray mass spectrometric assay of beta-adrenergic blockers: application to a pharmacokinetic study of sotalol in human plasma. *J Chromatogr B* 2003; 796: 45–54.

109. Gergov M, Robson JN, Ojanpera I, *et al*. Simultaneous screening and quantitation of 18 antihistamine drugs in blood by liquid chromatography ionspray tandem mass spectrometry. *Forensic Sci Int* 2001; 121: 108–115.

4

Method validation using LC-MS

Frank T. Peters

Introduction

During the last decade, the role and progress of liquid chromatography-(tandem) mass spectrometry [LC-MS(-MS)] in forensic and clinical toxicology has been assessed several times by leading experts in the field [1–11]. These review articles clearly show that this technique has left the development stage and is becoming increasingly important in routine toxicological analysis. However, despite the maturity of the technique itself, individual LC-MS(-MS) procedures must be validated before use to ensure their reliability and applicability for the intended purpose. This is of particular importance in forensic and clinical toxicology because reliable analytical data are a prerequisite for correct interpretation of toxicological findings. Unreliable results not only might be contested in court, but also could lead to unjustified legal consequences for the defendant or to wrong treatment of the patient. The importance of validation, at least of routine analytical methods, can therefore hardly be overestimated. This is especially true in the context of quality management and accreditation, which have become matters of increasing importance in analytical toxicology in recent years.

This chapter will give an overview on the state of the art in (bio)analytical method validation considering important validation guidelines. Experimental designs and statistical procedures for the estimation of validation parameters will be presented and discussed using examples from LC-MS(-MS) literature where appropriate. Readers not familiar with basic statistical procedures are referred to the *Handbook of Chemometrics and Qualimetrics* [12] for further reading.

Guidelines and literature on validation

Due to the importance of method validation in the whole field of analytical chemistry, a number of guidance documents on this subject have been issued by various international organisations or conferences. Two have been developed by the International Conference on Harmonization of Technical Requirements for Registration of Pharmaceuticals for Human Use (ICH), and approved by the regulatory agencies of the European Union, the US and Japan. The first, approved in 1994, concentrated on the theoretical background and definitions [13]; the second, approved in 1996, concentrated on methodology and practical issues concerning validation of methods used to acquire data for drug approval submissions [14]. Both can be downloaded from the ICH homepage free of charge (www.ich.org). The EURACHEM guide *The Fitness for Purpose of Analytical Methods* published in 1998 [15] is not limited to a certain field of analytical chemistry. It provides definitions of validation parameters as well as useful practical guidance on performance and evaluation of validation experiments. It is also available free of charge on the EURACHEM website (www.eurachem.ul.pt). More recently, the International Union of Pure and Applied Chemistry (IUPAC), the International Organization for Standardization (ISO) and the Association of Official Analytical Chemists (AOAC) International have developed a *Harmonized*

Guideline for Single-Laboratory Validation of Methods of Analysis (Harmonized Guide) [16]. It is also a general guideline applicable in many fields of analytical chemistry. It provides guidance on principles of method validation, but practical aspects like experimental designs are not included. Very recently, three review articles have been published on analytical method validation. The one published by Taverniers *et al.* [17] gives an excellent overview on the position of analytical method validation in the greater context of quality control, accreditation and proficiency testing. This paper includes a table listing definitions, expressions (calculations), requirements for acceptance and practical assessment (experimental design) for the most important validation parameters. The other two review articles focus on validation of qualitative analytical methods [18] and on validation of high-performance LC methods for analysis of pharmaceutical products [19].

While all of the papers mentioned in the previous paragraph are certainly important and potentially helpful for any method validation, they specifically address neither analysis of drugs, poisons and/or their metabolites in body fluids or tissues nor LC-MS(-MS)-based analysis. However, these topics are covered by the report on the conference on 'Analytical Methods Validation: Bioavailability, Bioequivalence and Pharmacokinetic Studies' held in Washington in 1990 (Conference Report) [20] and by the report on the follow-up conference in 2000 (Conference Report II) [21], in which experiences and progress since the first conference were discussed. Both of these reports were by Shah *et al.* and had an enormous impact on validation of bioanalytical methods in the pharmaceutical industry. The latter has even been used as a template for the respective regulatory guideline of the US Food and Drug Administration (FDA) [22]. Because of the close relationship to bioanalysis in the context of bioavailability, bioequivalence and pharmacokinetic studies, Conference Report II is probably also the most useful guidance paper for bioanalytical method validation in forensic and clinical toxicology. For comprehensive overviews on the validation of bioanalytical chromatographic methods, and on the implications of bioanalytical method validation in forensic and clinical toxicology, the reader is referred to the review articles by Hartmann *et al.* [23] and by Peters and Maurer [24], respectively.

It should be noted that different sets of terminology have been employed by different guidelines and authors in the literature on method validation, which may lead to confusion and misunderstandings. A detailed discussion of this topic is beyond the scope of this chapter, but can be found in the review by Hartmann *et al.* [23].

Validation parameters

Most of the LC-MS(-MS) methods in forensic and clinical toxicology are used for quantification of drugs, poisons and/or their metabolites in biological fluids or tissues [7, 8, 10, 11]. For quantitative bioanalytical procedures, there is a general agreement that at least the following validation parameters should be evaluated: selectivity, calibration model (linearity), stability, accuracy (bias), precision (repeatability, intermediate precision) and the lower limit of quantification (LLOQ). Additional parameters which might have to be evaluated include limit of detection (LOD), recovery, reproducibility and ruggedness (robustness) [20, 21, 23, 25–30].

Several authors have also used LC-MS(-MS) for systematic toxicological analysis [5, 9, 31–35]. A general validation guideline is currently not available for such qualitative procedures [18]. However, there seems to be agreement that at least selectivity and LOD should be evaluated, and that additional parameters like precision, recovery and ruggedness (robustness) might also be important [14, 18, 36, 37].

Of course, evaluation of possible matrix effects (ME), ion suppression or enhancements should be part of the validation of any LC-MS(-MS) method, particularly those employing electrospray ionisation (ESI) [10, 21, 38, 39]. This is of special importance for qualitative procedures because the LOD would be the first validation parameter to be negatively affected by signal suppression and because this negative effect could not even be compensated by a stable-isotope-labelled internal standard (IS).

Selectivity (specificity)

In Conference Report II [21], selectivity was defined as 'the ability of the bioanalytical method to measure unequivocally and to differentiate the analyte(s) in the presence of components, which may be expected to be present. Typically, these might include metabolites, impurities, degradants, matrix components, etc.'. The term specificity is often used interchangeably with selectivity, although in a strict sense specificity refers to methods that produce a response for a single analyte, whereas selectivity refers to methods that produce responses for a number of chemical entities, which may or may not be distinguished [40]. Selective multi-analyte methods (e.g. for the analysis of different drugs of abuse in blood) should of course be able to differentiate all interesting analytes from each other and from the matrix. One approach to establish method selectivity is to prove the lack of response in a blank matrix [17, 20, 21, 23, 25–30], i.e. that there are no signals interfering with the signal of the analyte(s) or the IS(s). The second approach is based on the assumption that, for merely quantitative procedures, small interferences can be accepted as long as accuracy (bias) and precision at the LLOQ remain within certain acceptance limits [15–17, 23, 29]. However, in forensic and clinical toxicology, analysis is often mainly performed to prove the intake of an (illicit) substance and qualitative data are, therefore, also important. Here, the approach to prove selectivity by absence of interfering signals seems much more reasonable [24].

Establishing selectivity by demonstrating absence of interfering signals

The requirement established by the Conference Report [20] to analyse at least six different sources of blank matrix has become state of the art during the last decade and this number has been used in the validation of most published bioanalytical LC-MS(-MS) methods. However, Hartmann et al. [23] stated from statistical considerations that with analysis of such a small number of matrix blanks relatively rare interferences will remain undetected with a rather high probability. For the same reason, Dadgar et al. [29] proposed to evaluate at least 10–20 sources of blank samples. In Conference Report II [21], however, the required number of matrix sources was even reduced to a single source for methods using hyphenated MS methods like LC-MS(-MS) for detection. This confinement does not seem reasonable for toxicological applications because of the great importance of selectivity in this field. Furthermore, it increases the risk of falsely assuming sufficient selectivity of the tested procedure. In this case, serious problems might be encountered during routine application because even relatively rare matrix interferences are not unlikely to occur if large numbers of samples are analysed. The number of blank matrix sources proposed by Dadgar et al. [29] seems to be a good compromise between reducing the workload during method validation and lowering the risk of unexpected interferences during routine application. Some working groups, including the author's, have taken such considerations into account and checked matrix samples from at least 10 or even 20 sources for the absence of interfering matrix peaks [41–45].

In contrast to samples from pharmacokinetic studies, where usually only a single or very limited number of drugs are applied under controlled conditions, samples from forensic or clinical toxicology cases often contain many different drugs, poisons and/or their metabolites. In this field it is, therefore, also important to check for possible interferences from other xenobiotics which may be expected to be present in authentic samples [24]. This can be accomplished by analysing blank samples spiked with possibly interfering compounds at their highest expectable concentrations. The number and spectrum of compounds to be used in such spiking experiments depend on the purpose of the procedure. For example, Pichini et al. [46] checked interference from various amphetamines, cannabinoids, benzodiazepines and antidepressants (altogether 23 compounds) during the selectivity experiments of an assay for determination of opiates and cocaine in meconium. In the validation of an assay for simultaneous quantification of the chemotherapeutic drugs cyclophosphamide, 4-hydroxycyclophosphamide, thiotepa and tepa in human plasma, De Jonge et al. [47] performed spiking experiments to investigate potential interference from four metabolites of these drugs, 17 other therapeutic drugs often used in co-medication, and caffeine. Crommentuyn et al. [48] studied possible interference even from over 30 often co-administered drugs in an assay for atazanavir and tipranavir, two human immunodeficiency virus (HIV) protease inhibitors used in the treatment of acquired immune deficiency syndrome (AIDS). On first sight, this may look exaggerated, but it is not in fact if one considers that treatment of AIDS usually involves at least three different anti-HIV drugs and co-administration of further drugs is the rule rather than an exception. In addition, spiking mixtures of possibly interfering compounds rather than single compounds allows effective simultaneous investigation of interference from many compounds, while the workload is kept at a minimum.

Another way to exclude interference from other drugs or their metabolites is to check authentic samples containing these, but not the analyte of interest. This approach is preferable if the possibly interfering substance is known to be extensively metabolised, as it also allows excluding interferences from such metabolites, which are usually not available as pure substances. An example can be found in Streit et al. [49] which describes an LC-MS-MS assay for the immunosuppressant mycophenolic acid. Here, the authors used 30 samples from transplant recipients not treated with mycophenolic acid, but with a number of other immunosuppressive drugs and other drugs commonly prescribed in transplant recipients. The working group of Maurer also included routine samples known to contain various drugs but not the analytes in their selectivity studies of several LC-MS procedures [42–44].

It must be noted that metabolites of the target analyte can also be an important source of interference. This is of particular interest in LC-MS(-MS) analysis because metabolites that are in principle amenable to this type of analysis, such as glucuronides or N-oxides, can convert back to the parent drug by in-source collision-induced dissociation (CID) and thus lead to cross-talk in the channel of the parent drug [50–53]. Similarly, labile pro-drugs may be converted to the active metabolite, in this case the target analyte, by in-source CID and lead to cross-talk in the metabolite channel. As such phenomena occur only when there is insufficient separation between the respective compounds, they can be avoided by optimising the chromatographic conditions. However, this type of interference must be detected in the first place to be avoided. If the metabolite is available, this can be achieved by spiking experiments as described above; if not, the only way to check for this type of interference is analysis of incurred samples, i.e. samples from persons (or animals) who have ingested the parent compound [52, 53]. A useful strategy for the performance of such studies is presented in [53].

Stable-isotope-labelled analogues of the target analytes are often used as ISs in LC-MS(-MS). Due to their similar physicochemical properties, they can ideally compensate for variability during sample preparation and measurement, but still be differentiated from the target analyte by mass spectrometric detection. However, an isotopically labelled compound may contain the non-labelled compound as an impurity or their mass spectra may sometimes contain fragment ions with the same mass-to-charge (m/z) ratios as the monitored ions of the target analyte. In both cases, the peak area of the analyte peak would be overestimated, thus compromising quantification. The absence of such interference caused by the IS can be checked by analysing so-called zero samples, i.e. blank samples spiked with the IS, as exemplified in the literature [42–44, 47, 51, 54]. The importance of this is underlined by the findings of Janda et al. [54], who found a low signal of non-labelled ethyl glucuronide in a blank hair sample spiked only with ethyl glucuronide-d_5, most probably caused by traces of the non-labelled drug in the stable-isotope-labelled standard. These authors stated that it may be useful to reduce the total amount of IS to eliminate this interference. In a similar way to that described above, the analyte might interfere with a stable-isotope-labelled IS. This even becomes a principal problem with deuterated analogues when the number of deuterium atoms of the analogue or one of its monitored fragments is three or less [55]. Blank samples spiked with the analyte at the upper limit of the calibration range, but without the IS, can be used to check for the absence of such interferences [51].

None of the above-mentioned experiments

requires quantification of the analytes. They can therefore be performed in the pre-validation phase, where it is still possible to switch back to the development phase without having spent a lot of time and expense on further validation experiments.

Establishing selectivity by acceptable accuracy (bias) and precision at the lower limit of quantification

This approach was preferred by Dadgar et al. [29] and Hartmann et al. [23]. Both authors proposed analysis of up to 20 different blank samples spiked with analyte at the LLOQ and with possibly interfering compounds at their highest likely concentrations, if available. In this approach, the method can be considered sufficiently selective if precision and accuracy (bias) data for these LLOQ samples are acceptable. Detailed accounts of experimental designs and statistical methods to establish selectivity are given by Dadgar et al. [29]. An important disadvantage of this approach is that it requires quantification and can therefore be performed only in the main validation phase. If the selectivity of the method is found to be insufficient at this late stage of validation, the method needs further development and all validation experiments have to be repeated.

In their spiking experiments discussed in the previous paragraph, Crommentuyn et al. [48] and de Jonge et al. [47] used an intermediate approach and defined the maximum acceptable area of interfering peaks to be 20% of the peak area of the analyte at the LLOQ. This semi-quantitative approach can be used in the pre-validation phase because it requires no calibration or precision estimate.

Calibration model (linearity)

The choice of an appropriate calibration model is necessary for reliable quantification. Therefore, the relationship between the concentration of analyte in the sample and the corresponding response (in bioanalytical methods mostly the area ratio of analyte versus IS) must be investigated. This can be done by analysing spiked calibration samples and plotting the resulting response versus the corresponding concentrations. A plot of an exemplary data set and the corresponding calibration line obtained from simple linear regression is shown in Figure 4.1(a). The resulting standard curves can then be further evaluated by graphical methods like residual plots or by mathematical methods; the latter also allow statistical evaluation of the response functions.

Residuals are the deviations of the observed values from the values predicted by the applied calibration model. They can be calculated according to equation (1). In residual plots, these residuals are plotted against the respective concentrations, which ideally results in a random distribution of residuals around zero. A residual plot corresponding to the calibration plot in Figure 4.1(a) is shown in Figure 4.1(b). The plotted residuals are clearly not ideally distributed; the reasons for this phenomenon will be discussed below.

$$e_{ij} = y_{ij} - \hat{y}_j \qquad (1)$$

where e_{ij} is the ith residual at the jth concentration level, y_{ij} is the ith observed value at the jth concentration level and $\hat{y}_j$ is the predicted value at the jth concentration level.

There is general agreement that for bioanalytical methods, calibrators should be matrix based, i.e. prepared by spiking of blank matrix, and that calibrator concentrations must cover the whole calibration range [21, 23, 25–28, 30]. However, recommendations on how many concentration levels should be studied with how many replicates per concentration level differ significantly in the literature on analytical method validation [15–17, 21, 23, 25–28, 30]. In Conference Report II [21], 'a sufficient number of standards to define adequately the relationship between concentration and response' was demanded. Furthermore, it was stated that at least five to eight concentration levels should be studied for linear and maybe more for non-linear relationships. However, no information was given on how many replicates should be analysed at each level. The guidelines established by the ICH [14] and those of the *Journal of Chromatography B* [14] also required at least five concentration levels, but again no specific requirements for the number of replicates

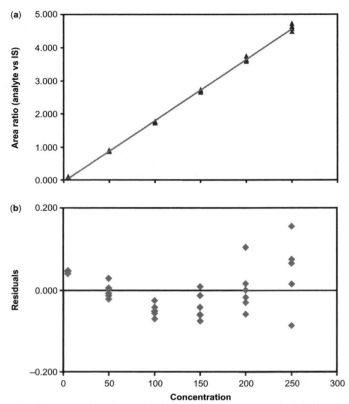

Figure 4.1 Regression plot of an exemplary data-set with the calibration curve calculated using simple linear regression (a) and the corresponding residual plot (b).

at each level were given. In the validation guide by EURACHEM [15] and in the Harmonized Guide [16], at least six concentration levels are requested. The latter has further specified that the concentration levels should be evenly spaced over the concentration range and that at least two (preferably three or more) replicates should be analysed at each concentration level [16]. Based on studies by Penninckx et al. [56], Hartmann et al. [23] proposed in their review rather to use fewer concentration levels with a greater number of replicates, e.g. four evenly spread levels with nine replicates, in order to achieve more reliable variance estimations while still having enough concentration levels for a preliminary indication of possible non-linearity. These different recommendations probably explain why almost every thinkable combination of concentration levels and replicates per level has been used for individual assays in the literature. However, two important points need to be considered: (i) with a decreasing number of concentration levels, the evaluation of non-linear models becomes increasingly unreliable and, thus, their evaluation requires more concentration levels than linear models [20, 21, 23, 56] and (ii) assessing the behaviour of variance across the calibration range becomes increasingly difficult with a decreasing number of replicates [56].

The initial step of studying the calibration function is to check for outliers. This can first be done visually, e.g. by evaluation of residual plots (Figure 4.1). In Figure 4.1, the highest data point of the second highest concentration level might be suspected to be an outlier. Such suspected outliers should be checked by appropriate statistical procedures like the Grubbs test and eliminated, if found to be significant. The suspected outlier in

Figure 4.1(b) was found to be not significant and was left in the data-set. More than two outliers in a complete data-set of a calibration model experiment may indicate serious problems with the method, which require further investigation [56].

The next step is to check for homogeneity of variance (homoscedasticity) over the calibration range. Again, this can first be done visually using residual plots. For example, in the residual plot in Figure 4.1a, one can clearly see that the scatter of the replicates increases with concentration indicating inhomogeneous variances (heteroscedasticity) over the calibration range. In such cases, homoscedasticity should be checked by an appropriate statistical procedure, e.g. by a simple one-sided F-test between the variances at the highest and the lowest concentration levels. For a more detailed account and alternative statistical procedures, see Penninckx et al. [56]. The evaluation of the behaviour of variance is very important for the choice of the correct regression model. Ordinary least squares regression models are applicable only for homoscedastic data sets, whereas in case of heteroscedasticity the data should mathematically be transformed or a weighted least squares model should be applied [15, 16, 21, 23, 26, 27, 56]. It should be noted that, for calibration ranges spanning more than one order of magnitude, heteroscedasticity is the rule rather than the exception. Regarding the usually wide concentration range of assays used in analytical toxicology or in the area of pharmacokinetics, it is not surprising that weighted regression models were used in the vast majority of the validated LC-MS(-MS) assays in literature. The weighting factors $1/x$ and $1/x^2$, i.e. the inverse of the concentration or the inverse of the squared concentration, respectively, are by far the most often used. Only few authors used $1/y$ or $1/y^2$, i.e. the inverse of the response or the inverse of the squared response, respectively. Theoretically, the suitability of a weighting factor can easily be checked by comparing the variances of the weighted residuals. Figure 4.2(a) shows the calibration line obtained by weighted linear regression with the weighting factor $1/x^2$ of the same data as in Figure 4.1. Figure 4.2(b) shows the corresponding plot of the weighted residuals. It can clearly be seen that the scatter of the weighted residuals is comparable at all concentration levels indicating that the weighting factor used was appropriate. Such a finding can again be tested statistically by performing an F-test on the variances of the weighted residuals at the highest and lowest concentration levels.

However, some authors chose the weighting factor in a more practical way [47, 57, 58]. They evaluated alternative weighting schemes and back-calculated the concentrations of the calibrators using the respective regression models. Then they added the relative deviations of these back-calculated values from the nominal values and chose the weighting scheme as the most appropriate with which the smallest total relative deviation had been obtained.

After significant outliers have been purged from the data-set and, if applicable, an appropriate weighting factor for regression has been found, a mathematical model has to be found that adequately describes the relationship between analyte concentration in the sample and response. Usually, linear models are preferable, but, if necessary, the use of non-linear models is not only acceptable but even recommended. Examples for the use of second-order (quadratic) [59–63] and more complicated non-linear regression models [64] have been described in the LC-MS(-MS) literature. The model-fit can again first be evaluated visually using residual plots. For example, the plot of weighted residuals in Figure 4.2(b) corresponds to a weighted linear regression model applied to the same data as in Figure 4.1. One can see that the residuals at the lowest and the two highest concentration levels are on one side of the zero line, while the majority of those at the intermediate concentration levels are on the opposite side. Such patterns are typically obtained when linear models are fitted into curved data-sets. This indicates that, in the presented example, a non-linear model might be more appropriate to describe the data. In any case, the model-fit should also be tested by appropriate statistical methods [14, 16, 20, 21, 23, 28, 56]. For example, the fit of simple regression models (homoscedastic data) can be tested by the analysis of variance (ANOVA) lack-of-fit test [23, 28, 56]. Testing the fit of weighted regression models is somewhat more complicated. For example, deviation from weighted linear model can be tested by evaluation of an alternative

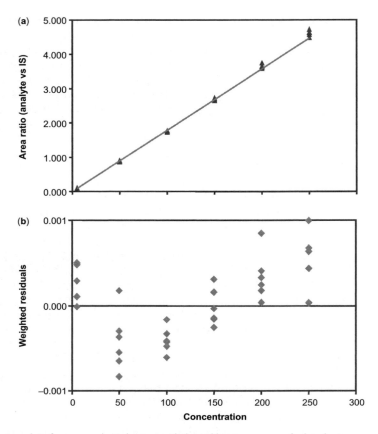

Figure 4.2 Regression plot of an exemplary data-set with the calibration curve calculated using a weighted ($1/x^2$) linear regression model (a) and the corresponding plot of the weighted residuals (b).

weighted second-order regression model. If the quadratic term describing the curvature of the second-order model is significantly different from zero, non-linearity has to be assumed. Such significant non-linearity was also found for the data in Figure 4.2. The regression line of the alternative second-order model and the corresponding plot of the weighted residuals are shown in Figure 4.3(a and b, respectively). It can be seen that the weighted residuals are now randomly distributed around zero indicating that the weighted second-order model adequately describes the data-set. A detailed discussion of alternative statistical procedures for testing the model-fit of both unweighted and weighted calibration models can be found in Penninckx *et al.* [56]. The widespread practice of evaluating a calibration model via its coefficients of correlation or determination is not acceptable from a statistical point of view [23].

One important point should be kept in mind when statistically testing the model-fit – the higher the precision of a method, the higher the probability of detecting a statistically significant deviation from the assumed calibration model [23, 28, 40]. Therefore, the practical relevance of the deviation from the assumed model should also be taken into account. If the accuracy (bias) and precision data are within the required acceptance limits or an alternative calibration model is not applicable, slight deviations from the assumed model may be neglected [23, 28].

Once a calibration model has been established, the calibration curves for other validation

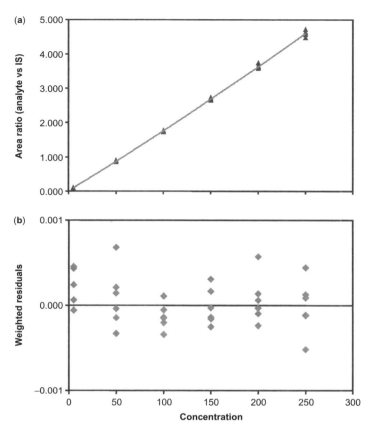

Figure 4.3 Regression plot of an exemplary data-set with the calibration curve calculated using a weighted ($1/x^2$) second-order regression model (a) and the corresponding plot of the weighted residuals (b).

experiments (precision, bias, stability, etc.) and for routine analysis can be prepared with fewer concentration levels and fewer or no replicates [23, 28].

Accuracy (bias)

In a strict sense, the accuracy of a method is affected by systematic (bias) as well as random (precision) error components [23, 65]. This fact has been taken into account in the definition of accuracy as established by the ISO [66]. However, it must be mentioned that accuracy is often used to describe only the systematic error component, i.e. in the sense of bias [13, 20, 21, 25–28, 40]. In the following, the term accuracy will be used in the sense of bias, which will be indicated in brackets.

According to the ISO, bias is 'the difference between the expectation of the test results and an accepted reference value' [66]. It may consist of more than one systematic error component. Bias can be measured as a percentage deviation from the accepted reference value. The term trueness expresses the deviation of the mean value of a large series of measurements from the accepted reference value. It can be expressed in terms of bias. Due to the high workload of analysing such large series, trueness is usually not determined during method validation, but rather from the results of a great number of quality control samples during routine application or in inter-laboratory studies. It should be noted that trueness and bias are sometimes reported in terms of (analytical) recovery of an accepted reference value [16]. This must not be confused with the validation parameter recovery in the sense of, for example, extraction efficiency (RE, *see below*).

Precision

According to the ICH, precision is 'the closeness of agreement (degree of scatter) between a series of measurements obtained from multiple sampling of the same homogeneous sample under the prescribed conditions and may be considered at three levels: repeatability, intermediate precision and reproducibility' [13]. Precision is usually measured in terms of imprecision expressed as an absolute or relative standard deviation (RSD) and does not relate to reference values.

Repeatability

'Repeatability expresses the precision under the same operating conditions over a short interval of time. Repeatability is also termed intra-assay precision' [13]. Within-run or within-day precision values are also often used to describe repeatability.

Intermediate precision

'Intermediate precision expresses within-laboratories variations: different days, different analysts, different equipment, etc.' [13]. The ISO definition used the term 'M-factor different intermediate precision', where the M-factor expresses how many and which factors (time, calibration, operator, equipment or combinations of those) differ between successive determinations [66]. In a strict sense intermediate precision is the total precision under varied conditions, whereas so-called inter-assay, between-run or between-day precision measure only the precision components caused by the respective factors (*see below*). However, the latter terms are not clearly defined and obviously often used interchangeably with each other and also with the term intermediate precision.

Reproducibility

'Reproducibility expresses the precision between laboratories (collaborative studies, usually applied to standardization of methodology)' [13]. Reproducibility has to be studied only if a method is supposed to be used in different laboratories.

Unfortunately, some authors also used the term reproducibility for within-laboratory studies at the level of intermediate precision [25, 27]. However, this should be avoided in order to prevent confusion.

Precision and bias experiments and calculations

As already mentioned above, precision and bias can be estimated from the analysis of quality control samples under specified conditions. As both precision and bias can vary substantially over the calibration range, it is necessary to evaluate these parameters at least at three concentration levels: low, medium and high relative to the calibration range [14, 17, 20, 21, 23, 40]. In Conference Report II [21], it was further defined that the concentration of the low quality control sample must be within three times LLOQ. The *Journal of Chromatography B* requirement is to study precision and bias at two concentration levels (low and high) [25], whereas Causon [27] suggested estimating precision and bias at four concentration levels. Four concentration levels (LLOQ, low, medium, high) were also studied in the experimental design proposed by Wieling *et al.* [28]. Due to the often higher concentration ranges in the fields of forensic and clinical toxicology, it might be reasonable also to validate the analysis of quality control samples containing concentrations above the highest calibration standard after dilution or after reduction of sample volumes, as has been described by Dadgar *et al.* [30]. These authors also described the use of quality control samples with concentrations below those of the lowest calibration standard using greater sample volumes. The Harmonized Guide [16] recommends studying precision at the extremes of the working range while the EURACHEM guide [15] recommends various concentrations across the working range without any further specification. None of these two guides contains any specific recommendations concerning the number of concentration levels for studying bias.

Several authors have specified acceptance limits for precision and/or accuracy (bias) [20, 21, 25–27]. Both Conference Reports [20, 21] required precision to be within 15% RSD except at the LLOQ, where 20% RSD was accepted. Bias was required to be within ±15% of the accepted true value, except at the LLOQ, where ±20% was accepted. These requirements have been subject to criticism in the analysis of the Conference Report by Hartmann et al. [65]. These authors concluded from statistical considerations that it is not realistic to apply the same acceptance criteria at different levels of precision (repeatability, reproducibility) as RSD under reproducibility conditions is usually considerably greater than under repeatability conditions. Furthermore, if precision and bias estimates are close to the acceptance limits, the probability to reject an actually acceptable method (β-error) is quite high. Causon proposed the same acceptance limits of 15% RSD for precision and ±15% for accuracy (bias) for all concentration levels [27].

The guidelines established by the *Journal of Chromatography B* [25] require precision to be within 10% RSD for the high-quality control samples and within 20% RSD for the low-quality control samples. Acceptance criteria for accuracy (bias) were not specified therein.

Again, the proposals on how many replicates at each concentration levels should be analysed vary considerably. The Conference Reports [20, 21] and *Journal of Chromatography B* [25] guidelines required at least five replicates at each concentration level. However, one would assume that these requirements apply only to repeatability studies; at least no specific recommendations were given for studies of intermediate precision or reproducibility.

The EURACHEM guide [15] recommends performing 10 independent determinations for each concentration level under each repeatability, M-factor-different intermediate precision and reproducibility conditions. In this straightforward approach, the corresponding precision data can simply be calculated as RSDs of the values obtained under these stipulated conditions.

Examples for this approach can be found in the literature [59, 67–69].

Other authors have proposed an approach where replicates are analysed on a number of different occasions (e.g. different runs, days, etc.) [16, 23, 27, 28]. In their experimental design, Wieling et al. [28] analysed three replicates at each of four concentration levels on each of 5 days. Similar approaches were suggested by Causon [27] (six replicates at each of four concentrations on each of four occasions), Hartmann et al. [23] (two replicates at each concentration level on each of 8 days) and the Harmonized Guide [16] (duplicates in a number of successive runs). Using one-way ANOVA with the varied factor (e.g. day) as the grouping variable, such experimental designs allow separate calculation of repeatability and of the precision component caused by the grouping variable from the same data-sets, as well as the calculation of factor-different intermediate precision as the combination of the previous two [70–72]. Thus, repeatability, expressed as percentage RSD, can be calculated from equation (2):

$$RSD_r\ (\%) = \frac{\sqrt{MS_{wg}}}{\overline{X}} \cdot 100 \qquad (2)$$

where RSD_r is the repeatability, within-group precision, expressed as percentage RSD, MS_{wg} is the mean square within groups, obtained from the ANOVA table and $\overline{X}$ is the grand mean of all observations.

The precision component caused by the varied factor or grouping variable corresponds to the standard deviation (SD) of the group means after subtraction of the contribution from within-group variability. Expressed as RSD, this precision component can be calculated from equation (3):

$$RSD_{bg}\ (\%) = \frac{\sqrt{\frac{MS_{bg} - MS_{wg}}{n}}}{\overline{X}} \cdot 100 \qquad (3)$$

set $RSD_{bg} = 0$, if $MS_{bg} < MS_{wg}$

where RSD_{bg} is the between-group precision (component), expressed as percentage RSD, MS_{bg} is the mean square between groups, obtained from the ANOVA table, MS_{wg} is the mean square within groups, obtained from the ANOVA table, n is the number of observations in each group and $\overline{X}$ is the grand mean of all observations.

From a strictly statistical point of view, this is the between-group precision (component). Unfortunately this term is often used for total precision or factor-different intermediate precision, which is not correct.

The factor-different intermediate precision or total precision can be calculated according to equation (4) as a combination of within- and between-group effects:

$$RSD_{I(F)} (\%) = \frac{\sqrt{\frac{MS_{bg} + (n-1) \cdot MS_{wg}}{n}}}{\overline{X}} \cdot 100 \quad (4)$$

where $RSD_{I(F)}$ is the factor-different intermediate precision, expressed as percentage RSD, MS_{bg} is the mean square between groups, obtained from the ANOVA table, MS_{wg} is the mean square within groups, obtained from the ANOVA table, n is the number of observations in each group and $\overline{X}$ is the grand mean of all observations.

This value corresponds to the intermediate precision estimate obtained with the previously mentioned alternative design proposed by EURACHEM [15]. However, in the author's opinion, the ANOVA approach should be preferred because more information can be derived from this approach with a comparable number of analyses. For example, in the case of unacceptably high intermediate precision values, the ANOVA approach allows easy comparison of the precision components. This can be very helpful in finding the cause of the problem.

The ANOVA approach has been used in a number of LC-MS(-MS) publications (e.g. [42–44, 53, 63, 73–79]). In the majority of these publications, four to six replicates were analysed on 3–4 days, i.e. a total number of 15–20 samples [53, 63, 73–79]. With such designs, repeatability is estimated with quite a large number of degrees of freedom (12–16) while there are only few degrees of freedom for the between-day component (two to three) making its estimation quite unreliable. In the designs proposed by Hartmann et al. [23] and in the Harmonized Guide [16], the degrees of freedom for both estimations are more balanced, e.g. eight for repeatability and seven for the between-day component in the case of the Hartmann design. For this reason the latter design is preferred by the author's working group [42–44]. In the information for authors of the *Clinical Chemistry* journal, an experimental design with two replicates per run, two runs per day over 20 days for each concentration level is recommended, which has been established by the National Committee for Clinical Laboratory Standards [72]. This not only allows estimation of within-run and between-run SDs, but also of within-day, between-day and total SDs, which are in fact all estimations of precision at different levels. A modified version of this design was applied in one of the earlier validation studies of the author's working group [80]. From this experience, it seems rather questionable if the additional information provided by this approach can justify the high workload and costs compared with the other experimental designs.

In many publications, a mixture of the above-mentioned approaches is used. The experimental design is the same as in the ANOVA approach, with replicates being analysed on several occasions, but calculations are performed as in the approach in the EURACHEM guide [15]. Only the replicates from the first occasion are considered when estimating repeatability, which is statistically correct, but still a waste of valuable information because the other occasions are not taken into account. For this reason, such a repeatability estimate is of course much less reliable than one derived from the whole data-set, as in the ANOVA approach. The calculation of intermediate precision as the (R)SD of all observations is even more problematic. This way of calculation treats all observations as independent, which in fact they are not because they were obtained in groups on several occasions. This is not only statistically incorrect, but leads to underestimation of intermediate precision [70]. For these reasons, the mixed approach should not be used in any validation study. Furthermore, publications describing validated assays should report not only the experimental design and results of precision studies, but also a detailed description on how the results were calculated. Otherwise comparison between precision data in different publications will not be possible.

Bias experiments usually include a number of replicate measurements from which a mean value is calculated. The EURACHEM guide [15] recommended 10 replicates. The mean value is compared with a certain reference value, e.g. the target value of a certified reference material, the results obtained for the same sample using a reference method or the theoretical concentration of a spiked sample [15–17, 21]. Bias is then usually calculated as the percentage deviation of the observed mean value from the respective reference value according to equation (5):

$$\text{Bias (\%)} = \frac{\overline{X} - \mu}{\mu} \cdot 100 \quad (5)$$

where $\overline{X}$ is the observed mean value and μ is the accepted reference value.

Bias experiments can be carried out together with the precision experiments, i.e. the same results are used for calculation of both. In the approach using separate experiments for evaluation of precision under repeatability and intermediate precision conditions, this will yield two bias values, one from the mean value of the repeatability experiment (within-day or within-run bias) and one from the mean of the factor-different intermediate precision experiment (between-day or between-run bias). Using the ANOVA approach, usually only a single bias value is obtained from the grand mean of all observations.

It is important to note that daily variations of the calibration curve can influence bias estimation. Therefore, bias estimations should be based on data calculated from several calibration curves [23]. In the experimental design of Wieling et al. [28], the results for quality control samples were calculated via daily calibration curves. Therefore, the overall means from these results at the different concentration levels reliably reflect the average bias of the method at the corresponding concentration level. Alternatively, as described in the same paper, the bias can be estimated using confidence limits around the calculated mean values at each concentration [28]. If the calculated confidence interval includes the accepted true value, one can assume the method to be free of bias at a given level of statistical significance. Another way to test the significance of the calculated bias is to perform a t-test against the accepted true value. However, even methods exhibiting a statistically significant bias can still be acceptable, if the calculated bias lies within previously established acceptance limits. Other methods for bias evaluation can also be found [23].

Limits

Lower limit of quantification

The LLOQ is the lowest amount of an analyte in a sample that can be quantitatively determined with suitable precision and accuracy (bias) [13, 21]. There are different approaches for the determination of LLOQ.

Lower limit of quantification based on precision and accuracy (bias) data

This is probably the most practical approach and defines the LLOQ as the lowest concentration of a sample that can still be quantified with acceptable precision and accuracy (bias) [13, 14, 20, 21, 23, 26, 27]. In the Conference Reports [20, 21], the acceptance criteria for these two parameters at LLOQ are 20% RSD for precision and ±20% for bias. Only Causon suggested 15% RSD and ±15%, respectively [27]. It should be pointed out, however, that these parameters must be determined using an LLOQ sample independent of the calibration curve. The advantage of this approach is the fact that the estimation of LLOQ is based on the same quantification procedure used for real samples. This approach is the most commonly used in validation of LC-MS(-MS) methods. In order to include the possible influence of matrix differences into the LLOQ evaluations, many authors used six to ten different sources of blank matrix to prepare quality control samples at the LLOQ [47, 60, 62, 63, 73, 78, 79].

Lower limit of quantification based on signal-to-noise ratio [14, 25]

This approach can be applied only if there is baseline noise, e.g. to chromatographic methods. The signal-to-noise (S/N) ratio can then be defined as the height of the analyte peak (signal) and the amplitude between the highest and lowest point of the baseline (noise) in a certain area around the analyte peak. For LLOQ, the S/N ratio is usually required to be ≥ 10.

Estimation of baseline noise can be quite difficult for bioanalytical methods if matrix peaks elute close to the analyte peak. Nevertheless, this approach for estimation of the LLOQ is quite popular in LC-MS(-MS) analysis [42–44, 59, 81–83], possibly because LC-MS-MS methods in particular often exhibit very low noise levels and few peaks besides the analyte peak.

Lower limit of quantification based on the standard deviation of the response from blank samples

Another definition of LLOQ is the concentration that corresponds to a response that is k times greater than the estimated SD of blank samples [14]. From the response, the LLOQ can be calculated using the slope of the calibration curve using equation (6) (for blank corrected signals):

$$\text{LLOQ} = k \cdot \frac{SD_{bl}}{S} \quad (6)$$

where k is a factor (usually 10), SD_{bl} is the standard deviation of the blank response and S is the slope of the calibration curve.

This approach is applicable only for methods where SD_{bl} can be estimated from replicate analysis of blank samples. It is therefore not applicable for most quantitative chromatographic methods, as here the response is usually measured in terms of peak area units, which can of course not be measured in a blank sample analysed with a selective method. This is probably the reason why this approach has rarely been used in LC-MS(-MS) analysis. An example can be found in Pichini *et al.* [45].

Lower limit of quantification based on a specific calibration curve in the range of the lower limit of quantification

In this approach, a specific calibration curve is established from samples containing the analyte in the range of LLOQ [14]. One must not use the calibration curve over the whole range of quantification for this determination because this may lead to overestimation of the LLOQ. The SD of the blank can then be estimated from the residual SD of the regression line or the SD of the y-intercept. This approach is also applicable for chromatographic methods, but has been used by only a few authors in LC-MS(-MS) analysis [54, 84–86], e.g. using special software such as VALISTAT (www.arvecon.de).

Upper limit of quantification

The upper LOQ (ULOQ) is the maximum analyte concentration of a sample that can be quantified with acceptable precision and accuracy (bias). In general, the ULOQ is identical to the concentration of the highest calibration standard [21].

Limit of detection

Quantification below the LLOQ is by definition not acceptable [13, 14, 20, 21, 23, 30]. Therefore, below this value a method can produce only semiquantitative or qualitative data. However, it can still be important to know the LOD of the method. According to the ICH, it is the lowest concentration of analyte in a sample that can be detected but not necessarily quantified as an exact value [13]. According to Conference Report II [21], it is the lowest concentration of an analyte in a sample that the bioanalytical procedure can reliably differentiate from background noise.

In most published LC-MS(-MS) methods, the LOD has not been evaluated due to their purely quantitative character. However, authors from the field of toxicology mostly reported the LOD of their LC-MS(-MS) procedures [36, 41–45, 54, 59, 61, 67, 68, 81, 84–86] because qualitative data can also be very important in this field. The approaches most often applied for estimation of the LOD are basically the same as those described for LLOQ with the exception of the approach using precision and accuracy data, which cannot be used here for obvious reasons. In contrast to the LLOQ determination, for LOD an S/N ratio or k-factor ≥ 3 is usually chosen [14–17, 23, 28, 40]. If the calibration curve approach is used for determination of the LOD, then only calibrators containing the analyte in the range of LOD must be used to avoid overestimation of the LOD. All these approaches evaluate only the pure response of the analytes. In toxicology, however, unambiguous identification of an analyte in a sample requires more complex acceptance criteria to be fulfilled. Such criteria have recently been reviewed by Rivier [37] and are described by the same author in Chapter 5 of this book. Especially in forensic toxicology and doping control, it would certainly be more appropriate to define the LOD as the lowest concentration of analyte in a sample, for which specific identification criteria can still be fulfilled.

Stability

The definition according to Conference Report II [21] was as follows: 'The chemical stability of an analyte in a given matrix under specific conditions for given time intervals.' Stability of the analyte during the whole analytical procedure is a prerequisite for reliable quantification. Therefore, full validation of a method must include stability experiments for the various stages of analysis including storage prior to analysis.

Long-term stability

The stability in the sample matrix should be established under storage conditions, i.e. in the same vessels, at the same temperature and over a storage period at least as long as the one expected for authentic samples [20, 21, 23, 25, 29, 30, 40].

Freeze/thaw stability

As samples are often frozen and thawed, e.g. for re-analysis, the stability of analyte during several freeze/thaw cycles should also be evaluated. The Conference Reports [20, 21] require a minimum of three cycles at two concentrations in triplicate, which has also been accepted by other authors [23, 28, 29].

In-process stability/bench-top stability

The stability of analyte under the conditions of sample preparation (e.g. ambient temperature over time needed for sample preparation) is evaluated here. There is general agreement that this type of stability should be evaluated to find out if preservatives have to be added to prevent degradation of analyte during sample preparation [21, 23, 29].

Processed sample stability

Instability can occur not only in the sample matrix, but also in processed samples. It is therefore important also to test the stability of an analyte in the prepared samples under conditions of analysis (e.g. auto-sampler conditions for the expected maximum time of an analytical run). One should also test the stability in prepared samples under storage conditions, e.g. refrigerator, in case prepared samples have to be stored prior to analysis [21, 23, 28–30].

Stability experiments and evaluations

A detailed account of experimental designs and statistical evaluations of stability experiments can be found elsewhere [23, 29, 30]. Stability can be tested by comparing the results of quality control samples analysed before (comparison samples) and after (stability samples) being exposed to the conditions for stability assessment. It has been recommended to perform stability experiments at at least two concentration levels (low and high) [23, 28–30]. For both comparison and stability samples, analysis of at least six replicates was recommended [23]. Ratios between comparison samples and stability samples of 90–110%, with 90% confidence intervals within 80–120 or 85–115% [29], were regarded as acceptable. Alternatively, the mean of the stability samples can be tested against a lower acceptance limit corresponding to 90% of the mean of the comparison samples [23, 27]. An alternative approach for statistically testing processed samples in the auto-sampler was used in the experimental design proposed by Wieling et al. [28]. A sample extract was injected repeatedly at certain intervals for a total time period corresponding to the time required to measure a batch of samples under routine conditions. The peak areas of the analytes were then plotted against the respective injection times. Subsequently, regression analysis was performed to check for a significantly negative slope, which would indicate instability.

The importance of stability is obviously widely accepted in the LC-MS(-MS) community, which can be deduced from the fact that the majority of LC-MS(-MS) papers published in recent years included descriptions of stability experiments. Due to the many different experimental designs used in these papers, a comprehensive overview of them all would be beyond the scope of this chapter. Exemplary stability studies can be found elsewhere [42–44, 63, 64, 74, 75, 81, 84, 87–90]. The papers of the group of Maurer *et al.* [42–44] describe studies on bench-top, freeze/thaw, long-term and auto-sampler stability including statistics and acceptance criteria. In particular the latter are missing in most other papers. The papers by the group of Jemal *et al.* [63, 74, 75] describe stability studies including the effect of modifications introduced to stabilise labile analytes. Egge-Jacobsen *et al.* [64] performed a comprehensive stability study of various anti-HIV drugs including stability under conditions usually applied for HIV inactivation in blood samples. Skopp and Poetsch [84] studied the stability of the labile 11-nor-Δ^9-carboxy-tetrahydrocannabinol glucuronide in plasma and urine at various temperatures ranging from –20 to 49°C. Lin *et al.* [87] described a comprehensive stability study including the stability of stock solutions at –20°C, which should also be of special interest for forensic and clinical toxicologists because this paper deals with the analysis of cocaine and benzoylecgonine, two potentially unstable compounds relevant in these fields of toxicology. In another interesting paper, Scheidweiler and Huestis [81] studied hydrolysis of cocaine and 6-monoacetylmorphine during sample preparation of an LC-MS(-MS) assay for quantification of opiates, cocaine and metabolites in hair. They spiked blank hair with cocaine and 6-monoacetylmorphine, and estimated their hydrolysis by quantification of the degradation products benzoylecgonine and morphine after sample preparation. All these examples show that besides the above-mentioned types of stability, which should be performed in any validation study, additional experiments might be necessary for certain analytes.

A serious problem encountered during stability testing for bioanalytical methods in forensic and clinical toxicology is the fact that there are many different sampling vessels. Furthermore, the anticoagulants used also differ. Both facts make it difficult to assess long-term stability, as the workload to analyse all possible combinations of vessels and anticoagulants is of course far too great. However, for some analytes relevant for forensic and clinical toxicology (e.g. cocaine) stability problems with different sampling vessels have been reported [91], which shows that at least for some analytes such extensive studies may have to be considered.

Recovery

As already mentioned above, recovery is not among the validation parameters regarded as essential by the Conference Reports [20, 21]. Most authors agree that the value for recovery is not important, as long as the data for LLOQ (LOD), precision and accuracy (bias) are acceptable [21, 23, 26, 27, 30, 40]. Nevertheless, the guidelines of the *Journal of Chromatography B* require the determination of the recovery for analyte and IS at high and low concentrations [25]. Recovery can usually be calculated as the percentage of the analyte response after sample work-up compared with that of a solution containing the analyte at the theoretical maximum concentration. However, one must be aware that in LC-MS(-MS) analysis a different experimental design must be used to determine recovery because part of the change of the response in prepared samples in comparison to respective standard solutions might be attributable to ME. For this reason it is more appropriate to perform the recovery experiments together with ion suppression/enhancement experiments as described below.

Ruggedness (robustness)

Ruggedness is a measure for the susceptibility of a method to small changes that might occur during routine analysis, e.g. small changes of pH values, mobile-phase composition, temperature, etc. Full validation must not necessarily include ruggedness testing; however, it can be very helpful during the method development/pre-validation phase as problems that may occur during validation are often detected in advance. Ruggedness should be tested if a method is supposed to be transferred to another laboratory [13,

14, 23, 40, 92]. A detailed account and helpful guidance on experimental designs and evaluation of ruggedness/robustness tests can be found in Vander-Heyden *et al.* [92].

Matrix effects (ion suppression/enhancement)

Suppression or enhancement of analyte ionisation by co-eluting compounds is a well-known phenomenon in LC-MS(-MS) analysis, mainly depending on the sample matrix, sample preparation procedure, quality of chromatographic separation, mobile-phase additives and ionisation type [38, 39, 50, 52, 93–98]. While ESI has been reported to be much more prone to such effects, they may also occur with atmospheric pressure chemical ionisation (APCI) [38, 39, 50, 93–96, 98]. It is obvious that ion suppression as well as ion enhancement may affect validation parameters such as LOD, LLOQ, linearity, precision and/or bias. Sojo *et al.* [99] demonstrated that mutual suppression of analytes and their stable-isotope-labelled ISs in the ESI mode should not affect quantification, but they were concerned about a negative influence on the LOD if too high concentrations of IS were used. In principle, these findings were confirmed by Liang *et al.* [95], who also reported mutual suppression of analytes and IS in the ESI mode resulting in negative effects on LOD and LLOQ. They also confirmed the findings concerning linearity, but only for certain IS concentrations. For the APCI mode, they reported mutual ionisation enhancement for part of the analytes and their ISs, possibly improving LOD and LLOQ data. Furthermore, the calibration curves were linear when an appropriate concentration of the IS was chosen. From these two studies, one might conclude that, while the limits might be affected by ion suppression/enhancement, quantification would not as long as a stable-isotope-labelled IS was used. However, this is not always true, as could be demonstrated by Jemal *et al.* [77], who reported on certain batches of urine ME that lead to changes in the response ratio (analyte versus IS), which could have a negative influence on quantification. Worse problems might have to be expected when no stable-isotope-labelled IS is available.

All this clearly shows that studies of ion suppression/enhancement should be an integral part of the validation of any LC-MS(-MS) method. This is also in line with Conference Report II [21], which in the case of LC-MS(-MS) procedures explicitly recommends studies on possible ME 'to ensure that precision, selectivity and sensitivity will not be compromised'. However, no specific experimental design was proposed in this document. In the literature, two approaches have been used extensively to study ion suppression/enhancements. In the first approach, a solution of the analyte is constantly infused into the eluent from the column via a post-column T-connection using a syringe pump. The continuous post-column infusion leads to a constant signal in the detector, unless compounds that elute from the column suppress or enhance ionisation, which would lead to a decreased or increased detector response, respectively. Thus, monitoring of the detector response after injection of blank matrix extracts can be used to check for possible ion suppression/enhancement of blank matrix compounds and for their retention times. Applications of this approach can be found in the literature [49, 82, 83, 93–97]. Strategies for the second approach, which are based on their previous work [100], were recently published by Matuszewski *et al.* [39]. This paper provides excellent guidance on how to perform and evaluate studies on ME in LC-MS(-MS) analysis. The principal approach involves determination of peak areas of analyte in three different sets of samples, one consisting of neat standards (set 1), one prepared in blank matrix extracts from different sources and spiked after extraction (set 2), and one prepared in blank matrix from the same sources, but spiked before extraction (set 3). From these data, one can then calculate the ME (ion suppression/enhancement), RE and process efficiencies (PE) according to equations (7) and (8), respectively [39].

$$\text{ME (\%)} = \frac{B}{A} \cdot 100 \qquad (7)$$

where ME is the matrix effect (ion suppression/enhancement), B is the peak area in sample from set 2 (spiked blank matrix extract) and A is the peak area in the sample from set 1 (neat standard).

$$\text{RE (\%)} = \frac{C}{B} \cdot 100 \qquad (8)$$

where RE is the extraction efficiency (extraction recovery), C is the peak area in the sample from set 3 (extract of spiked blank matrix) and B is the peak area in the sample from set 2 (spiked blank matrix extracts).

$$\text{PE (\%)} = \frac{C}{A} \cdot 100 = \frac{\text{ME} \cdot \text{RE}}{100} \qquad (9)$$

where PE is the process efficiency, C is the peak area in sample from set 3 (extract of spiked blank matrix), A is the peak area in the sample from set 1 (neat standard), ME is the matrix effect (ion suppression/enhancement) and RE is the extraction efficiency (extraction recovery).

The full experimental design proposed in Matuszewski et al. [39] involves analysis of 105 samples (three sets with seven concentration levels and $n = 5$ per level). Thus, a tremendous amount of data is acquired which can provide valuable information concerning the performance of the studied assay. As analysis of such large numbers of samples is also time-consuming and expensive, a number of simplified experimental designs are also described in this paper [39]. Other examples using the same or very similar approaches for evaluation of ME can be found in [39, 45, 48, 57, 59–61, 69, 77, 81, 83, 89, 100].

With two well-established procedures for studying MEs, the question arises of which one of them is best suited for validation studies. In the author's opinion, the post-column infusion experiments is very useful during method development because it provides information on the retention times where ion suppression/enhancement has to be expected, which can then be avoided by optimising the separation system. For a validation study, the alternative approach seems to be more suitable because it yields a quantitative estimation of MEs and their variability, and is thus more objective. However, no matter which approach is used in validation studies, it is essential to evaluate several sources of blank matrix [38, 39], just as it has been described for the selectivity experiments. Otherwise, MEs caused by less-frequently occurring matrix compounds might be overlooked. Unfortunately, this important point has not been taken into account in many publications so far [39, 49, 77, 89, 90, 93, 100]. Finally, possible mutual suppression or enhancement by co-eluting analytes has to be considered, which is particularly important in the field of toxicology, where multi-analyte procedures are often used. An example for mutual ion suppression by co-eluting analytes has been reported in Egge-Jacobsen et al. [64].

Experimental design for method validation

In the following, a rational experimental design for method validation using LC-MS(-MS) will be described, which is summarized in Table 4.1. It is based on the experimental design proposed by Wieling et al. [28], the modifications introduced by the author of this chapter [101] and, of course, the considerations presented above. Similar experimental designs have been used in several LC-MS publications from the group of Maurer et al. [42–44].

It is recommended to start the validation studies with the selectivity and ion-suppression/enhancement experiments because if any of these two parameters is not acceptable, major changes of the method might be required. For the reasons outlined above, ten instead of only six sources of blank matrix are evaluated to exclude interference from the blank matrix. The zero samples (blank + IS) are analysed to exclude interference from the IS. If appropriate, spiking experiments with compounds likely to be present in real samples should also be performed. If no interferences are detected during these selectivity experiments, the next step is to check for a possible ME, i.e. ion suppression or enhancement. For the reasons given above, an approach similar to the one used by Matuszewski et al. [39] should be used. However, a simplified design should be sufficient as long as it includes analysis of different sources of blank matrix. In the design proposed in Table 4.1, five sources of blank matrix are analysed, and extracts of spiked samples are also included to allow simultaneous determination of the ME, RE and PE using the calculations described above. If there is no

Table 4.1 Experimental design for the validation of LC-MS(-MS) methods

Run	Selectivity	ME/RE/PE	Processed sample stability	Linearity
	10 different sources of blank matrix, two zero samples, x spiked samples	30 samples (five neat standards, five spiked blank extracts and five extracts of spiked blanks at each of two concentrations)	16 injections of extracts (at certain time intervals, eight at each of two concentrations)	36 samples (six concentration levels, six replicates each)
Total	94 (+x)			

Run	Calibration samples (six levels)	Validation samples						Optional				Total
		Low		Medium	High			LLOQ	Dilution		
		Precision and accuracy	Freeze/thaw stability	Precision and accuracy	Precision and accuracy		Freeze/thaw stability		Precision and accuracy	Precision and accuracy		
1	6	2	6	2	2		6		2	2		24 (+4)
2	6	2		2	2				2	2		12 (+4)
3	6	2		2	2				2	2		12 (+4)
4	6	2	6	2	2		6		2	2		24 (+4)
5	6	2		2	2				2	2		12 (+4)
6	6	2		2	2				2	2		12 (+4)
7	6	2		2	2				2	2		12 (+4)
8	6	2		2	2				2	2		12 (+4)
Total	120 (+32)											

For details, see text.

relevant ME, the processed sample stability should be assessed to ensure stability of processed samples under the conditions on the auto-sampler tray during analysis of large batches of samples. For this purpose, 10 extracts each at low and high concentration can be pooled, thoroughly mixed and divided again into aliquots which are then injected at certain time intervals until the maximum expected run-time is reached [101]. After plotting the observed peak areas versus injection time, stability can be checked by regression analysis as described above. If the processed samples are stable, one can proceed to the linearity experiments; if not, one must try to stabilise the processed samples, e.g. by cooling the auto-sampler tray, or further optimise the method. For the evaluation of the calibration model (linearity experiments) six concentration levels are analysed in replicates of six. This allows one to check for outliers and to study the behaviour of variance across the calibration range as well as the evaluation of non-linear models, at least in a preliminary fashion. After an appropriate calibration model has been established, the early validation phase is complete and one can proceed to the main validation phase.

During the main validation phase, bias and precision as well as freeze/thaw stability are evaluated. For determination of bias and precision, duplicate quality control samples are analysed at a minimum of three concentration levels on each of 8 days. Where applicable, a quality control sample at very high concentrations can be analysed in the same fashion after appropriate dilution to check if reliable quantification should be possible after dilution of real samples containing very high concentrations. In addition, quality control samples corresponding to the LLOQ can be analysed in the same way to investigate if precision and bias are acceptable at such low concentrations. From the data obtained during these experiments, the precision components and intermediate precision can be calculated using one-way ANOVA as described above. The bias values can be calculated as the percentage deviations of the observed mean values from the nominal concentrations of the respective quality control samples. It should be noted that a calibration curve is run on each of the 8 days in order to include variability of daily calibration curves into the precision and bias estimates. Performing the freeze/thaw experiments together with the precision and bias experiments allows using the daily calibration curves also for the stability evaluations. Freshly prepared control samples are analysed on the first day to obtain the control value. Samples from the same pools are then frozen and thawed three times before they are analysed based on the calibration curve of the fourth day. The initial mean values and those obtained after storage are then compared as described above. If the stability samples are kept at room temperature for the time needed to prepare a batch of samples before they are frozen again, the freeze/thaw stability evaluation also includes bench-top stability. As mentioned above, a detailed description of a similar experimental design can be found in Peters *et al.* [101].

Conclusions

LC-MS(-MS) methods are becoming increasingly important in the field of forensic and clinical toxicology. Before their integration into routine analysis, they have to be validated to prove their suitability for the intended purpose. Apart from the general recommendations of (bio)analytical method validation, particular aspects have to be considered in the validation of LC-MS(-MS) procedures in forensic and clinical toxicology. One is the need to study possible MEs (ion suppression/enhancement), which are unique to the LC-MS(-MS) technique, particularly in the ESI mode. The other is the special importance of the validation parameters' selectivity, LOD, LLOQ and stability for analytical methods to be applied in forensic and/or clinical toxicology. The presented experimental design allows determination of the most important validation parameters, but additional validation experiments may be necessary for certain analytical problems.

References

1. Hoja H, Marquet P, Verneuil B, *et al.* Applications of liquid chromatography-mass spectrometry in

analytical toxicology: a review. *J Anal Toxicol* 1997; 21: 116–126.

2. Maurer HH. Liquid chromatography-mass spectrometry in forensic and clinical toxicology [review]. *J Chromatogr B* 1998; 713: 3–25.

3. Bogusz M. Hyphenated liquid chromatographic techniques in forensic toxicology [review]. *J Chromatogr B* 1999; 733: 65–91.

4. Marquet P, Lachatre G. Liquid chromatography-mass spectrometry: potential in forensic and clinical toxicology [review]. *J Chromatogr B* 1999; 733: 93–118.

5. Polettini A. Systematic toxicological analysis of drugs and poisons in biosamples by hyphenated chromatographic and spectroscopic techniques [review]. *J Chromatogr B* 1999; 733: 47–63.

6. Maurer HH. Screening procedures for simultaneous detection of several drug classes used in the high throughput toxicological analysis and doping control [review]. *Comb Chem High Throughput Screen* 2000; 3: 461–474.

7. Van Bocxlaer JF, Clauwaert KM, Lambert WE, *et al.* Liquid chromatography-mass spectrometry in forensic toxicology [review]. *Mass Spectrom Rev* 2000; 19: 165–214.

8. Marquet P. Progress of LC-MS in clinical and forensic toxicology [review]. *Ther Drug Monit* 2002; 24: 255–276.

9. Maurer HH. Position of chromatographic techniques in screening for detection of drugs or poisons in clinical and forensic toxicology and/or doping control [review]. *Clin Chem Lab Med* 2004; 42: 1310–1324.

10. Maurer HH. Advances in analytical toxicology: the current role of liquid chromatography-mass spectrometry for drug quantification in blood and oral fluid [review]. *Anal Bioanal Chem* 2005; 381: 110–118.

11. Maurer HH. Multi-analyte procedures for screening for and quantification of drugs in blood, plasma or serum by liquid chromatography-single stage or tandem mass spectrometry (LC-MS or LC-MS/MS) relevant to clinical and forensic toxicology [review]. *Clin Biochem* 2005; 38: 310–318.

12. Massart DL, Vandeginste BGM, Buydens LMC, *et al. Handbook of Chemometrics and Qualimetrics: Part A*, 1st edn. Amsterdam: Elsevier, 1997.

13. International Conference on Harmonization. *Validation of Analytical Methods: Definitions and Terminology ICH Q2 A*. Geneva: ICH, 1994.

14. International Conference on Harmonization. *Validation of Analytical Methods: Methodology ICH Q2 B*. Geneva: ICH, 1996.

15. EURACHEM. *The Fitness for Purpose of analytical Methods – A Laboratory Guide To Method Validation and Related Topics*. Teddington: EURACHEM, 1998.

16. Thompson M, Ellison SLR, Wood R. Harmonized guidelines for single-laboratory validation of methods of analysis [IUPAC technical report]. *Pure Appl Chem* 2002; 74: 835–855.

17. Taverniers I, De Loose M, Van Bockstaele E. Trends in quality in the analytical laboratory. II. Analytical method validation and quality assurance. *Trends Anal Chem* 2004; 23: 535–552.

18. Trullols E, Ruisanchez I, Rius FX. Validation of qualitative analytical methods. *Trends Anal Chem* 2004; 23: 137–145.

19. Epshtein NA. Validation of HPLC techniques for pharmaceutical analysis. *Pharm Chem J* (translation of *Khimiko-Farmatsevticheskii Zhurnal*) 2004; 38: 212–228.

20. Shah VP, Midha KK, Dighe S, *et al.* Analytical methods validation: bioavailability, bioequivalence and pharmacokinetic studies. Conference report. *Pharm Res* 1992; 9: 588–592.

21. Shah VP, Midha KK, Findlay JW, *et al.* Bioanalytical method validation – a revisit with a decade of progress. *Pharm Res* 2000; 17: 1551–1557.

22. US Department of Health and Human Services. *Guidance for Industry. Bioanalytical Method Validation*. Rockville, MD: FDA, 2001. www.fda.gov/cder/guidance/index.htm (accessed 6 June 2005).

23. Hartmann C, Smeyers-Verbeke J, Massart DL, *et al.* Validation of bioanalytical chromatographic methods. *J Pharm Biomed Anal* 1998; 17: 193–218.

24. Peters FT, Maurer HH. Bioanalytical method validation and its implications for forensic and clinical toxicology – a review [review]. *Accred Qual Assur* 2002; 7: 441–449.

25. Lindner W, Wainer IW. Requirements for initial assay validation and publication in *J Chromatogr B* [editorial]. *J Chromatogr B* 1998; 707: 1–2.

26. Bressolle F, Bromet PM, Audran M. Validation of liquid chromatographic and gas chromatographic

methods. Applications to pharmacokinetics. *J Chromatogr B* 1996; 686: 3–10.

27. Causon R. Validation of chromatographic methods in biomedical analysis. Viewpoint and discussion. *J Chromatogr B* 1997; 689: 175–180.

28. Wieling J, Hendriks G, Tamminga WJ, *et al.* Rational experimental design for bioanalytical methods validation. Illustration using an assay method for total captopril in plasma. *J Chromatogr A* 1996; 730: 381–394.

29. Dadgar D, Burnett PE. Issues in evaluation of bioanalytical method selectivity and drug stability. *J Pharm Biomed Anal* 1995; 14: 23–31.

30. Dadgar D, Burnett PE, Choc MG, *et al.* Application issues in bioanalytical method validation, sample analysis and data reporting. *J Pharm Biomed Anal* 1995; 13: 89–97.

31. Marquet P, Saint-Marcoux F, Gamble TN, *et al.* Comparison of a preliminary procedure for the general unknown screening of drugs and toxic compounds using a quadrupole-linear ion-trap mass spectrometer with a liquid chromatography-mass spectrometry reference technique. *J Chromatogr B* 2003; 789: 9–18.

32. Marquet P. Is LC-MS suitable for a comprehensive screening of drugs and poisons in clinical toxicology? *Ther Drug Monit* 2002; 24: 125–133.

33. Pelander A, Ojanpera I, Laks S, *et al.* Toxicological screening with formula-based metabolite identification by liquid chromatography/time-of-flight mass spectrometry. *Anal Chem* 2003; 75: 5710–5718.

34. Thieme D, Grosse J, Lang R, *et al.* Screening, confirmation and quantification of diuretics in urine for doping control analysis by high-performance liquid chromatography-atmospheric pressure ionisation tandem mass spectrometry. *J Chromatogr B* 2001; 757: 49–57.

35. Weinmann W, Wiedemann A, Eppinger B, *et al.* Screening for drugs in serum by electrospray ionization/collision-induced dissociation and library searching. *J Am Soc Mass Spectrom* 1999; 10: 1028–1037.

36. Jimenez C, Ventura R, Segura J. Validation of qualitative chromatographic methods: strategy in antidoping control laboratories. *J Chromatogr B* 2002; 767: 341–351.

37. Rivier L. Criteria for the identification of compounds by liquid chromatography-mass spectrometry and liquid chromatography-multiple mass spectrometry in forensic toxicology and doping analysis. *Anal Chim Acta* 2003; 492: 69–82.

38. Annesley TM. Ion suppression in mass spectrometry. *Clin Chem* 2003; 49: 1041–1044.

39. Matuszewski BK, Constanzer ML, Chavez-Eng CM. Strategies for the assessment of matrix effect in quantitative bioanalytical methods based on HPLC-MS/MS. *Anal Chem* 2003; 75: 3019–3030.

40. Karnes HT, Shiu G, Shah VP. Validation of bioanalytical methods. *Pharm Res* 1991; 8: 421–426.

41. Sottani C, Bettinelli M, Lorena FM, *et al.* Analytical method for the quantitative determination of urinary ethylenethiourea by liquid chromatography/electrospray ionization tandem mass spectrometry. *Rapid Commun Mass Spectrom* 2003; 17: 2253–2259.

42. Kratzsch C, Weber AA, Peters FT, *et al.* Screening, library-assisted identification and validated quantification of fifteen neuroleptics and three of their metabolites in plasma by liquid chromatography/mass spectrometry with atmospheric pressure chemical ionization. *J Mass Spectrom* 2003; 38: 283–295.

43. Kratzsch C, Tenberken O, Peters FT, *et al.* Screening, library-assisted identification and validated quantification of twenty-three benzodiazepines, flumazenil, zaleplone, zolpidem and zopiclone in plasma by liquid chromatography/mass spectrometry with atmospheric pressure chemical ionization. *J Mass Spectrom* 2004; 39: 856–872.

44. Maurer HH, Tenberken O, Kratzsch C, *et al.* Screening for, library-assisted identification and fully validated quantification of twenty-two beta-blockers in blood plasma by liquid chromatography-mass spectrometry with atmospheric pressure chemical ionization. *J Chromatogr A* 2004; 1058: 169–181.

45. Pichini S, Pacifici R, Pellegrini M, *et al.* Development and validation of a liquid chromatography-mass spectrometry assay for the determination of opiates and cocaine in meconium. *J Chromatogr B* 2003; 794: 281–292.

46. Alexander MS, Kiser MM, Culley T, *et al.* Measurement of paclitaxel in biological matrices: high-throughput liquid chromatographic-tandem mass spectrometric quantification of paclitaxel and metabolites in human and dog plasma. *J Chromatogr B* 2003; 785: 253–261.

47. De Jonge ME, Van Dam SM, Hillebrand MJX, et al. Simultaneous quantification of cyclophosphamide, 4-hydroxycyclophosphamide, N,N',N''-triethylenethiophosphoramide (thiotepa) and N,N',N''-triethylenephosphoramide (tepa) in human plasma by high-performance liquid chromatography coupled with electrospray ionization tandem mass spectrometry. *J Mass Spectrom* 2004; 39: 262–271.

48. Crommentuyn KML, Rosing H, Hillebrand MJX, et al. Simultaneous quantification of the new HIV protease inhibitors atazanavir and tipranavir in human plasma by high-performance liquid chromatography coupled with electrospray ionization tandem mass spectrometry. *J Chromatogr B* 2004; 804: 359–367.

49. Streit F, Shipkova M, Armstrong VW, et al. Validation of a rapid and sensitive liquid chromatography-tandem mass spectrometry method for free and total mycophenolic acid. *Clin Chem* 2004; 50: 152–159.

50. Ackermann BL, Berna MJ, Murphy AT. Recent advances in use of LC/MS/MS for quantitative high-throughput bioanalytical support of drug discovery. *Curr Top Med Chem* 2002; 2: 53–66.

51. Zhang N, Yang A, Rogers JD, et al. Quantitative analysis of simvastatin and its beta-hydroxy acid in human plasma using automated liquid-liquid extraction based on 96-well plate format and liquid chromatography-tandem mass spectrometry. *J Pharm Biomed Anal* 2004; 34: 175–187.

52. Sun H, Naidong W. Narrowing the gap between validation of bioanalytical LC-MS-MS and the analysis of incurred samples. *Pharm Technol* 2003; 27: 74–86.

53. Jemal M, Zheng OY, Powell ML. A strategy for a post-method-validation use of incurred biological samples for establishing the acceptability of a liquid chromatography/tandem mass-spectrometric method for quantitation of drugs in biological samples. *Rapid Commun Mass Spectrom* 2002; 16: 1538–1547.

54. Janda I, Weinmann W, Kuehnle T, et al. Determination of ethyl glucuronide in human hair by SPE and LC-MS/MS. *Forensic Sci Int* 2002; 128: 59–65.

55. Bogusz MJ. Large amounts of drugs may considerably influence the peak areas of their coinjected deuterated analogues measured with APCI-LC-MS [letter]. *J Anal Toxicol* 1997; 21: 246–247.

56. Penninckx W, Hartmann C, Massart DL, et al. Validation of the calibration procedure in atomic absorption spectrometric methods. *J Anal At Spectrom* 1996; 11: 237–246.

57. Breda M, Basileo G, James CA. Simultaneous determination of estramustine phosphate and its four metabolites in human plasma by liquid chromatography-ionspray mass spectrometry. *Biomed Chromatogr* 2004; 18: 293–301.

58. Pereira AS, Kenney KB, Cohen MS, et al. Simultaneous determination of lamivudine and zidovudine concentrations in human seminal plasma using high-performance liquid chromatography and tandem mass spectrometry. *J Chromatogr B* 2000; 742: 173–183.

59. Mortier KA, Maudens KE, Lambert WE, et al. Simultaneous, quantitative determination of opiates, amphetamines, cocaine and benzoylecgonine in oral fluid by liquid chromatography quadrupole-time-of-flight mass spectrometry. *J Chromatogr B* 2002; 779: 321–330.

60. Li AC, Sabo AM, McCormick T, et al. Quantitative analysis of squalamine, a self-ionization-suppressing aminosterol sulfate, in human plasma by LC-MS/MS. *J Pharm Biomed Anal* 2004; 34: 631–641.

61. Mortier KA, Dams R, Lambert WE, et al. Determination of paramethoxyamphetamine and other amphetamine-related designer drugs by liquid chromatography/sonic spray ionization mass spectrometry. *Rapid Commun Mass Spectrom* 2002; 16: 865–870.

62. Chang SY, Whigan D, Shyu WC, et al. Sensitive triple-quadrupole mass spectrometric assay for the determination of BMS-181885, a 5-HT1 agonist, in human plasma following solid phase extraction. *Biomed Chromatogr* 1999; 13: 425–430.

63. Jemal M, Ouyang Z, Chen BC, et al. Quantitation of the acid and lactone forms of atorvastatin and its biotransformation products in human serum by high-performance liquid chromatography with electrospray tandem mass spectrometry. *Rapid Commun Mass Spectrom* 1999; 13: 1003–1015.

64. Egge-Jacobsen W, Unger M, Niemann CU, et al. Automated, fast, and sensitive quantification of drugs in human plasma by LC/LC-MS: quantification of 6 protease inhibitors and 3 nonnucleoside transcriptase inhibitors. *Ther Drug Monit* 2004; 26: 546–562.

65. Hartmann C, Massart DL, McDowall RD. An

analysis of the Washington Conference Report on bioanalytical method validation. *J Pharm Biomed Anal* 1994; 12: 1337–1343.

66. International Organization for Standardization. *Accuracy (Trueness and Precision) of Measurement Methods and Results (ISO/DIS 5725–1 to 5725–3).* Geneva: ISO, 1994.

67. Tracqui A, Kintz P, Ludes B, *et al.* High-performance liquid chromatography coupled to ion spray mass spectrometry for the determination of colchicine at ppb levels in human biofluids. *J Chromatogr B* 1996; 675: 235–242.

68. Nordgren HK, Beck O. Multicomponent screening for drugs of abuse: direct analysis of urine by LC-MS-MS. *Ther Drug Monit* 2004; 26: 90–97.

69. He X, Batchelor TT, Grossman S, *et al.* Determination of procarbazine in human plasma by liquid chromatography with electrospray ionization mass spectrometry. *J Chromatogr B* 2004; 799: 281–291.

70. Krouwer JS, Rabinowitz R. How to improve estimates of imprecision. *Clin Chem* 1984; 30: 290–292.

71. Massart DL, Vandeginste BGM, Buydens LMC, *et al.* Internal method validation. In: Vandeginste BGM, Rutan SC, eds, *Handbook of Chemometrics and Qualimetrics: Part A*, 1st edn. Amsterdam: Elsevier, 1997: 379–440.

72. National Committee for Clinical Laboratory Standards. *Evaluation of Precision Performance of Clinical Chemistry Devices: Approved Guideline. Document EP5-A.* Wayne, PA: NCCLS, 1999.

73. Bennett PK, Li YT, Edom R, *et al.* Quantitative determination of Orlistat (tetrahydrolipostatin, Ro 18–0647) in human plasma by high-performance liquid chromatography coupled with ion spray tandem mass spectrometry. *J Mass Spectrom* 1997; 32: 739–749.

74. Jemal M, Mulvana DE. Liquid chromatographic-electrospray tandem mass spectrometric method for the simultaneous quantitation of the prodrug fosinopril and the active drug fosinoprilat in human serum. *J Chromatogr B* 2000; 739: 255–271.

75. Jemal M, Ouyang Z, Powell ML. Direct-injection LC-MS-MS method for high-throughput simultaneous quantitation of simvastatin and simvastatin acid in human plasma. *J Pharm Biomed Anal* 2000; 23: 323–340.

76. Xue YJ, Turner KC, Meeker JB, *et al.* Quantitative determination of pioglitazone in human serum by direct-injection high-performance liquid chromatography mass spectrometry and its application to a bioequivalence study. *J Chromatogr B* 2003; 795: 215–226.

77. Jemal M, Schuster A, Whigan DB. Liquid chromatography/tandem mass spectrometry methods for quantitation of mevalonic acid in human plasma and urine: method validation, demonstration of using a surrogate analyte, and demonstration of unacceptable matrix effect in spite of use of a stable isotope analog internal standard. *Rapid Commun Mass Spectrom* 2003; 17: 1723–1734.

78. Xue Y-J, Pursley J, Arnold ME. A simple 96-well liquid–liquid extraction with a mixture of acetonitrile and methyl *t*-butyl ether for the determination of a drug in human plasma by high-performance liquid chromatography with tandem mass spectrometry. *J Pharm Biomed Anal* 2004; 34: 369–378.

79. Yao M, Shah VR, Shyu WC, *et al.* Sensitive liquid chromatographic-mass spectrometric assay for the simultaneous quantitation of nefazodone and its metabolites hydroxynefazodone m-chlorophenylpiperazine and triazole-dione in human plasma using single-ion monitoring. *J Chromatogr B* 1998; 718: 77–85.

80. Maurer HH, Kratzsch C, Weber AA, *et al.* Validated assay for quantification of oxcarbazepine and its active dihydro metabolite 10-hydroxy carbazepine in plasma by atmospheric pressure chemical ionization liquid chromatography/mass spectrometry. *J Mass Spectrom* 2002; 37: 687–692.

81. Scheidweiler KB, Huestis MA. Simultaneous quantification of opiates, cocaine, and metabolites in hair by LC-APCI-MS/MS. *Anal Chem* 2004; 76: 4358–4363.

82. Fan B, Bartlett MG, Stewart JT. Determination of lamivudine/stavudine/efavirenz in human serum using liquid chromatography/electrospray tandem mass spectrometry with ionization polarity switch. *Biomed Chromatogr* 2002; 16: 383–389.

83. Naidong W, Bu H, Chen YL, *et al.* Simultaneous development of six LC-MS-MS methods for the determination of multiple analytes in human plasma. *J Pharm Biomed Anal* 2002; 28: 1115–1126.

84. Skopp G, Potsch L. Stability of 11-nor-delta(9)-carboxy-tetrahydrocannabinol glucuronide in plasma and urine assessed by liquid chromatography-tandem mass spectrometry. *Clin Chem* 2002; 48: 301–306.

85. Weinmann W, Schaefer P, Thierauf A, *et al.* Confirmatory analysis of ethylglucuronide in urine by liquid-chromatography/electrospray ionization/tandem mass spectrometry according to forensic guidelines. *J Amer Soc Mass Spectrom* 2004; 15: 188–193.

86. Weinmann W, Goerner M, Vogt S, *et al.* Fast confirmation of 11-nor-9-carboxy-delta(9)-tetrahydrocannabinol (THC-COOH) in urine by LC/MS/MS using negative atmospheric-pressure chemical ionisation (APCI). *Forensic Sci Int* 2001; 121: 103–107.

87. Lin SN, Moody DE, Bigelow GE, *et al.* A validated liquid chromatography-atmospheric pressure chemical ionization-tandem mass spectrometry method for quantitation of cocaine and benzoylecgonine in human plasma. *J Anal Toxicol* 2001; 25: 497–503.

88. Naidong W, Eerkes A. Development and validation of a hydrophilic interaction liquid chromatography-tandem mass spectrometric method for the analysis of paroxetine in human plasma. *Biomed Chromatogr* 2004; 18: 28–36.

89. Ramakrishna NVS, Vishwottam KN, Puran S, *et al.* Quantitation of tadalafil in human plasma by liquid chromatography-tandem mass spectrometry with electrospray ionization. *J Chromatogr B* 2004; 809: 243–249.

90. Shou WZ, Jiang X, Naidong W. Development and validation of a high-sensitivity liquid chromatography/tandem mass spectrometry (LC/MS/MS) method with chemical derivatization for the determination of ethinyl estradiol in human plasma. *Biomed Chromatogr* 2004; 18: 414–421.

91. Toennes SW, Kauert GF. Importance of vacutainer selection in forensic toxicological analysis of drugs of abuse. *J Anal Toxicol* 2001; 25: 339–343.

92. Vander-Heyden Y, Nijhuis A, Smeyers-Verbeke J, *et al.* Guidance for robustness/ruggedness tests in method validation. *J Pharm Biomed Anal* 2001; 24: 723–753.

93. Souverain S, Rudaz S, Veuthey J-L. Matrix effect in LC-ESI-MS and LC-APCI-MS with off-line and on-line extraction procedures. *J Chromatogr A* 2004; 1058: 61–66.

94. Mallet CR, Lu Z, Mazzeo JR. A study of ion suppression effects in electrospray ionization from mobile phase additives and solid-phase extracts. *Rapid Commun Mass Spectrom* 2004; 18: 49–58.

95. Liang HR, Foltz RL, Meng M, *et al.* Ionization enhancement in atmospheric pressure chemical ionization and suppression in electrospray ionization between target drugs and stable-isotope-labeled internal standards in quantitative liquid chromatography/tandem mass spectrometry. *Rapid Commun Mass Spectrom* 2003; 17: 2815–2821.

96. Dams R, Huestis MA, Lambert WE, *et al.* Matrix effect in bioanalysis of illicit drugs with LC-MS/MS: influence of ionization type, sample preparation, and biofluid. *J Am Soc Mass Spectrom* 2003; 14: 1290–1294.

97. Muller C, Schafer P, Stortzel M, *et al.* Ion suppression effects in liquid chromatography-electrospray-ionisation transport-region collision induced dissociation mass spectrometry with different serum extraction methods for systematic toxicological analysis with mass spectra libraries. *J Chromatogr B* 2002; 773: 47–52.

98. King R, Bonfiglio R, Fernandez-Metzler C, *et al.* Mechanistic investigation of ionization suppression in electrospray ionization. *J Am Soc Mass Spectrom* 2000; 11: 942–950.

99. Sojo LE, Lum G, Chee P. Internal standard signal suppression by co-eluting analyte in isotope dilution LC-ESI-MS. *Analyst* 2003; 128: 51–54.

100. Matuszewski BK, Constanzer ML, Chavez EC. Matrix effect in quantitative LC/MS/MS analyses of biological fluids: a method for determination of finasteride in human plasma at picogram per milliliter concentrations. *Anal Chem* 1998; 70: 882–889.

101. Peters FT, Schaefer S, Staack RF, *et al.* Screening for and validated quantification of amphetamines and of amphetamine- and piperazine-derived designer drugs in human blood plasma by gas chromatography/mass spectrometry. *J Mass Spectrom* 2003; 38: 659–676.

5

Identification and confirmation criteria for LC-MS

Laurent Rivier

Introduction

Demands for proper identification in clinical and forensic toxicology are growing in accordance with the implementation of new quality assurance schemes and the forever changing drug scene. Most of the time, the analytical toxicologist will not know exactly what kind of chemicals (poisons, therapeutic drugs, drugs of abuse, pollutants, etc.) to look for, but he or she has to bring unequivocal evidence of the presence or absence of chemically well-defined substances in specified biological matrices [1]. There is a clear need for implementing some basic rules concerning the performance of analytical methods and the interpretation of results in order to obtain the necessary and correct evidence from the laboratory. As a result of advances in analytical chemistry, the concept of routine and reference methods has been superseded by the criteria approach, in which performance criteria and procedures for validation of screening, identification and confirmation methods have been established [2].

By definition, identification is obtained by a rather complex procedure. It is considered to take place when the analytical results collected from a number of tests on a substance in a sample are compared with reference data on all relevant substances that may come into consideration. When the results obtained for the substance fit with those of a single and unique molecule, identification is achieved. This process does not make any presumptions on the presence of any possible compounds in that sample. However, when investigating relevant analytical problems in clinical or forensic toxicology, many toxins, poisons, natural or synthetic drugs and related xenobiotics can be present. Endogenous organic substances belonging to each biological matrix also have to be included as well as artefacts and impurities contained in reagents. Some drugs are metabolised very quickly (e.g. heroin) and are not detectable at all in the blood. Metabolites of parent drugs may remain in the body for a longer period, especially in urine. Most of the toxicologically relevant drugs are small, stable organic compounds, ranging from 70 to 700 amu in molecular weight. The geographical area where the sample has been collected may also play a certain role in terms of the accessibility to therapeutic drugs as this varies from country to country. Similarly, different vegetal or animal poisons may be encountered at different latitudes. Therefore, it is imperative to remember that only a portion of the 6 billion or more chemically defined substances in existence should be considered at any one time.

Each laboratory has to carefully define the size of that portion to be included in their identification/confirmation procedures [3]; the importance or relevance of what is included is nevertheless open to discussion within the toxicological community [4].

However, confirmation presumes the presence of a substance in a sample, based on initial tests or prior information. The presence of the substance in the sample can then be confirmed by further tests. Moreover, a confirmation test does not always correspond to the unambiguous identification of the substance because it may indicate only that the result of the test does not contradict the first presumption (other compounds might be able to give the same test results as the presumed substance). This is quite different from the process of identifying a substance where only one compound unambiguously remains as the only possible candidate when all (relevant) substances have been excluded in the matrix [5]. Consequently, confirmation can be considered as a part of the whole identification process.

The identification process is a key step in forensic and clinical toxicology because any result coming out of the laboratory should be as close as possible to the reality of the sample's composition. Errors are not permitted and the analyst should bring considerable effort and skill to ensure proper reporting. If any doubt in identification remains, it should be clearly indicated in the reporting of results.

Procedures for the identification of organic compounds in biological matrices are numerous and novel technologies or latest instrument developments add significant benefits to the overall efficacy of the analytical strategy. In short, the acceptability of the analytical results released by the toxicological laboratory depends on the quality or performance capabilities of the methods used. Due to the high requirements for quality, forensic and clinical toxicology laboratories have to use highly specific and selective methods, which also need to be repeatable and reproducible, and hence a degree of method harmonisation must be implemented. Mass spectrometry (MS) linked to a suitable separation device is often considered as the best tool for achieving the formal unambiguous identification of low amounts of organic molecules in biosamples. This is true to an extent, but only if the technique is applied properly and knowledgeably. Formal education in fundamental MS is nowadays very rare and toxicologists are trained most of the time 'on the spot'. This situation may lead to dangerous interpretations taken directly from the instrument output, whereas it is vital to critically assess the analytical data and put these into context. This situation is becoming even more acute with the introduction of new and diversified instrument designs in liquid chromatography (LC)-MS.

In contrast to gas chromatography (GC)-MS, for which ionisation/fragmentation processes of an organic molecule can be very stable when standardised, LC-MS exhibits a high degree of complexity as several instrument parameters may influence the ionisation process and the fragmentation pattern. It is therefore very difficult or even impossible to rely on any previously acquired collection of reference data. This is amplified by the fact that different instrument designs from different manufacturers will not provide readily comparable results.

As in most instances the identification procedure will be contested and has to stand up to the scrutiny of the courts, the analyst has to rely on sound directives and have confidence in his or her methods. Several guidelines have been released that are very helpful when applied in concrete cases. The Society of Forensic Toxicologists (SOFT) together with the American Academy of Forensic Sciences (AAFS) has published a set of *Forensic Laboratory Guidelines* to that effect [6]. In essence, the guidelines rest on the principle of accumulating a finite quantity of information about the unknown molecule and the corresponding reference substance. However, even when clear rules have been openly disclosed in professional circles, they do not remain mandatory for all. Scientists do not feel obliged to apply such guidelines, and this invariably leads to the release of erroneous, contradictory and/or confusing results. It is advisable that these laboratories adopt identification criteria, and include those in their reports in order to gain a credible standing within the community and in the courts

Identification

The general criteria for identification procedures are simple to express, but actually complex to achieve. The method has to be able to distinguish between the analyte and all known interfering materials (i.e. other drugs with closely related structures, metabolites, etc.) that may possibly occur in the matrix, and the physical and chemical behaviour of the substance during the analysis has to be indistinguishable from that of the corresponding reference substance in the corresponding matrix. Hence, detailed information about the molecular structure of the analyte is essential.

In practice, identification is achieved by applying step-by-step extraction, isolation and chromatographic or other separations, which allow the scientist to safely discriminate the substance of interest from any possible interferences.

Each analytical technique possesses a certain discrimination power of its own and a definite number of identification points (IPs) can be attributed to each measurement [7]. The total number of IPs obtained is the sum of each IP derived from each individual analytical step of the method, ranging from partition coefficients to fragmentation patterns in MS. This procedure is called the '*ad hoc* confirmation package' in the veterinary area [8].

It is important in LC-MS to differentiate among the many possible instrumental configurations (e.g. single MS versus tandem MS [MS-MS], low-mass versus high-mass resolution) as IP earnings can differ strongly from one another. Examples of IP earnings by each ion on LC-MS are illustrated in Table 5.1.

If several ions are monitored, the identification value for each of them cannot simply be added to the other, as these ions do not necessarily originate from independent sources (Table 5.2).

Thus, the total IP obtained is the sum of each piece of information or IP derived from each individual analytical step of the method(s). Several official controlling bodies have agreed on a minimum of three IPs [6, 8, 9] and recent publications referring to real cases have shown that this is feasible in practice (*see below*). For example, in selected-ion monitoring (SIM) identification, the three IPs need the following strict criteria to be met:

- exact correspondence of the three characteristic ions
- relative abundances of ions correct
- retention times the same as standards
- samples processed through selective procedure designed to remove interferences
- peak intensities are at least three times larger than surrounding background noise.

Identification can also be achieved by many other LC-MS approaches. On the basis of a minimum of three IPs, the best method (or combination of methods) for LC-MS identification can be designed. LC-MS-MS and LC coupled with high-resolution (HR)-MS are the most effective and rapid methods to achieve this minimum of three IPs.

Generally speaking, the actual identification process starts when the differentiation/detection phase is finished. This is done by database retrieval where the analytical (molecular)

Table 5.2 Examples of the number of IPs earned for confirmation of identity by a range of LC-MS instrumental configurations and operation modes and combinations thereof (N = an integer) (adapted from Rivier [3])

Technique(s)	No. of ions	IP
LC-MS	N	N
LC-MS-MS	one precursor and two daughters	4
LC-MS-MS	two precursor ions, each with one daughter	5
LC-MS-MS-MS	one precursor, one daughter and two granddaughters	5.5
HR-MS	N	2N
LC-MS and GC-MS	2 + 2	4
LC-MS and HR-MS	2 + 1	4

Table 5.1 Relationship between a range of classes of mass fragments and IPs earned for confirmation of identity using different LC-MS configurations (adapted from Rivier [3])

LC-MS technique	Resolution	IP earned per ion
MS	low	1.0
MS-MS precursor ion	low	1.0
MS-MS transition products	low	1.5
MS	high	2.0
MS-MS precursor ion	high	2.0
MS-MS transition products	high	2.5

Each ion may be counted only once. Electrospray ionisation LC-MS might be regarded as a different technique from atmospheric pressure chemical ionisation LC-MS. Different metabolites belonging to the same parent compound can be used to increase the number of IPs. Transition products include both daughter and granddaughter products.

properties of the analyte are compared with the properties of reference substances collected in one or several relevant databases. As indicated earlier, the nature and the quality of these databases are of course crucial for the validation of the procedure. In principle, if the database content represents the panel of substances to be possibly found, identification is achieved when the properties of the unknown substance adequately match those of only one single reference substance. However, in practice, the matching process will often result in a list of substances whose properties correspond more or less to those of the unknown. The identification process must then continue (i.e. adding more differentiation techniques) until only one single substance remains on the list [1].

When this approach is applied, the following controls should be used in order to validate it:

- What are the requirements for suitable relevant databases?
- How are these analytical properties compared?
- What should be considered as 'adequate match'?
- What is the probability of correctness of identification?
- What are the criteria to reject substances?

Most of the time, instrument providers offer ready-to-use library searching systems without specifying details about the items listed above. The analyst must then apply some critical judgement before reporting any result.

Mass spectrometers provide different types of information. The usefulness depends on the selectivity of the ions monitored and on the quality of the chromatographic separation. In this respect, it should be mentioned that LC requires a more specific detection setting for achieving proper discrimination power as it is typically less discriminative than capillary GC [10].

In addition to the total ion current (TIC), full-scan and SIM spectra data, extraction ion profiles can be displayed. The amounts of information contained in each of these differ significantly. After collecting scan data, the recorded raw data can be searched for the presence of specific ions and the reconstructed ion current (RIC) or 'mass chromatogram' of these ions can be displayed individually. This type of data treatment can identify chromatographic peaks that exhibit the ions selected. These extracted-ion profiles [or selected-ion retrieval (SIR)] differ from SIM data in that they are retrieved from collected scan data and do not offer the enhancement in sensitivity available with SIM on quadrupole or magnetic sector MS instruments.

Confirmation of identity

In forensic toxicology, drug identity confirmation is based upon the use of (at least) two different techniques, each confirming the results of the other [1]. Until recently, most of the identifications were obtained by combining results from the simple combination of separation and detection devices like thin-layer chromatography (TLC), GC with flame ionisation detection or LC coupled with spectrophotometry or fluorimetry and, finally, GC-MS. The selectivity of each of these techniques is of course not identical and any chromatographic tool coupled with MS detection has long since been recognised as offering the best discrimination power between the ranges of all of them. In this respect, LC-MS can be valued at the same level as GC-MS and adds a lot of confidence in the accuracy of the results. This should be done following specific guidelines or rules as the IP values for each LC-MS technique are different and may not simply be additive. For example, confirmation of identity of cannabinoids detected in cannabis products could be done following the Table 5.2 criteria for LC-MS-MS analysis [11]. In a completely different area, the expected detection of endogenous boldenone in an entire male horse was confirmed by using LC-MS-MS with quadrupole-time of flight (QTOF) HR mass measurements on two daughter ions with a deviation of less than 5 mDa from the calculated expected masses (IP > 3) [12]

Value and role of mass spectral library searching in the identification of organic compounds

The discriminate and critical use of any ready-made mass spectral libraries is paramount.

Various mathematical algorithms have been developed for automatic spectra comparison (so-called 'library searching'), providing identification probabilities in percentage form. This can easily breed complacency in the analyst, with too much reliance on software and instilling over-confidence in the reliability of results. Some of the pitfalls inherent in such systems are as follows:

- a typical library does not always contain complete mass spectra and only a few characteristic ions are retained
- the mass range for data acquisition may differ from the mass range of the library spectra, possibly resulting in mismatching (e.g. amphetamine ions acquired starting from mass-to-charge [m/z] ratio 30 when the library spectrum starts from m/z 50)
- some libraries are not free of gross mistakes
- instrumental differences may account for different relative intensities of ions which can result in poor matching
- impurities co-eluting with the compound of interest may give mismatching
- numbers and types of substances present in libraries are finite.

The search algorithms used by the different commercially available systems are not identical and may give very different answers. Any retrieval system tends to give poor results when the unknown substance is not in the reference library. Another more valuable approach is the use of structure-oriented databases and mathematical procedures able to interpret unknown mass spectra. HR mass spectra, based on exact mass determination, can be used in such a way as they offer high selectivity or IP numbers [13]. Although such identification processes are based on exact molecular and/or ion mass measurements, the actual paradigm for full identification in forensic toxicology of analytes still requires direct comparison with authentic reference compounds. Sometimes *ad hoc* synthesis of new compounds has to be considered to cover the specific identification requirements of forensic and clinical toxicology [14].

Spectral data collected by scanning the mass range during the chromatographic run can be encoded through the use of an appropriate algorithm and then used to search in-house library databases. Most of these libraries use reversed-searching capabilities with the probability-based matching algorithm (PBM) [15, 16] because this approach is more robust against contaminations often encountered in biosamples. Most of the time, no definitive answers are obtained as several candidates are presented on a list. Other approaches are linked more to structure-related similarities or pattern recognition techniques [17]. However, the interpretation of the results is most often complex and (to the author's knowledge) shows limited use in practice. It is important to note that all of these very powerful tools have associated pitfalls and often lack sufficient critical examination by the operator which may then lead to the erroneous interpretation of results. It is thus important to understand the principles involved in these basic operations and always consider results in the context of common sense. Finally, it should be remembered that not all analytical problems can or should be addressed by MS alone.

Interpretation of mass spectra for compound identification may also be attempted based on the rules of fragmentation of organic ions in the gas phase [18]. Unfortunately, this approach is frequently slow and tedious, and generally requires considerable experience and training. It is often difficult to sustain the necessary accurate deductive reasoning process to make the correct identification and this is coupled with the fact that the processes that occur in the fragmentation of organic ions in LC-MS have not yet been completely elucidated. Nevertheless, some interpretative power can be obtained from some search algorithm results in that the hit-list structures are often similar to the structure of the query (e.g. [17, 19]). If the unknown is not present in the library, spectra-similarity hit lists may be less useful or even misleading, bringing the analyst to the wrong conclusions [14].

Together with the characteristics of the reference library and of the search algorithm, correct MS identification obviously requires the optimisation of the whole analytical procedure used to acquire MS data to be submitted to the search. In particular, care should be taken to avoid the presence of interfering ions or distortion of the spectrum (e.g. applying a proper purification of the sample and chromatographic separation, and checking for background contribution and for detector saturation).

In the past, the size of reference databases and the complexity of algorithms used strongly limited the search speed, but these limitations do not exist with modern computer systems. The most serious limitations of LC-MS database searching are:

- size and the quality of the database (e.g. mass spectra are not representative because they have been more often registered using one solvent mixture without chromatographic separation)
- difficulty of updating the database (e.g. adding new MS editions, running several MS collections, changing names of the compounds)
- very limited searching capabilities usually offered (e.g. only a reversed-search algorithm is proposed).

In general, when examining the mass spectrum to be identified, the following key determinants should be localised:

- a molecular (parent) ion may be present, but adduct ions or losses of water are quite frequent in LC-MS
- in some cases, it is possible to deduce structural information from the ion clusters close to the molecular ion (this is especially the case when halogens are present in the molecule)
- atmospheric pressure ionisation (API) is a low-energy ionisation technique and therefore high mass ions are most frequently produced; for preliminary and rapid molecule identification, these high mass ions have the edge over fragmentation ions although these are also very useful
- retention time is often misevaluated because it is difficult to reproduce it from run to run and therefore relative retention time is a more appropriate measurement
- the presence in the biological sample of metabolites originating from the same parent molecule(s) is, of course, a strong additional indication of the exposure to that drug(s).

It is obvious that a substance that is not present in the database cannot be found by database retrieval (false negative) and that a reference substance having a mass spectrum similar to the unknown may give a false identification. Depending on the area of interest, the size and the scope of the database may be different: in doping analysis, for example, the relevant substances are those that are banned in sports and listed by official organisations (e.g. the World Anti-Doping Agency [WADA] and the International Olympic Committee [IOC]), and comprise substances such as anabolic steroids, diuretics or stimulants, to mention just a few.

Generally, reference MS databases should be as large as possible, and must include parent compounds as well as metabolites, decomposition products, interferents (matrix components, plasticisers, antioxidants, etc.) and analytical artefacts. The selection of compounds should be as broad as possible and the list kept up to date [1].

For substances where no reference standards are available, some specialists consider correct identification to be impossible (e.g. [1, 3]). In any case, when no criteria have ever been defined, it is the experience of the specialists that becomes crucial for setting a correct identification procedure. Last, but not least, mass spectra published in the scientific literature should be used only with extreme caution, even for preliminary identification, as they are not always correct!

Recommendations of regulatory bodies

During the last few years several recommendations have been published concerning analytical areas related to clinical and forensic toxicology [3]. They describe, among other things, detailed requirements that the analyst has to consider when working with LC-MS identification and/or confirmation. However, criticism has been raised recently because of the lack of formal validation of these criteria [4], although quite a few recently published examples [11, 20–22] taken from real cases demonstrate their suitable applicability.

All existing recommendations in clinical and forensic toxicology are based on the ISO/IEC 17025 guidelines [2]. In order to improve quality, it is accepted that any conscientious toxicologist should apply them without delay in his or her practice even if clients do not directly request them. These guidelines are very general and look difficult to apply in specific fields. Further details, explanation and guidance are provided by numerous scientific societies like the American Association of Clinical Chemistry (AACC), SOFT, AAFS, the International Association of Forensic Toxicologists (TIAFT), Gesellschaft für Toxikologische und Forensiche Chemie (GFTCh), etc., the references for which are given throughout this chapter.

One of the most useful sets of rules concerns the identification of toxic residues in meat by the US Food and Drug Administration (FDA) [8]. Practically, they can be well applied to clinical and forensic toxicology after harmonising their nomenclature as indicated below.

MS identification/confirmation criteria vary depending on the acquisition mode:

(i) MS full scan or partial scan. The mass spectrum should include at least three structure specific ions. All these ions should appear in the scan range. Matching with the contemporaneous (e.g. introduced in the same analytical system at the same batch of samples) standard is done visually. Since full-scan spectra may include hundreds of significant data points for comparison, strict numerical criteria need not be applied (matching of relative intensities within ±20% arithmetic difference on major ions is a useful rule). Library-search algorithms should not be used to confirm identity. The next criteria apply also when scan data are used:
   (a) all structure-specific ions indicated in the standard operating procedures (SOPs) are present above a specified relative abundance
   (b) there is general correspondence between relative abundances or ranked abundances obtained for sample and standard
   (c) ions other than from the target analyte can be explained (e.g. present in controls, blanks, etc.)
   (d) if background subtraction is used, this should be specified in the SOPs (the range used as background should always be indicated on the chromatogram).

(ii) SIM or MS scan acquisition; SIM treatment:
   (a) relative abundance for three structure specific ions should match the comparison standard within ±10% (arithmetic difference, not relative difference), i.e. at 50% relative abundance, the matching window would be in the range 40–60%, not 45–55%
   (b) if not achievable, relative abundance of four or more unique, structure specific ions should match the comparison standard within ±15%
   (c) otherwise, relative abundances for more than three ions, which include ions due to isotopes or loss of water, should match the comparison standard within ±10%.

(iii) MS-MS and MS^n full scan or partial scan:
   (a) the spectrum obtained from the suspect compound should visually match the spectrum obtained from a contemporaneous standard; since full-scan data may include hundreds of significant data points for comparison, strict numerical criteria do not need to be applied
   (b) the criteria set for MS scan should also be applied here
   (c) furthermore, if a structure-specific precursor ion dissociates completely to product ions after MS-MS or MS^n, the appearance of at least two additionally structure specific product ions in the MS^n + 1 spectrum is required.
   (d) ions produced from sources other than the target analyte should be explained (e.g. present in controls, blanks, etc.)
   (e) if background subtraction is used, the range used as background should be specified.

(iv) MS-MS and MS^n multiple-reaction monitoring (MRM) or MS-MS and MS^n scan acquisition, MRM treatment:
   (a) if the parent ion selected by MS-MS or MS^n is completely dissociated and only two structure-specific product ions are monitored in MS^n + 1, the relative abundance ratio should match the comparison standard within ±10%

Table 5.3 Maximum tolerance windows for relative ion intensities to ensure appropriate certainty in LC-MS and LC-MS-MS identification (adapted from WADA [9])

Relative abundance (% of base peak)	LC-MS; LC-MS-MS
≥50	±15% (absolute)
<50 and ≥25	±25% (relative)
<25	±10% (absolute)

For example, ±15% absolute difference for a 50% abundance gives a tolerance window of 35–65%; ±25% relative difference for a 50% relative abundance gives a tolerance window of 37.5–62.5%.

(b) if three or more structure specific product ions are monitored, the relative abundance ratios should match the comparison standard within ±20%.

Other sets of rules can vary slightly and may include retention time requirements that also represent a very helpful addition for gathering more IPs rapidly. Requirements for LC-MS identification in the doping control area are given in Table 5.3.

Specific difficulties in constructing an in-house LC-MS system

Under a given set of experimental conditions, the mass spectrum of a molecule is like a fingerprint. Both a liquid chromatograph and a mass spectrometer are relatively simple instruments conceptually, and the analytical data that each produces are easily understood and used. When these two instruments are directly combined into one LC-MS system, the capabilities of that system are modulated by the interface where the ionisation processes are taking place. First, the solvent of the effluent should evaporate and the remaining organic molecules ionise at the same time as the pressure is reduced. Several interfaces have been designed by manufacturers with no comparative performance tests available except for those on sensitivity.

With the introduction of API sources, LC-MS has become a powerful tool because of its capacity to ionise polar and thermally unstable compounds. However, the high selectivity and sensitivity of the MS detector has brought some misconceptions about the speeding up or even elimination of sample preparation and chromatographic separation. In fact, in many reports, it has been demonstrated that matrix effects (MEs) and adduct formation are of major concern because they can invalidate both qualitative and quantitative results [18]. Several adduct ions such as $[M + Na]^+$, $[M + K]^+$ and $[M + NH_4]^+$ are also frequently reported. Sodium and potassium originate from the biological matrix or from the glassware. Ammonium comes from ammonium acetate (or formate) added to the LC solvent. The process of adduct formation is unfortunately poorly reproducible and its contribution should be minimised. Practically speaking, ME and adduct ion production can be reduced by carefully selecting the extraction procedure, the LC solvents and the ionisation mode [23].

Thus, the following aspects should be taken into consideration when setting up LC-MS conditions for constructing a mass spectral database:

- influence of the mobile phase (solvents, additives and their concentration, flow rate)
- influence of sodium and potassium on the formation of adducts
- influence of the curtain or sweep gas flow
- reproducibility of mass spectra upon retuning (over more than 10 days)
- influence of ion suppression
- running the same sample at HR or LC-MS-MS for verification of the correct m/z value of the selected ions
- possibly independent data obtained by GC-MS (electron impact and/or chemical ionisation) to further verify the correct m/z value of the selected ions if LC-MS-MS not available.

Nature of databases for identification in LC-MS

The nature of the LC-MS spectra and the quality of the libraries using peak intensity ratios are discussed in Chapter 6. In the context of the identification proper, new, original approaches are emerging that tend to avoid ion intensity variations and still need to be fully confronted in real cases.

One of these approaches sets the abundance of significant daughter ions arbitrarily between 100 and 50%. Such a special simplified LC-MS-MS library makes use of chemical ionisation combined with collision-induced dissociation (CID) fragmentation techniques and offers very stable simplified mass spectra patterns on the most intense ions for about 200 substances over 'universal instrumentation coverage' [24, 25]. However, it has the disadvantage of never differentiating between isomers.

The second promising approach is based on the elemental composition of selected drugs by LC and HR-LC-TOFMS. In Table 5.2, one can see that the greatest value for IPs is gathered by the determination of the elemental composition of a chosen ion in the mass spectrum. This is obtained by determining the accurate mass of that unique ion. The reference library is simply built by entering the elemental formula of a targeted substance and calculating its respective mono-isotopic molecular mass. Modern LC-HR-MS and LC-TOFMS operate routinely with sufficient scan speed and dynamic resolving power of up to 10 000. If accurate mass measurements under true LC-MS conditions are guaranteed, a satisfactory reference HR mass spectrum can be recorded. For example, the mass accuracy in an LC-MS Q-trap instrument (approximately 20 ppm for pure compounds at m/z 500) is already much better than in conventional triple quadrupoles (full width at half maximum [FWHM] = 0.7 Da), but not as good as that of LC-TOFMS or LC-QTOFMS instruments (approximately 5 ppm) [13]. Ion traps (ITs) may offer new alternatives for similar screening approaches [26]. Therefore, methods using LC coupled with IT-MS show a promising future in forensic toxicology to identify a wide range of drugs (e.g. [12, 26, 27]).

Requirement for pure standards

As clearly indicated previously, the most convincing evidence for identification is obtained by the examination of pure standards corresponding to suspected sample components. However, this depends entirely on the availability of the pure standard (particularly in the case of metabolites), and the additional cost and time required to examine it. In addition, it is not always possible to predict which compounds will be found in real cases. There are also many very practical limitations imposed on the development and maintenance of a large collection of pure standards. The same limitation in the availability of standards also affects calibrated concentration measurements.

The question remains of whether or not reliable identification can be achieved without reference standards. One possible criterion for reliable identification is a quantitative measure of the exactness of match between an experimental mass spectrum and a spectrum obtained from the literature or a computer database. Sometimes called the similarity index (SI), the number typically ranges from 0 to 1. Even a reasonable match of 0.75 may sometime suggest 'identification', but more often than not this is not the case. Quite a number of compounds are not uniquely characterised by their mass spectrum and position isomers or members of homologous series of compounds may give very similar mass spectra. In some cases, e.g. for the discrimination of stereoisomers, full characterisation is based only on significant differences in ion abundance measurements. As previously stated, this implies that proper calibration of the ion abundance settings of the instrument is performed beforehand, especially around the level of the limit of detection (LOD).

Decision limits

There are relevant differences between forensic and clinical toxicology – diagnostic, sensitive, rapid and cost-effective testing in clinical toxicology differs from forensic testing where absolute rigorous identification is required. This is especially true when trace amounts have to be identified. For magnetic sector or quadrupole MS instruments, it is often true that scanning through a very large mass range will not provide workable results because the intensities of the background ions are too close to those of the ions relevant for the identification. A scan limited over the most significant mass values (e.g. around the molecular weight and most important fragments m/z values) provides a better signal-to-noise (S/N) ratio than a scan over the whole range (e.g. m/z 40–750), the reason being that statistically truer values for the intensity of ions are obtained the longer the collecting time is above each m/z value. In magnetic sector or quadrupole SIM experiments where three or four significant ions are measured, the best S/N ratio is usually obtained. Thus, at the LOD in the SIM mode, the further key determinants for proper identification are:

- Careful consideration must be given to choose proper ions for monitoring. The priority must be given to ions specific to the compound. They should be the most abundant ions as, obviously, detection limits are improved. The molecular ion is often chosen, if sufficiently abundant, as well as the base peak. If several high-intensity ions are available, then the ones providing the highest S/N ratio should be privileged in order to reduce background contribution.
- The monitoring of several ions (three or four) from different fragmentation origins reflects the need to achieve sufficient selectivity of detection.
- For each compound, all ions should peak at the same retention time, and have the same peak shape. Relative abundances of the ions monitored should correspond to those of the reference standard examined under identical conditions within a given tolerance window (see Table 5.3).
- Clean extracts should be prepared as the possibility of interference increases when measurements are made at lower detection limits.
- Background subtractions should be applied uniformly throughout the batch of samples and, if applied, it should be clearly indicated.

The repeatability of ion intensities on both pure solution and extracts obtained from various matrices should be measured at the LOD. At that level, the smallest amount should give a single result which, with a stated probability of 2σ standard deviation (commonly 95% probability), can be distinguished from a suitable blank. The overall batch acceptance criteria, being similar to those of the control material within a 10–20% tolerance [2, 5, 6], are:

- chromatography similarity (peak shape, symmetry, integration peak and baseline resolution)
- retention time within 0.1% of the extracted reference compound
- mass ion ratios within 10–25% of the extracted ions depending on their origin and their relative intensity

This LOD can be calculated daily using calibration sample data. The ruggedness of the assay assumes a normal Gaussian frequency distribution of readings at a definite concentration, α and β representing the error probability levels to (falsely) classify the analyte as present or not present, respectively (Figure 5.1).

The unreliability region of the assay can be defined as the interval of concentration of the analyte in which errors of identification take place (required qualifying parameters are not met) [28]. Most of the time this occurs at the LOD. The LOD can thus be defined as a concentration at which the analyte is classified as present or not present with equal probability (α and β are equal). In ordinary practice, threshold or cut-off concentrations are used for the LOD. The limit of identification (LOI) should now be regarded as the lowest concentration at which the less advantageous identification criteria can be met, e.g. to the lowest abundant targeted ion. In clinical toxicology, the requirements for α and β values are large. In forensic toxicology, a more robust approach is used to avoid the occurrence of false positives and/or false negatives.

For LC-MS methods referring to a single ion such as HR-MS or MS-MS, the limit of identification can be calculated on that very ion. For quadrupole instruments where unit mass resolution is standard, a more lengthy approach can be applied as the detection of several ions is required to identify an analyte (Table 5.1).

Usually, the most intense ions are chosen as target ions. The corresponding accumulation of IP depends on the availability and quality of the next qualifying ions (Table 5.2).

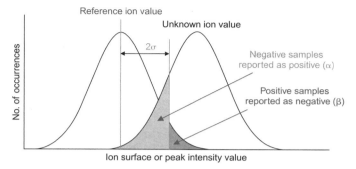

Figure 5.1 Statistical decision rules for identification criteria based on mass ion surface or peak intensities. Here the repartitions of ion intensity are represented for the reference standard ('reference ion') and the analyte to be identified ('unknown ion'). When enough measurements are repeated, measured intensities spread over a target value according to Gaussian curves. If these are overlapping to the predefined statistical degree (e.g. within a mean abundance value of the unknown situated at ≤2σ standard deviation of the maximum of the mean reference ion abundance), criteria for collecting IPs are met.

Thus, in order to properly identify an analyte, the abundance of the least intensive of the remaining qualifying ions must also be recorded within the 2σ standard deviation error margin (Figure 5.1).

Consequently, the experimental data showing the uncertainty on peak area measurements of the ions on the least-intensive ion monitored need to be used to statistically calculate the LOD. This is in contrast to the method for determining the limit of quantification where the most intensive ion is used for calculation [10] and should prevent the so-called 'ion shopping' (i.e. focussing only on the most suitable ions for the purpose and neglecting the others) [4].

Final considerations

Clinical and forensic toxicology deal with the detection and identification of a very large array of compounds. It is certain that the dedicated laboratory has to use several analytical instruments to cover it. Noticeably, LC-MS can significantly add to the expansion of the array of compounds already detected by GC-MS, including large, polar and/or thermolabile substances. As MS is considered the less likely identification technique to be challenged in court, identification of a molecule can be best based on its mass spectrum. This is commonly done by analysing, on the same day, the corresponding reference compound or standard through the same analytical procedure and instrumentation, and then comparing both sets of data.

When reference standards are not readily available, the unknown mass spectra are compared directly with a library of reference mass spectra. These data collections can be obtained from commercial sources or built within the laboratory. However, this should be done with caution as, even when the ionisation energies are the same, LC-MS mass spectra are most of the time specific to the laboratory and instrument. Thus, in forensic and clinical toxicology, quality criteria regarding LC-MS and LC-MS-MS compound identification are best established by comparing mass spectra obtained within the same batch of runs, measured on the same day, on the same instrument.

Confirmation of identity should be objective and reliable, and not depend on the sole interpretation of the operator. External experts to the laboratory should be available for reviewing and critically appraising all data [6]. This might be difficult to achieve when the procedure is highly automated. Statistically, it is important to agree on the level of uncertainty to be tolerated in these highly regulated fields.

Several recent official documents propose various rules and strict quality assurance criteria for the confirmation of identity of organic molecules in biosamples taking into account the latest advances in LC-MS. They might appear confusing or even contradictory. The commonality in all these is the necessary comparison of the mass spectrum of the unknown to that of a reference substance obtained side by side in the same batch. Within the spectra themselves, relative intensities of a minimum of three well-chosen ions have to match.

Conclusions

It is to be expected that LC-MS will play a greater role for trace organic identification in biological matrices, and particularly in clinical and forensic toxicology. In reality, full confirmation of identity can often only be attained by repeating several analytical runs, independently of each other, until a minimum numbers of IPs are obtained or a predefined threshold of converging data is reached. The more selective and specific the technique, the more IPs are readily accumulated. Complex LC-MS-MS techniques or LC-HR-MS present the highest potential for gathering IPs. For more simple instrumentation, like single-quadrupole LC-MS, a similar approach is also possible, but it requires the monitoring of several ion fragments. The nature of the ions to be monitored is important because they have to be as structurally close to the analyte as possible. IP earning tables presented here can give useful indications for choosing the appropriate approach. Depending on the analyte to be identified, it is essentially the analyst's responsibility to decide how and when the minimum requirement for identity confirmation is reached. Last, but not least, the chosen criteria should always be made transparent when communicating the final results together with the corresponding conclusions.

It is the author's conviction that all published rules or recommendations are rather simple in order to be followed by any kind of laboratory. By doing so, laboratories can considerably add to their credibility. However, it should be remembered that good rules (as detailed as they can be) are just not enough. In the critical fields of medical or medico-legal expertise, expert knowledge on the extraction, isolation and identification of organic substances as well as knowledge of the pharmacology and metabolism of drugs are indispensable.

Acknowledgements

The author wishes to thank Dr Einar Jensen (Department of Pharmacy, University of Tromsø, Norway) for his contribution to the literature survey, and great interest and continuous support through many constructive discussions that we shared during the various steps of writing this chapter.

References

1. De Zeeuw RA, Franke JP. 'General unknown' analysis. In: Bogusz MJ, ed., *Handbook of Analytical Separations*. Vol 2: *Forensic Science*. Amsterdam: Elsevier, 2000: 567–599.

2. International Organization for Standardization. *General Requirements for the Competence of Testing and Calibration Laboratories (ISO/IEC 17025)*. Geneva: ISO, 1999.

3. Rivier L. Criteria for the identification of compounds by liquid chromatography-mass spectrometry and liquid chromatography-multiple mass spectrometry in forensic and doping analysis. *Anal Chim Acta* 2003; 492: 69–82.

4. De Zeeuw RA. Substance identification: the weak link in analytical toxicology. *J Chromatogr B* 2004; 811: 3–12.

5. De Zeeuw RA. Laboratory guidelines in analytical toxicology: how to approach qualitative analysis. *J Forensic Sci* 1992; 37: 1437–1442.

6. Society of Forensic Toxicologists/American Academy of Forensic Sciences. *Forensic Toxicology Laboratory Guidelines*. Mesa, AZ: SOFT/AAFS, 2002: www.soft-tox.org/docs/Guidelines.2002.final.pdf (accessed 8 June 2005).

7. Moffat AC, Owen P, Brown C. Evaluation of weighted discrimination power calculations as an

aid to the selection of chromatographic systems for the analysis of drugs. *J Chromatogr* 1978; 161: 179–185.

8. US Food and Drug Administration. *Guidance for Industry: Mass Spectrometry for Confirmation of the Identity of Animal Drug Residues. Draft Guidance 118.* Rockville, MD: FDA, 2001.

9. World Anti-Doping Agency. *International Standard for Laboratories (Version 2)*. Montreal: WADA, 2003. www.wada-ama.org (accessed 8 June 2005).

10. Aderjan RE. Aspects of quality assurance in forensic toxicology. In: Bogusz MJ, ed., *Handbook of Analytical Separations*. Vol 2: *Forensic Science*. Amsterdam: Elsevier, 2000: 489–530.

11. Stocker AM, van Schoonhoven J, de Vries AJ, *et al.* Determination of cannabinoids in cannabis products using liquid chromatography-ion trap mass spectrometry. *J Chromatogr A* 2004; 1058: 143–151.

12. Ho ENM, Yiu KCH, Tang FPW, *et al.* Detection of endogenous boldenone in the entire male horses. *J Chromatogr B* 2004; 808: 287–294.

13. Gergov M, Boucher B, Ojanperä I, *et al.* Toxicological screening of urine for drugs by liquid chromatography/time-of-flight mass spectrometry with automated target library search based on elemental formulas. *Rapid Commun Mass Spectrom* 2001; 15: 521–526.

14. Aebi B, Bernhard W. Advances in the use of mass spectral libraries for forensic toxicology. *J Anal Toxicol* 2002; 26: 149–156.

15. McLafferty FW. Computer identification of mass spectra. probability based matching of mass spectra. Rapid identification of specific compounds in mixtures. *Org Mass Spectrom* 1974; 9: 690.

16. Staufer DB, McLafferty FW, Ellis RD, *et al.* Adding forward searching capabilities to a reverse search algorithm for unknown mass spectra. *Anal Chem* 1985; 57: 771–773.

17. MassLib. *Mass Spectral Database System*. Zollikofen: MassLib, 2003. www.msp.ch (accessed 8 June 2005).

18. Lambert W. Pitfalls in LC-MS(-MS) analysis. *Bull TIAFT* 2004; 34: 59–61.

19. McLafferty FW. *Mass spectrometry of Organic Ions*. New York: Academic Press, 1963.

20. Johansen SS, Jensen JL. Liquid chromatography-tandem mass spectrometry determination of LSD, iso-LSD and the main metabolite 2-oxo-3-hydroxy-LSD in forensic samples and application in a forensic case. *J Chromatogr B* 2005; 825: 21–28.

21. Maurer HH, Tenberken O, Kraatzsch C, *et al.* Screening for library-assisted identification and fully validated quantification of 22 beta-blockers in blood plasma by liquid chromatography-mass spectrometry with atmospheric pressure chemical ionization. *J Chromatogr A* 2004; 1058: 169–181.

22. Maralikova B, Weinmann W. Confirmatory analysis for drugs of abuse in plasma and urine by high-performance liquid chromatography-tandem mass spectrometry with respect to criteria for compound identification. *J Chromatogr B* 2004; 811: 21–30.

23. Souverain S, Rudaz S, Veuthey JL. Matrix effects in LC-ESI-MS and LC-APCI-MS with off-line and on-line extraction procedure. *J Chromatogr A* 2004; 1058: 61–66.

24. Kienhuis PGM, Geerdink RB. A mass spectral library based on chemical ionization and collision-induced dissociation. *J Chromatogr A* 2002; 974: 161–168.

25. Gergov M. Library search-based drug analysis in forensic toxicology by liquid chromatography-mass spectrometry. PhD Thesis. Helsinki University of Technology, 2004.

26. Fitzgerald RI, Rivera JD, Herold DA. Broad spectrum drug identification directly from urine, using liquid chromatography-tandem mass spectrometry. *Clin Chem* 1999; 45: 1224–1234.

27. Saint-Marcoux F, Lachâtre G, Marquet P. Evaluation of an improved general unknown screening procedure using liquid chromatography-electrospray-mass spectrometry by comparison with gas chromatography and high-performance liquid chromatography-diode array detection. *J Am Soc Mass Spectrom* 2003; 14: 14–22.

28. Simonet BM, Rios A, Valcarcel M. Unreliability of screening methods. *Anal Chim Acta* 2004; 516: 67–74.

6

Systematic toxicological analysis with LC-MS

Pierre Marquet

Introduction

Systematic toxicological analysis (STA), also called 'general unknown screening (GUS) of drugs and toxic compounds' or 'comprehensive screening of drugs and toxic compounds', comprises analytical methods or combinations of analytical methods aimed at detecting and identifying *a priori* unknown xenobiotics (i.e. compounds or substances foreign to the body, mainly therapeutic drugs, drugs of abuse and toxic compounds) in biological fluids [1]. Indeed, in some cases, the substances involved in intoxication cases are not known with certainty and must be identified before quantification (e.g. in cases of unexplained death or of severe intoxication where the patient is unconscious, or simply because the information provided by patients or their relatives about the drugs taken is often unreliable). It is also sometimes important to exclude toxic hypotheses raised by clinical findings. With that aim, the laboratory is generally provided with blood and/or urine samples. Urine is preferred for first-line STA by most laboratories because it generally contains xenobiotics, or more often their metabolites, at rather elevated concentrations [2] and several immunoassays have been developed for this matrix. However, for cultural or specific reasons, urine is not always collected and blood may be used (in the living). Blood, plasma and serum can often be interchanged in most methods, although postmortem blood may present problems due to its high viscosity [3]. In some postmortem cases, only exudation fluids, hair or tissues are available.

As not all substances can be detected with a single screening method [3] STA generally involves a combination of methods based on different technologies:

- immunoassays for the most common drugs in urine (morphine and derivatives, cocaine and metabolites, amphetamine and a few derivatives, cannabinoids, methadone, propoxyphene, benzodiazepines, etc.), or in serum or plasma (paracetamol, salicylates, highly dosed benzodiazepines, barbiturates, tricyclic antidepressants, etc.)
- chromatographic techniques, ideally coupled to specific methods of detection such as mass spectrometry (MS) or UV-diode array detection (DAD).

However, it is generally recognised that this strategy may fail to detect different types, families, or individual drugs or toxic compounds that may be responsible for serious or even lethal intoxication cases. Indeed, even gas chromatography (GC)-MS, universally applied for STA over recent decades due to its specificity, sensitivity and the availability of very large libraries of standardised spectra [4], may fail to detect polar or thermally labile compounds, or high-molecular-weight molecules. Up till now, LC-DAD has been the most widely used complement to GC-MS for STA because it is compatible with large, non-volatile and thermolabile molecules; however, the separation power of LC is much less than that of capillary GC; DAD is not as specific as MS and it can fail to detect molecules with no or little UV absorbance.

The coupling of MS with liquid chromatography (LC) has therefore been expected to increase the range of compounds amenable to MS. However, the mass information brought by modern systems, based on atmospheric pressure ionisation (API) sources, is very different from that of electron ionisation (EI), in that the different ionisation sources induce soft ionisation of the samples with little fragmentation. Of these, electrospray ionisation (ESI) and atmospheric pressure chemical ionisation (APCI) sources have been mainly used for STA in clinical and forensic toxicology, whereas, to the best of the authors' knowledge, no paper involving atmospheric pressure photo-ionisation (APPI), not to mention the seldom used sonic-spray or laser-spray sources [5], has been published so far. Different systems and concepts have thus been investigated to obtain rich enough mass information that could be stored in libraries to be used for the specific identification of the compounds when detected in biological matrices. This chapter gives an overview of the different concepts that are employed for such LC-MS STA techniques, based on single-stage or tandem MS (MS-MS), LC methods and extraction procedures developed for this purpose, as well as the suitability and performance of such approaches compared with GC-MS and LC-DAD techniques.

Molecular identification using LC-MS

Types of sources/interfaces used

LC-MS ionisation sources compatible with electron ionisation

As EI at 70 eV produces universally reproducible mass spectra, this ionisation mode is the gold standard for the specificity of MS detection. As far as LC-MS is concerned, only moving-belt and particle beam interfaces (PBI) are compatible with EI sources [6], but both are no longer commercialised and are probably very seldom used today. In 1998, Spratt and Vallaro proposed a STA technique using LC coupled on-line to both DAD and PBI-MS [7]. Mass spectra generated by EI at 70 eV were acquired in the 50–400 amu range, and then searched in different commercial and in-house libraries. With this procedure, they analysed about 150 compounds, mainly drugs, with a limit of detection (LOD) varying from 0.02 to more than 1 mg/L in whole blood. Interestingly, many compounds with the highest LODs were polar compounds, such as morphine, benzoylecgonine and ibuprofen. Indeed, PBI involves a volatilisation step by heating (like moving belts) and is thus not suited to polar, non-volatile or thermally labile compounds, which are also those generally missed by GC-MS.

Atmospheric pressure ionisation sources

ESI-type and APCI sources have been used for a large majority of the applications of LC-MS in forensic and clinical toxicology, mainly quantitative procedures, published over the last decade [8]. Pneumatically assisted ESI sources allow the analysis of moderately non-polar to highly polar, thermally labile or high-mass compounds. Fewer applications in toxicology were reported with APCI sources, probably because of their more limited polarity range (partly overlapping with that of GC-MS for less-polar molecules), their relative incompatibility with thermally labile compounds and the higher background noise produced compared with ESI. However, APCI is less prone to the phenomenon of ion suppression than ESI, which might be a non-negligible advantage when analysing non-selective extraction products of complex matrices such as urine or blood.

Neither ESI nor APCI sources are compatible with EI. Both involve a soft ionisation process, mainly giving rise, in the positive-ion mode, to protonated molecules or adducts with cations from the mobile phase and, in the negative-ion mode, to pseudo-molecular ions or adducts with anions. Mass information is thus limited, with virtually no fragment and a mass resolution of only about 0.7 amu full width at half maximum (FWHM) for the most commonly employed quadrupole analysers [5]. The first limitation can be overcome by using collision-induced dissociation (CID), resulting in efficient fragmentation of most compounds. CID consists of accelerating the ions generated and making them collide with molecules of a neutral gas, either in a specialised 'collision cell' or in the intermediate pressure part of the mass spectrometer, between the atmospheric pressure source and the high vacuum of the mass analyser (so-called 'in-source CID').

In-source CID relies on acceleration of the ions produced in the API source by an electrical field directed towards the mass spectrometer, inducing dissociation of these ions through collision with the residual solvent and gas molecules present in the source or the intermediate pressure region between ambience and high vacuum. Alternatively, ion dissociation can be obtained by electromagnetic excitation in ion traps (ITs).

Single-stage mass spectrometry

The fragments produced by in-source CID are generally the same as those produced by conventional CID in the collision cell of a triple-quadrupole (QqQ) instrument, but not necessarily with the same intensity.

Systematic toxicological analysis procedures based on single-quadrupole instruments

Principles and practices of in-source collision-induced dissociation

Principle In-source CID can be induced by increasing the skimmer nozzle, cone, capillary or orifice voltage, depending on the source and terms used by the manufacturers [9].

Selection of fragmentation voltages An in-source CID MS technique in which the fragmentation voltage was linearly increased ('ramped') during each scan gave encouraging preliminary results with a few test compounds [9]. With another type of instrument, this process was found to give poorer results than alternated, fixed fragmentation voltages [10]. All the other teams used fixed fragmentation conditions at one [11], two [11, 12] or three [13–16] different voltages in the positive mode; or four, continuously alternated positive and negative fragmentation voltages: two in the positive-ion mode (+20 and +80 eV) and two in the negative-ion mode (–20 and –80 eV) [10]. When more than one fragmentation voltage was chosen, the authors selected mild and strong fragmentation conditions in order to obtain, for a majority of their test compounds, molecular information (i.e. molecular or pseudo-molecular ions or adducts of the molecules with ions from the mobile phase) in the former condition and structural information (informative fragment ions) in the latter.

Spectral entries in the libraries The spectra of a single compound acquired at different fragmentation voltages were either stored individually in a MS library [12, 13–19] or summed to give reference spectra showing both fragment ions and the pseudo-molecular ion for most compounds [10, 11, 20, 21].

Mass resolution and mass range The mass ranges usually scanned cover the range of molecular weights of most therapeutic drugs, drugs of abuse or pesticides and their major mass fragments, i.e. from 50–100 to 500–600 amu [9, 13, 17–19]. Only a few teams extended this range, e.g. up to 750 [22] or 1100 amu [10, 20, 21], to cover molecules with higher molecular weights, e.g. digoxin or tacrolimus (molecular weight 781.9 and 821.5, respectively). However, it is worth noting that none of the reported experimental conditions was designed to detect peptides or proteins, e.g. heparins, insulins or erythropoietins. Also, the electronics of some LC-MS systems allow a scanning step size down to 0.1 amu, i.e. better mass accuracy than mass unit (as is traditional for GC-MS), even though the mass resolution of the quadrupoles used is generally not better than 0.7 amu FWHM. Some teams took advantage of this fact, using step sizes of 0.2 [10, 20, 21], 0.25 [9] or 0.5 amu [13].

Comparison of in-source CID with LC-MS-MS and EIS [10] The reconstructed CID MS spectra obtained by adding up the low and high fragmentation spectra of each compound compared favourably with LC-MS-MS product-ion spectra in terms of number and intensity of ions in the spectrum. Most were even as rich as EI spectra and had the advantage of including the pseudo-molecular ion, which is structurally more informative than any fragment. Moreover, the use of both positive- and negative-ion modes in the same run provided significantly more information than either of them alone. Less than half the molecules were ionised in the negative mode (mainly acidic compounds, hardly amenable to GC-MS without convenient derivatisation), but some were detected only using this polarity.

Several other authors pointed out that analysis should be conducted in the negative-ion mode for acidic compounds [9, 13].

Reproducibility and robustness of in-source CID mass spectra

Influence of the mobile phase Several authors [9, 13, 23, 24] confirmed that the mobile-phase composition in terms of ionic strength, pH or organic solvent content had little or no influence on fragmentation. The optimal formate concentration for the intensity of the mass spectrometric signal was claimed to be 2 [10] to 2.5 mmol/L [24]. It was verified that such mass spectra were not dependent on compound concentration within a wide range, e.g. 10 µg/L–10 mg/L [25], except when saturation of the channel electron multiplier occurred [13].

Reproducibility with time and standardisation In-source CID fragmentation is dependent on ion kinetic energy and on the distance that the accelerated ions have to cross before entering the high vacuum region [25]. This, in turn, is partly dependent on the geometry of the transition zone between pumping stages, which at least for some instruments can be modified by adjusting the distance between the ionisation needle and the mass spectrometer entrance orifice [10]. Several teams [10, 13] showed that fragmentation tuning based on one or several reference compounds ensured reproducible in-source CID spectra, at least using the same type of instrument. Haloperidol [26], glafenine [10, 20, 21, 27], [D_3]trimipramine [17, 19, 29] and reserpine [30], among others, were proposed as tuning compounds. Of note, all are small molecules, generally giving rise to several intense fragment ions. Intra- and inter-assay coefficients of variation (CV%) of fragment-to-parent ion ratios in reconstructed mass spectra were found to be below 18 and 20%, respectively, in the positive-ion mode (and most often less than 10%), and below 18 and 22%, respectively, in the negative-ion mode [10]; the lower the ion ratio, the higher the CV%. This variability, including that of the chromatographic background noise, was found to be acceptable for correct identification of compounds, inasmuch as a low weight could be attributed to intensity ratios in the 'library search' procedure. Moreover, visual inspection of the spectra recorded showed a very satisfactory reproducibility of the spectrum pattern, including mass-to-charge (m/z) ratios of low intensity [10]. Gergov et al. [11] found reproducibility CVs of 0.9% for the reconstructed spectra of 17 β-blocking drugs over a 1-week period [11], although no attempt was apparently made to standardise the in-source CID fragmentation energy.

Inter-instrument and inter-laboratory robustness Some of the spectra obtained with one brand of instruments in one study [12] looked very different from those obtained in the others, most of which employed systems of a single, different brand. Indeed, the former showed very intense adduct ions and virtually no protonated molecular ion, even at high fragmentation energy (as shown for carbamazepine) [12]. This was a first hint of limited inter-instrument robustness of in-source CID STA procedures. An inter-laboratory study with three different instruments of the same brand reported poor reproducibility of in-source CID mass spectra with both ESI and APCI sources, but they used no fragmentation tuning [23]. Interestingly, they found satisfactory short- and long-term intra-laboratory reproducibility when using ESI sources. Based on a previously published technique [10], Rivier tried to standardise in-source CID using glafenine as a tuning compound on several instruments: API 150 (Applied Biosystems) with a Turbospray source; MSD 1100 (Agilent); AQA, TSQ Quantum and LCQ Deca XP (Finnigan); ZQ 2000 (Waters) [27]. He found that with each instrument it was possible to find a couple of fragmentation conditions (i.e. voltages) in order to obtain the positive reference spectrum of glafenine previously obtained with an Applied Biosystem-Sciex API 100 instrument with an Ionspray interface [10], but that, under these standardised conditions, the ion ratios of diazepam could be very different following the geometry of the ESI sources. This was confirmed by another team that showed that the fragmentation voltages giving similar fragmentation on an Applied Biosystems API 365 used in the single-quadrupole mode and an Agilent MSD 1100 SL instrument were different according to the compounds analysed [31].

Bristow et al. [30] proposed two different standardisation techniques prior to comparing the CID mass spectra generated with LC-MS systems of different types and brands. The first involved a tuning compound, i.e. reserpine, and concerned five different instruments from four manufacturers, operated at low and medium CID energy levels. At low fragmentation energy, the $[M + H]^+$, $[M + Na]^+$ or $[M + K]^+$ could be obtained as base peaks for single compounds, depending on the instrument, whereas at medium fragmentation energy, not only the abundance of the ions formed varied, but also sometimes the content of the spectra (Figure 6.1).

In a second approach, Bristow et al. [29] tried to standardise in-source CID by adjusting the fragmentation energy in order to reduce the $[M + H]^+$ ion to 50% of its original abundance. However, this did not provide sufficient reproducibility either. In one instance, both the pseudo-molecular and fragment ions obtained with two systems were different, one giving the $[M + H]^+$ ion and corresponding fragments, and the other the $[M + NH_4]^+$ adduct and, most surprisingly, the ammonium adducts of the fragments (Figure 6.2). The authors concluded that both methods appeared not to be universally transferable for all compounds within their test group across all instrument types.

Published libraries

Several teams built up MS libraries for drugs, pesticides and other toxicants or explosives. These libraries contained spectra from a few therapeutic drugs to hundreds of various xenobiotics: three libraries for 22 benzodiazepines and 16 sulfonylurea herbicides, one for each fragmentation level [14]; a library of 70 spectra (35 compounds) in the positive mode [12]; a library of reconstructed CID MS spectra obtained by the addition of spectra recorded respectively at +25 and +90 V, from about 400 therapeutic or illicit drugs [11]; another library of about 400 compounds, including three spectra obtained at +20, +50 and +80 V orifice voltages for each compound [26], etc. One library containing about 1100 reconstructed mass spectra was built in the positive-ion mode and another library containing about 500 reconstructed spectra was built in the negative-ion mode (combined spectra at two fragmentation voltages for each polarity) [10]. These two libraries together included spectra from a total of about 1300 therapeutic drugs, drugs of abuse, pesticides, plants, and industrial and domestic toxicants.

Positive identification criteria

Depending on the authors, the definition of positive identification of a compound could vary greatly, including: a similar search result with high confidence for at least two of the three spectra generated at three fragmentation energies [14]; a search result with a reverse fit (comparison of the reference mass spectrum with the unknown) of more than 60% [21], with a low weight attributed to ion intensity ratios; and similarity of the five most abundant ions (20% fit each) [30]. Several authors stressed that the relative intensity of fragments generated by in-source CID should not be given too much attention (or weight) and that similarity assessment should mainly rely on the nature of the ions present in the mass spectrum [20, 21, 30]. Anyway, in-source CID mass spectral identification criteria should be standardised, at least to some extent, before this technique can spread out in toxicological laboratories. Further details on LC-MS identification criteria are available in Chapter 5.

Automation of library searching

Looking through rich and noisy LC-MS chromatograms to detect a lot of peaks, drawing the corresponding mass spectra (often after background subtraction), searching them against MS libraries and, finally, reporting the results is a long and cumbersome task that can be partially alleviated by automated processes. Depending on the manufacturers, some computer programs, similar to their GC-MS counterparts, can do that, while others cannot. In cooperation with the manufacturer of the LC-MS instrument they employed, Marquet et al. developed a computer program to automatically detect chromatographic peaks, reconstruct positive and negative spectra by adding up one spectrum at low fragmentation voltage and one at high fragmentation voltage in each polarity, and compare these positive or negative spectra together with their relative retention time (under optional,

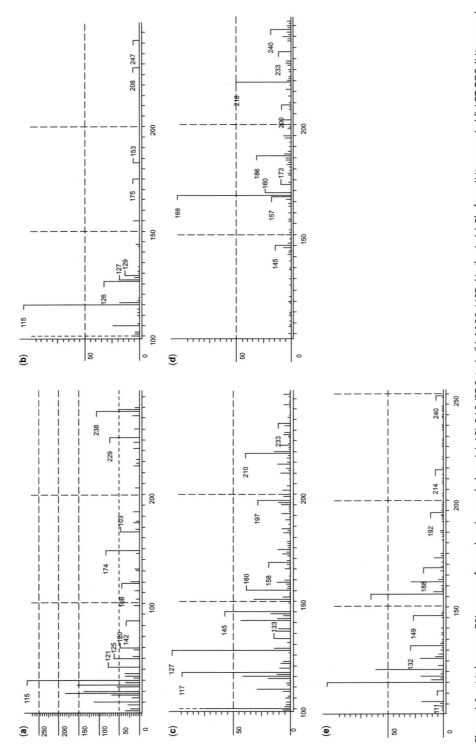

Figure 6.1 Mid-energy ESI spectra for carbaryl recorded on (a) API 365 (PE-Sciex), (b) 1100 MSD (Agilent), (c) Platform (Micromass), (d) LCT TOF (Micromass) and (e) Quattro Ultima (Micromass) instruments. (Reproduced from [30] with permission from John Wiley & Sons.)

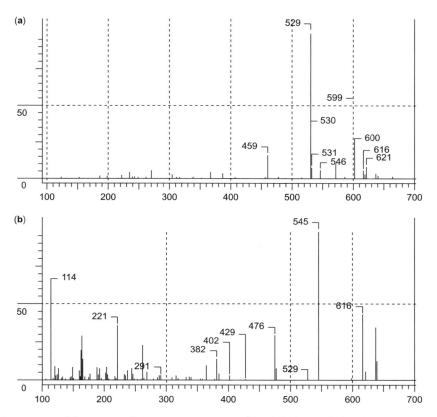

Figure 6.2 Comparison of the spectra of a proprietary compound (Kodak 3) recorded on (a) LCT TOF (Micromass) and (b) Quattro Ultima (Micromass) instruments tuned using 50% molecular ion attenuation. (Reproduced from [30] with permission from John Wiley & Sons.)

standardised chromatographic conditions) to those in the corresponding libraries [20, 21].

Systematic toxicological analysis procedures based on time-of-flight mass spectrometers

A single team [22] used LC-ESI-TOFMS for a STA procedure, based on the higher mass resolution of time-of-flight (TOF) detectors over quadrupoles. They used a heated, pneumatically assisted ESI source in the positive-ion mode and scanned a mass range of 100–750 amu with an acquisition time of 2 s/spectrum. They built a library of 433 toxicologically relevant substances and metabolites, simply by entering the elemental formulas of the compounds in the instrument software for the calculation of their respective mono-isotopic masses. Identification was based on the exact masses and mono-isotopic ratios of the compounds with respect to those of library entries. Using instrument mass calibration with Jeffamine D-230 before every sample injection, the method provided a 5–10 ppm mass accuracy for most compounds identified in urine samples. The authors stated that, due to the limited size of their library, an accuracy of 20 ppm was still adequate for drug identification, but this would probably not apply to larger libraries. One limitation of this technique is that, using solid-phase extraction (SPE) of basic, neutral and acidic drugs, and a combination of extracts, the authors obtained very noisy total ion chromatograms (TICs), in which the peaks of the compounds of interest were undetectable. This led them to back-search library entries in the TIC by automatically generating extracted ion chromatograms centred on the mass of a given library entry with a 'large' mass window of ±200 ppm, detecting chromatographic peaks and comparing the corresponding mass spectrum with that of the library entry. Again, this approach could be unrealistic if libraries including thousands of compounds were generated.

Detection of the signal in background noise and signal processing

As in-source CID is purposely a non-selective technique and background noise is generally high in the TIC after non-selective extraction of such complex media as urine or blood, different signal-processing procedures have been employed to try to extract signal from noise.

Manual selective-ion extraction was proposed to locate suspected compounds in very noisy TICs [13, 15]. However, only *a priori* suspected compounds can be searched for in this way, which is not compatible with a real STA. Also, when interfering compounds or background noise impeded compound identification, the authors used MS-MS fragmentation of selected ions and MS-MS library searching [15, 16], which required a second injection of the extract. Although the combination of in-source CID MS and MS-MS library searching is obviously a valuable tool, it is a rather time-consuming and cumbersome procedure for it to be used routinely.

As previously mentioned, an automated version of selected-ion extraction was proposed together with an LC-ESI-TOFMS method, where extracted-ion chromatograms were drawn for each pseudo-molecular mass of 344 library entries, the peaks detected and the corresponding spectra compared with the spectrum of the entry [22].

The computer programs employed with the LC-MS instruments generally include signal-processing algorithms such as background subtraction, filters, signal smoothing, three-dimensional representation of time, m/z and intensity ('eagle's view'), etc. However, only a few authors reported testing such possibilities [10, 20, 21]. After comparing all these possibilities, they claimed that the best results were obtained using an 'enhance' algorithm that eliminated the contributions of solvent and buffer ions to the acquired mass spectra, recognised and removed noise spikes caused by experimental variations, and extracted weak chromatographic peaks containing significant ions from the overall TIC trace, based on the premise that the occurrence of background ions is more frequent than that of ions due to analytes. An example of the efficiency of this algorithm is presented in Figure 6.3, in which the 'enhanced' mass chromatogram is superimposed over the raw mass chromatogram. This algorithm could also resolve partly co-eluted chromatographic peaks.

Tandem mass spectrometry

The radiofrequency-only daughter-ion-scan mode

The radiofrequency-only daughter ion scan (RFD) mode was proposed by a single team for screening *a priori* unsuspected compounds in surface and waste water using LC-APCI-MS-MS. In this mode, the first quadrupole is not used as a mass filter, but as an ion transmission path to the collision cell [32]. A cut-off mass of m/z 70, i.e. above the highest masses of the main mobile-phase ions and clusters, was applied. With this setting, all ions produced in the APCI source were admitted and fragmented in the collision cell, and the resulting fragments analysed in the third quadrupole. The authors applied low- and high-collision offset voltages in the collision cell to obtain declustered parents and fragments, respectively. They compared this RFD mode with in-source CID-Q3 MS and found that the signal-to-noise (S/N) ratios were comparable, but that, due to the cut-off mass, a much higher background of solvent ions was found with in-source CID [32]. In summary, this approach is very close to that of in-source CID MS, with the corresponding advantages and drawbacks, except that clusters of eluent ions with parent compounds or even fragments are avoided (which can also be obtained with a curtain gas between the API source and primary vacuum on some LC-MS instruments).

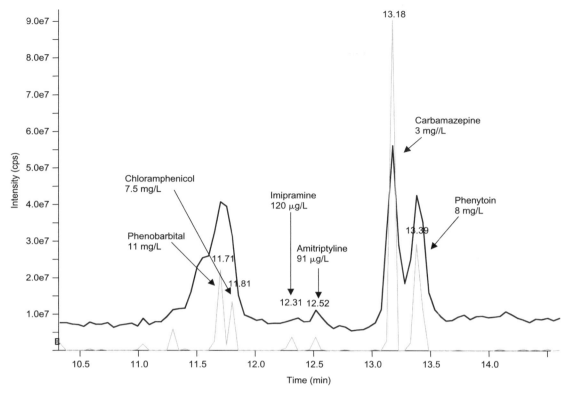

Figure 6.3 Efficacy of signal-processing software (the so-called 'enhance' procedure) for finding and individualising chromatographic peaks in a TIC (extract of a commercial quality control for therapeutic drug monitoring). (Reproduced from [1] with permission from Lippincott Williams & Wilkins.).

The selected (or multiple)-reaction monitoring mode

In this acquisition mode, universally used for quantification purposes, selected transitions between a parent ion and a single fragment are monitored. As a result of improvements in the electronics (scanning speed) and sensitivity of the most recent instruments, it is now possible to select tens or even hundreds of such ionic transitions in a single run, with an acceptable diminution of sensitivity and chromatographic resolution. One paper reported a method for screening 98 pesticides in crops, following two transitions per compound [33], and another one a qualitative screening for 238 drugs in blood, following only one transition per drug [34]. In the latter, fragmentation energy was adjusted for each drug and the total cycle time was 6 s. The LODs were less than 0.02 mg/L for about half the molecules up to 10 mg/L for the pesticide methylparathion. As far as therapeutic drugs were concerned, these LODs were within or below the therapeutic concentration range, except in only seven cases. The limitations of this approach are mainly the risk of cross-talk and the limited number of compounds screened for (even if this number is elevated). Cross-talk in the multiple-reaction monitoring (MRM) mode happens when an artefact transition is found, due to two co-eluting parent compounds with iso-baric fragment ions, where one can be artificially detected although absent, because the fragment issued from the other is still present in the collision cell. The extent of this problem depends on chromatographic resolution, speed with which the fragments exit the collision cell, and time given for each selected parent for fragmentation and transition monitoring. One can use a pause time between the transitions, but this causes 'dead time', making the scan cycle time significantly longer, which leads to fewer data points during a chromatographic peak [34]. Also, some instrument makers propose collision cells with variable quadrupole arrangements or electronic devices to accelerate fragment exit from the cell. In the above-mentioned study, the authors identified four pairs of compounds (out of 238) with the same retention time and fragment ion, and they took care to avoid this phenomenon in other cases using careful arrangement of the order of the transitions, i.e. by intercalating other transitions between two parents giving the same fragment and respecting a difference of at least 5 amu between consecutively monitored fragment ions [34]. However, the main limitation of this approach is indeed the incompatibility between a large number of compounds monitored, high sensitivity and good chromatographic resolution. The time spent monitoring each transition directly affects both the S/N ratio and chromatographic resolution, in opposite directions. For this reason, and as far as mass identification is based on time filtering, it is not possible to monitor more than a few hundred compounds and the paper discussed above probably presents what best can be done with this strategy. Also, contrary to full-scan approaches, it does not allow for detection of metabolites based on common fragments or fragmentation patterns with the parents.

The product (or daughter)-ion-scan approach

In this mode, selected parent ions (generally the protonated molecules in the positive-ion mode or the molecular ions in the negative-ion mode) are selected in the first quadrupole of tandem mass spectrometers, fragmented in the collision cell and the fragments analysed in the scan mode in the second mass filter (generally a quadrupole, but sometimes also a TOF mass filter). The resulting spectrum is thus supposed to be specific for the selected parent and devoid of chemical noise or interferences that may affect the quality of in-source CID spectra. However, fragmentation energy in the collision cell should be optimised in order to obtain rich and reproducible spectra, preferably including the mass of the parent ion, for a large panel of compounds.

Several teams applied this strategy with QqQ instruments to build libraries of 400 or 500 compounds. Four different positive-product-ion spectra were recorded in the library for each compound, corresponding to four different collision energies [26], or the spectra were obtained using a collision energy of 35 eV, with additional spectra acquired at 20 or 50 eV for those compounds giving no informative spectrum at 35 eV [11].

Baumann *et al.* [35] built a MS-MS library of 517 spectra using an IT mass spectrometer (LCQ; Thermo Electron) and either an ESI or an APCI source operated in the positive-ion mode. In order to obtain rich product ion spectra, resonance excitation at 20 amu below the parent ion selected was used to further dissociate the $[M + H - H_2O]^+$ ions generally produced by the rather soft fragmentation process involved in IT MS. Moreover, as the energy required for fragmentation decreases linearly when the molecular mass increases, a mass-dependent correction was automatically applied to the collision energy. As a result of both of these improvements, sufficiently specific MS-MS spectra of different drugs, as well as endogenous compounds, could be recorded.

Although theoretically interesting, the MS-MS libraries developed are of limited use in forensic or clinical toxicology. Indeed, the necessary selection of a limited number of ions in the first quadrupole implies that only a limited number of compounds can be screened for each time an extract is injected. As previously mentioned, the MRM mode can identify (based on a couple of specific transitions and the compound relative retention time) and possibly quantify a much larger number of suspected compounds. For the detection of the compounds of interest, the authors used a first injection in the single-quadrupole selected-ion monitoring (SIM) mode [11] or in the triple-quadrupole MRM mode [36], and, if any of the parent ions or transitions selected gave a chromatographic peak at their expected retention time, an automatic procedure created new experimental conditions for a second sample injection, where the parent ions detected were selected, fragmented and analysed in the product-ion-scan mode. This was an in-house version of the 'data- or information-dependent acquisition (DDA/IDA)' procedures described below.

LC-MS-MS with data- or information-dependent acquisition

DDA/IDA first uses two-stage MS instruments in the full-scan, single-stage MS mode to select the parent ions of interest, totally unexpected by definition, before dissociating them and monitoring their fragments in the daughter-ion full-scan mode. This approach, which can be summarised as 'self-adaptive MS-MS', was first employed in a preliminary STA procedure by Decaestecker et al. [37], using a two-stage instrument associating a quadrupole and a TOF mass analyser (QqTOF), operated in the positive ESI mode. During the chromatographic run, the quadrupole initially transmitted all masses in the 50–450 amu range to the TOF (pass-band mode), until one or more ions reached a predefined threshold. Instantly, the quadrupole selectively transmitted these high-intensity ions (maximum of four ions) to the collision cell operated with a single fragmentation energy and the resulting fragments were transmitted to the TOF mass analyser. Then, the instrument switched back to the initial conditions after 4 s, except that a 2-min refractory period was applied to the last selected ions. Seventeen common drugs were used to optimise the whole procedure, particularly fragmentation energy. This procedure was applied to the analysis of urine samples from toxicology cases in which all the compounds previously detected by the enzyme-multiplied immuno-technique (EMIT) and LC-DAD were also identified by LC-MS. However, no details were given about the mass spectra library used (or the conditions under which such a library was built). The same authors further refined this technique by applying two alternated collision energies (+15 and +25 eV), reducing the time allocated to MS-MS acquisition to an acceptable minimum (which is favoured by the use of a TOF mass analyser that does not scan, but continuously monitors the full mass range), and optimising SPE and LC conditions to the advantage of the S/N ratio and 'on-the-fly' selection of relevant parent ions for fragmentation. A library of mass spectra acquired at five different collision energies was built for more than 300 drugs and toxicants (1600 MS-MS spectra). The comparison of this approach to classic screening procedures including immunoassays, LC-DAD, GC-MS and GC with nitrogen phosphorous-selective detection (NPD) for the analysis of 20 blood samples from drug addicts or autopsies showed that, in three cases, bromazepam or diazepam was missed by LC-MS-MS, not considering caffeine or cotinine at low levels. However, a number of drugs were detected by the IDA approach and not by the other techniques, including ranitidine, propranolol, piroxicam, papaverine, atropine, midazolam, alprazolam, etc. [38].

DDA was employed in another preliminary study by Fitzgerald et al. [39] who used a particular instrumental setting, coupling the sample preparation/chromatographic separation parts of the REMEDi™ HS instrument from Bio-Rad Diagnostic to an IT mass spectrometer (LCQ) via an ESI source. After direct injection of urine samples in the column-switching system, the compounds of interest were analysed in the single-MS full-scan mode between 50 and 500 amu. When any ions exceeded a pre-set threshold, they were selected, fragmented by CID in the IT (no precise CID conditions reported) and the resulting fragments recorded in the product-ion-scan mode. Finally, the mass spectrometer reverted back to the full-scan mode. This procedure was tested on urine samples spiked with only 17 drugs again, but no fit values between the CID spectra obtained from these spiked samples and those from pure standards were given.

Marquet et al. compared their previously mentioned in-source CID STA technique with a preliminary procedure involving a triple quadrupole-linear IT (QqQ_{LIT}) mass spectrometer operated in the positive- and negative-ion IDA modes [40], with alternated low and high fragmentation energy in the collision cell. Despite optimisation of the intensity threshold for the detection of tiny chromatographic peaks, some peaks apparent on the survey scan chromatogram were not selected for the so-called enhanced product-ion-scan mode (Figure 6.4).

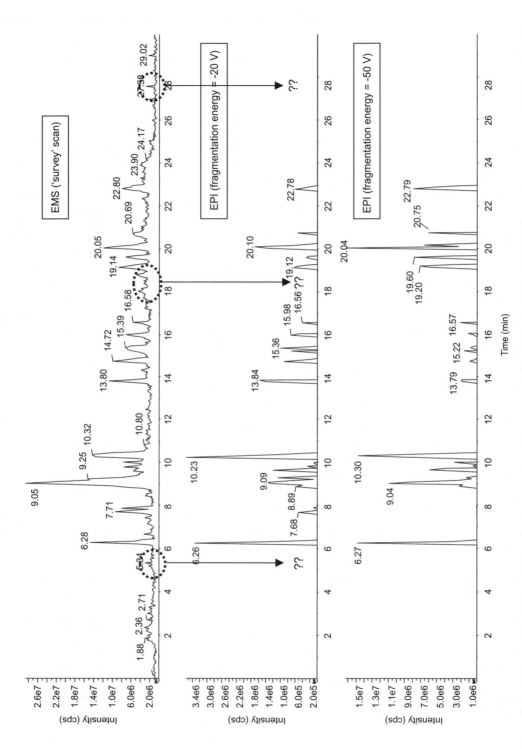

Figure 6.4 Influence of the IDA threshold on the detection of tiny chromatographic peaks (threshold set at 200 000 cps in the present example). (Reproduced from [40] with permission from Elsevier.)

However, five out of the eight compounds spiked to blank serum samples (milrinone, lorazepam, fluometuron, piretanide and warfarin), representing a wide range of lipid solubility and polarity, could be unambiguously identified at the concentration of 0.1 mg/L in serum, in the positive or negative modes, or in both, versus only two by in-source CID LC-MS. All of them could be identified at 1 mg/L by both techniques. Above all, a higher S/N ratio was obtained with the LIT instrument than with the reference technique, resulting in mass spectra devoid of contaminant ions and at least as informative as the reconstructed single-MS spectra. Lately, this technique has been further refined as a result of new software developments allowing automatically alternated collision energy during trap filling (which resulted in rich mass spectra without the need for adding spectra acquired at different collision energies), automatic on-the-fly background noise subtraction before IDA, resulting in enhanced detection of tiny chromatographic peaks, and automatic peak detection, library searching and reporting of results (personal communication).

LITs would be less prone to space-charging effects than traditional ITs, thus enabling a higher number of ions to be accumulated in the source and, hence, higher sensitivity [5]. Although further studies are needed to really evaluate their performance, they might represent efficient tools for LC-MS-MS STA despite a lower mass resolution than TOF instruments.

The major advantage of STA techniques relying on IDA MS-MS is their high specificity and selectivity, as the spectra recorded come from a single parent ion. However, their main drawback is probably that the setting of a given intensity threshold is difficult, due to the intense and, above all, highly variable background noise produced by extracts of real samples (in particular postmortem samples) or by gradient chromatographic elution. If this threshold is given a high value, it results in poor 'sensitivity' (i.e. detection efficiency); if it is given a very low value, it results in an overwhelming amount of information. However, even in the latter case, compounds of interest may be missed due to co-eluting compounds (or interferents) with more intense ions. Finally, processing the huge data files generated for a single injection (from several megabytes up to several gigabytes), due to the number of mass spectra drawn, the high mass resolution or both, may represent a new limitation to the technique.

Inter-instrument and inter-laboratory reproducibility of MS-MS spectra

Gergov et al. carried out a long-term reproducibility study of the MS-MS spectra of 30 test compounds acquired over 4 years on a single instrument [41]. The proprietary software criteria Fit, Reverse Fit and Purity were all excellent, despite the highest weight given to fragment ion intensity ratios: average Fit 95% (range 86–98%), Reverse Fit 94% (89–98%) and Purity 95% (83–98%).

As early as 1983, a round robin was undertaken on the reproducibility of MS-MS spectra obtained using different QqQ spectrometers, coupled either to GC and operated in the positive-ion chemical ionisation mode or to LC by means of APCI sources [42]. The results showed that the fragment ion ratios were in excellent agreement for a particular type of instrument, but different between different types of instruments, and that these differences were merely dependent on collision energy and/or collision gas nature (nitrogen or argon) and pressure (so-called 'thickness'). The authors finally concluded that 'it would be advantageous to set the standard [collision] energy arbitrarily by reference to some suitable ratio of peaks in a test spectrum. The requirement is for a parent which gives two daughters whose yield ratios depend largely on collision energy and two other daughters whose yield ratio depends largely on the amount of target gas' [42].

In 2000, two teams reported the results of an inter-laboratory comparison of MS-MS spectra using library searching with a combination of the libraries of 400 spectra each, built up by each team with the same model of instrument. A few test compounds were analysed as pure standards in methanol using three different QqQ instruments from the same manufacturer. When identical gas pressure and collision energy were applied, the results showed good inter-instrument fragmentation reproducibility. The libraries built using slightly different conditions (fragmentation energies of 20, 30, 40 and 50 eV for one team, 20, 35 and 50 eV for the other) could be used almost indistinctly [43]. Later, the same two laboratories further compared their respective MS-MS libraries by searching the spectra obtained for 30 test compounds in laboratory 1 against the library of laboratory 2 and vice versa. The average Purity score at low and high energies ranged between 85 and 96% for both directions of comparison. Although the medium collision energies were not exactly the same in both laboratories (35 eV in laboratory 1 versus 30 and 40 eV in laboratory 2), the average Purity score ranged between 63 and 88%, the spectra at 35 eV in laboratory 1 fitting better with the 30-eV than the 40-eV spectra of laboratory 2 [41].

Finally, four laboratories (including the previous two) compared the MS-MS spectra obtained on two similar instruments of brand A and two different models from brand B. Fragmentation energy was standardised in terms of gas pressure and collision voltage using a tuning compound, and rather complex steps of data exportation and spectral reduction, including enhancement of high mass peaks and removal of peaks of lowest intensity, were undertaken to allow spectral comparisons [41]. The spectra acquired for the same 30 test compounds with the instruments of brand B yielded Fit scores between 79 and 85% when compared with the mass spectral libraries of brand A instruments. Fit scores were lower than 70% for only 3 of the 30 compounds. However, the authors stressed that peak centroiding, generally used to reduce data before library searching, had a marked effect on the final Fit score: the smaller the merge, the better the distance. Similarly, the less the weight given to fragment ion intensity ratios, the greater the risk for false-positive results [41].

Bristow *et al.* compared product-ion MS-MS spectra recorded on three IT mass spectrometers, a QqQ and a Fourier-transform ion cyclotron resonance (FTICR) mass spectrometer [44]. Standardisation consisted only of selecting fragmentation energy resulting in attenuation of the abundance of the $[M + H]^+$ ion to 10–50% of its original abundance. The degree of similarity was calculated by comparing the five most abundant ions from the spectra and attributing them 20% fit each. Between 64 and 89% of spectra gave acceptable fit (60% or above), depending on the pair of instruments compared. When 20 selected spectra (five test compounds by four instruments) were compared with all 334 library entries (i.e. 6729 spectral comparisons), no false-positive result at the 60% fit threshold was obtained; 323 matches gave a 20% similarity and only 6 gave a 40% similarity [44]. However, such data reduction would probably result in a non-negligible number of false-positive results if the number of library entries and comparisons were increased, and it is our feeling, together with Gergov *et al.* [41], that mass spectral information should be kept as complete as possible for the sake of specificity.

Recently, Jansen *et al.* compared the spectra of 13 test compounds obtained with a QqQ_{LIT} instrument operated in the 'normal' quadrupole mode with those obtained in the LIT configuration, as well as with those of classic QqQ mass spectrometers of two different brands [45]. After tentative standardisation of ion fragmentation and transmission in the positive- and negative-ion modes using a tuning compound, the relative intensities of the ions present in the spectra obtained with the different settings and instruments in either polarity were statistically compared. Some fragments were absent in certain spectra, but no unexpected or unique fragments showed up. The two possible configurations of the LIT MS-MS instrument resulted in significant differences in relative ion intensities in both polarities and so did the other two instruments. However, the comparison criteria were very strict and, although significant differences were found, in most cases the spectra looked almost similar on visual inspection [45]. This suggests that library-searching algorithms should allow users to assign a variable weight to the relative intensity of *m/z* ratios, depending on the provenance of the libraries (in-house database or libraries from another laboratory, another instrument or of commercial origin, etc.).

Chromatographic separation

Several years ago, when LC-MS-MS was just arriving in therapeutic drug monitoring (TDM)/toxicology laboratories, Weinmann *et al.*

explored a very simple approach for the simultaneous quantitative screening of four illicit drugs and their respective deuterated internal standards (ISs) in serum and urine, by combining SPE and flow injection of the extracts in the ESI source of a QqQ instrument with no chromatographic step [46]. Detection was performed in the MRM mode, following one transition per compound and IS. It is now quite clear that, without chromatographic separation, the risks of ion suppression dependent on the ionic content of the matrix are very important and that a single transition per compound is probably not enough for sufficient specificity. Very few such methods have been proposed in the field of toxicology, since they could suffer from interferents, resulting in either false-positive or -negative results.

LC represents a limitation to LC-MS(-MS) STA due to its limited separation power with respect to capillary GC, and is limited itself by the relative incompatibility of MS with non-volatile salts, modifiers and ion-pairing agents in significant amounts that diminish the efficiency of the ionisation process, crucial in the ESI mode [47].

Most authors chose to use chromatographic columns with an internal diameter (i.d.) of 2 [12, 13, 15, 16] or 2.1 mm [7, 11, 35, 37], with a mobile-phase flow rate of 200 µL/min, generally without splitting. However, the lower the column internal diameter and mobile-phase flow rate, the better the LOD or limit of quantification (LOQ) [6]: narrow-bore columns of around 1 mm i.d. with flow rates of 40–50 µL/min generally give the best results [10, 20, 21], inasmuch as the sensitivity of ESI-type sources seems to be related to the concentration of the compounds of interest in the chromatographic effluent, rather than to the amount per unit time (mass flow rate) admitted in the source. This is attributed to a higher evaporation rate of the mobile phase at low flow rates and, thus, to better transmission of the ions formed towards the mass spectrometer [48]. However, this limitation can be partially compensated for through assisted evaporation of the spray by means of an orthogonal flux of heated nitrogen, such as in the TurboIonspray or equivalent sources. Another point is that chromatographic time should not be shortened too much, because efficient chromatographic separation is necessary for good selectivity (no interference), good sensitivity (no ion suppression) and, as far as in-source CID MS techniques are concerned, reproducible fragmentation (fragmentation efficiency being dependent on the ion density in the transition zone [48]). Finally, chromatographic separation should also be suited to compounds in a large polarity range, from the polar substances particularly targeted with LC-MS to more apolar drugs, if the technique is to be used as a complement to, and confirmation technique of, GC-MS and LC-DAD.

Sample preparation

Nordgren et al. reported a method for the 'screening' and determination of 23 drugs (phenetylamines, hypnotics and N-benzylpiperazine), in which urine samples were directly injected on-column after dilution (1/10) with water. They used a 6-min, gradient elution, reversed-phase LC separation, APCI and detection in the positive-ion MRM mode following one transition per compound [49]. Indeed, APCI is less prone to the phenomenon of ion suppression, which could be particularly important in this case because of the absence of an extraction/purification step. However, out of 529 authentic urine samples analysed, the authors found 6.6% positive, of which 2.2% could not be confirmed by another LC-MS-MS technique including SPE and should thus be regarded as false-positive results. They concluded that this technique, which is obviously not really STA, was (only) a valuable complement to immunochemical screening analysis [49].

One of the main problems when using LC-API-MS for STA is to detect even small signals against a high background noise. The critical point indeed is the signal-to-background noise ratio, mainly determined by the purity of the extracts injected. The extraction procedure is thus involved in terms of recovery and selectivity.

Filtration and injection, or protein precipitation with acetonitrile (ACN) and injection of the supernatant, can provide a direct means of introducing samples into LC. However, the lack of a concentration step may limit the detection of some of the most potent drugs [3] and favour ion suppression, above all when using an ESI source.

The molecules mainly targeted by LC-MS(-MS) STA procedures are generally those not amenable to GC-MS, i.e. mainly polar, acidic, thermally labile or hydrophilic compounds, and the extraction procedure should be chosen accordingly. A few authors used two liquid–liquid extraction (LLE) procedures in parallel, one for acidic and one for basic compounds [7, 9], while others used LLE in alkaline conditions only [37].

SPE was employed by many authors, either based on mixed-mode phases [11, 12, 15, 16] or on classic, hydrophobic C_{18}-bonded phases [12, 16]. Decaestecker et al. compared the extraction yields obtained for 18 basic and neutral compounds with 12 different SPE sorbents: seven apolar, three mixed mode and two polymeric packings [50]. For each type of extraction cartridge, the extraction procedure followed the instructions given by the manufacturer. They found that the Isolute™ C_8 and Oasis™ MCX (mixed-mode) phases had the highest clean-up potential, and that the former gave the best extraction yields. In a second step, they optimised the extraction procedure on the C_8 sorbent using a smart experimental design, which resulted in a higher recovery for all drugs, except benzoylecgonine and morphine, whose recovery was decreased by about 40% [51]. However, it is worth noting that no apolar or acidic substance was included in the test compounds.

After a similar comparative study, Saint-Marcoux et al. selected the Oasis™ MCX mixed-mode cartridges, allying a hydrophobic polymer and strong cationic exchanger properties, based on a higher S/N ratio than the other LLE and SPE procedures tested. They even combined the SPE dry extract with the supernatant of serum protein precipitation in order to recover the most hydrophilic compounds [20]. Maurer et al. described a rather comprehensive LLE procedure for both GC-MS and LC-MS STA [2]. One millilitre of plasma is extracted with 5 mL of a mixture of diethyl ether:ethyl acetate (1:1, v/v) and 5 mL saturated sodium sulphate solution. The organic phase is evaporated, while the aqueous phase is basified with sodium hydroxide and extracted again with the same solvent mixture. Finally, the combined dry residues are dissolved: in methanol for GC-MS; in ACN [17, 29] or ethanol [18] for LC-MS analysis. However, this procedure was not satisfactory for the identification and determination of 22 β-blockers and had to be replaced by a SPE procedure based on a mixed-mode, C_8 and cation-exchanger sorbent [19].

SPE based on either mildly hydrophobic or mixed-mode sorbents can probably cope with compounds in the largest polarity range. Although hardly addressed in the literature, emphasis should also probably be placed on the very last step of sample preparation, i.e. the nature of the solvent used to reconstitute dry extracts, as the solubility of polar compounds can be poor in pure organic solvents.

Practicality and efficiency of comprehensive LC-MS(-MS) systematic toxicology analysis procedures

Not all the papers cited here presented a global STA procedure, including extraction and chromatographic separation. When they did, they seldom included an evaluation of the procedure for the detection of unknown compounds, as compared to 'reference' STA techniques including or combining various immunoassays, or GC, GC-MS and/or LC-DAD techniques.

An LC-ESI-CID-MS procedure with optimised extraction and chromatographic separation, four ionisation/fragmentation conditions (two fragmentation energies × two polarities), and automatic data processing and reporting [10] were compared with GC-MS and LC-DAD for the analysis of 51 serum samples from routine clinical toxicology, using a common non-selective SPE on Oasis™ MCX [20]. It was found that LC-ESI-CID-MS was able to detect 63 out of 84 compounds detected (75%) versus 55 (65.5%) for GC-MS and 60 (71.4%) for LC-DAD. About 8.3% of all the compounds were detected only by LC-MS versus 8.3% for GC-MS and 9.5% for LC-DAD. Due to the small number of samples analysed, these results should only be regarded as a first indication of the efficiency of LC-MS as a screening procedure in toxicology, in combination with GC-MS and/or LC-DAD.

Gergov *et al.* evaluated the performance of their LC-MS-MS method in the MRM mode by analysing 71 postmortem blood samples and comparing the results with those obtained by a combination of methods (capillary GC with a dual-column system for blood samples, and thin-layer chromatography and over-pressured layer chromatography for urine or liver samples), taken as a reference [34]. If the results obtained by the MRM screening and the routine techniques were in disagreement, GC-MS in the SIM mode served as a confirmation technique. Of the findings by MRM screening, 256 (92%) were consistent with those obtained by the routine techniques; 76 findings were made only by MRM, of which 28 were metabolites at concentrations lower than the LOD of the reference methods. GC-MS confirmed 30 of the remaining findings and 18 were apparently false positives (although some may not have been detected by the GC-MS procedure if present).

More generally, assessing the performance of a new technique with a broader detection range can hardly rely on a single or combination of older technique(s), as the false-negative results of the latter will inevitably be counted as false-positive results for the former. In order to really validate LC-MS-MS STA procedures, further comparative studies based on validated, sensitive and specific LC-MS-MS methods in the MRM mode are warranted.

Conclusion

STA based on single-quadrupole LC-ESI-MS, although instrument specific, is now credible enough to be used as a complement to GC-MS and/or LC-DAD. However, complete, standardised, automated and routinely applicable procedures are still needed. Although IT MS has apparently not been employed in the single-MS mode in this purpose, maybe because of variable mass spectra depending on the ionic density in the trap, this problem could be overcome in the near future by means of LIT instruments, which do not suffer from this limitation.

As a general rule, MS-MS in toxicology brings higher sensitivity, specificity and selectivity (higher S/N ratio), as well as more structural information, when an unknown chromatographic peak has to be explored. However, the first step for STA is to detect unexpected compounds, which is not compatible with the classic MRM or product-ion-scan modes of mass spectrometers. IDA might be a solution to this problem and deserves to be further explored. Moreover, other major improvements now come from the MS part of the coupling: TOF mass spectrometers offer higher mass accuracy, and LIT ones offer increased S/N and MS^3 capabilities, which might greatly facilitate identification of unknown compounds. This, combined with software improvements to detect tiny chemical signals in the chromatographic noise or to rapidly compare all acquired spectra with those in libraries, will probably help to design even better methods, thus truly combining the relative universality of both LC and mass spectrometry.

Finally, widespread application of LC-MS(-MS) to STA will rely on its capability to detect those drugs not readily detectable at toxicological levels using GC-MS or LC-DAD screening procedures, such as Δ^9-tetrahydrocannabinol and metabolites, morphine and metabolites, buprenorphine, benzoylecgonine, lysergic acid diethylamide (LSD) and metabolites, low-dose benzodiazepines and analogues, neuroleptics, fentanyl and derivatives, a number of cardiovascular drugs, peptidic drugs, chloral hydrate and metabolites, anabolic steroids, coumarin-based anticoagulants, colchicines, α- and β-amanitins, etc. [3]. Experimental evidence is expected in the near future.

References

1. Marquet P. Is LC-MS suitable for a comprehensive screening of drugs and poisons in clinical toxicology? *Ther Drug Monit* 2002; 24: 125–133.

2. Maurer HH. Position of chromatographic techniques in screening for detection of drugs or poisons in clinical and forensic toxicology and/or doping control. *Clin Chem Lab Med* 2004; 42: 1310–1324.

3. Drummer OH. Chromatographic screening techniques in systematic toxicological analysis. *J Chromatogr B* 1999; 733: 27–45.

4. Pfleger K, Maurer HH, Weber AA. *Mass Spectral Library of Drugs, Poisons, Pesticides, Pollutants and their Metabolites*, 3rd edn. Palo Alto, CA: Agilent Technologies, 2000.

5. Niessen WMA. Progress in liquid chromatography-mass spectrometry instrumentation and its impact on high-throughput screening. *J Chromatogr A* 2003; 1000: 413–436.

6. Marquet P, Lachâtre G. Liquid chromatography-mass spectrometry: potential in forensic and clinical toxicology. *J Chromatogr B* 1999; 733: 93–118.

7. Spratt E, Vallaro GM. LC/MS with a particle beam interface in forensic toxicology. *Clin Lab Med* 1998; 18: 651–663.

8. Marquet P. Progress of liquid chromatography-mass spectrometry in clinical and forensic toxicology. *Ther Drug Monit* 2002; 24: 255–276.

9. Lips AGAM, Lameijer W, Fokkens RH, et al. Methodology for the development of a drug library based upon collision-induced fragmentation for the identification of toxicologically relevant drugs in plasma samples. *J Chromatogr B* 2001; 759: 191–297.

10. Marquet P, Venisse N, Lacassie E, et al. In-source CID mass spectral libraries for the 'general unknown' screening of drugs and toxicants. *Analusis* 2000; 28: 41–50.

11. Gergov M, Robson JN, Duchoslav E, et al. Automated liquid chromatographic/tandem mass spectrometric method for screening β-blocking drugs in urine. *J Mass Spectrom* 2000; 35: 912–918.

12. Rittner M, Pragst F, Neumann J. Screening method for seventy psychoactive drugs or drug metabolites in serum based on high-performance liquid chromatography-electrospray ionization mass spectrometry. *J Anal Toxicol* 2001; 25: 115–124.

13. Weinmann W, Wiedemann A, Eppinger B, et al. Screening for drugs in serum by electrospray ionization/collision-induced dissociation and library searching. *J Am Soc Mass Spectrom* 1999; 10: 1028–1037.

14. Hough JM, Haney CA, Voyksner RD, et al. Evaluation of electrospray transport CID for the generation of searchable libraries. *Anal Chem* 2000; 72: 2265–2270.

15. Müller C, Vogt S, Goerke R, et al. Identification of selected psychopharmaceuticals and their metabolites in hair by LC/ESI-CID/MS and LC/MS/MS. *Forensic Sci Int* 2000; 113: 415–421.

16. Weinmann W, Lehman N, Müller C, et al. Identification of lorazepam and sildenafil as examples for the application of LC/ionspray-MS and MS-MS with mass spectra library searching in forensic toxicology. *Forensic Sci Int* 2000; 113: 339–344.

17. Maurer HH, Kratzsch C, Kraemer T, et al. Screening, library-assisted identification and validated quantification of oral antidiabetics of the sulfonylurea-type in plasma by atmospheric pressure chemical ionization liquid chromatography-mass spectrometry. *J Chromatogr B* 2002; 773: 63–73.

18. Kratzsch C, Tenberken O, Peters FT, et al. Screening, library-assisted identification and validated quantification of 23 benzodiazepines, flumazenil, zaleplone, zolpidem and zopiclone in plasma by liquid chromatography/mass spectrometry with atmospheric pressure chemical ionization. *J Mass Spectrom* 2004; 39: 856–872.

19. Maurer HH, Tenberken O, Kratzsch C, et al. Screening for library-assisted identification and fully validated quantification of 22 beta-blockers in blood plasma by liquid chromatography-mass spectrometry with atmospheric pressure chemical ionization. *J Chromatogr A* 2004; 1058: 169–181.

20. Saint-Marcoux F, Lachâtre G, Marquet P. Evaluation of an improved general unknown screening procedure using liquid chromatography-electrospray-mass spectrometry by comparison with gas chromatography-mass spectrometry and high-performance liquid chromatography-diode array detection. *J Am Soc Mass Spectrom* 2003; 14: 14–22.

21. Venisse N, Marquet P, Duchoslav E, et al. A general unknown screening procedure for drugs and toxic compounds in serum using liquid chromatography-electrospray ionization mass spectrometry. *J Anal Toxicol* 2003; 27: 7–14.

22. Gergov M, Boucher B, Ojanperä I, et al. Toxicological screening of urine for drugs by liquid chromatography/time-of-flight mass spectrometry with automated target library search based on elemental formulas. *Rapid Commun Mass Spectrom* 2001; 15: 521–526.

23. Bogusz MJ, Maier RD, Krüger KD, et al. Poor reproducibility of in-source collisional atmospheric pressure ionization mass spectra in toxicologically relevant drugs. *J Chromatogr A* 1999; 844: 409–418.

24. Schreiber A, Efer J, Engewald W. Application of spectral libraries for high performance liquid chromatography-atmospheric pressure ionisation mass spectrometry to the analysis of pesticide and

explosive residues in environmental samples. *J Chromatogr A* 2000; 869: 411–425.

25. Marquet P, Venisse N, Gaulier JM, et al. 'General unknown' screening of xenobiotics in serum by LC-ES-MS. Abstract presented at *35th TIAFT Meeting*, Krakow, 1999.

26. Weinmann W, Lehmann N, Renz M, et al. Screening for drugs in serum and urine by LC/ESI/CID-MS and MS/MS with library searching. *Prob Forensic Sci* 2000; 22: 202–208.

27. Rivier L. Robustness evaluation for automatic identification of general unknown by reference ESI mass spectra comparison in LC-MS toxicological screening. Paper presented at *37th TIAFT Meeting*, Prague, 2001.

28. Rivier L. Criteria for the identification or compounds by liquid chromatography-mass spectrometry in forensic toxicology and doping analysis. *Anal Chim Acta* 2003; 492: 69–82.

29. Maurer HH, Kraemer T, Kratzsch C, et al. Negative ion chemical ionization gas chromatography-mass spectrometry and atmospheric pressure chemical ionization liquid chromatography-mass spectrometry of low-dosed and/drugs in plasma. *Ther Drug Monit* 2002; 24: 117–124.

30. Bristow AWT, Nichols WF, Webb KS, et al. Evaluation of protocols for reproducible electrospray in-source collisionally induced dissociation on various liquid chromatography/mass spectrometry instruments and the development of spectral libraries. *Rapid Commun Mass Spectrom* 2002; 16: 2374–2386.

31. Weinmann W, Stoertzel M, Vogt S, et al. Tune compounds for electrospray ionisation/in-source collision-induced dissociation with mass spectral library searching. *J Chromatogr A* 2001; 926: 199–209.

32. Kienhuis PGM, Geerdink RB. Liquid chromatography-tandem mass spectrometric analysis of surface and waste water with atmospheric pressure chemical ionisation. *Trends Anal Chem* 2000; 19: 460–474.

33. Klein J, Alder L. Applicability of gradient liquid chromatography with tandem mass spectrometry to the simultaneous screening for about 100 pesticides in crops. *J AOAC Int* 2002; 86: 1015–1037.

34. Gergov M, Ojanperä I, Vuori E. Simultaneous screening of 238 drugs in blood by liquid chromatography-ion spray tandem mass spectrometry with multiple-reaction monitoring. *J Chromatogr B* 2003; 795: 41–53.

35. Baumann C, Cintora MA, Eichler M, et al. A library of atmospheric pressure ionization daughter ion mass spectra based on wideband excitation in an ion trap mass spectrometer. *Rapid Commun Mass Spectrom* 2000; 14: 349–356.

36. Gergov M, Robson JN, Ojanperä I, et al. Simultaneous screening and quantitation of 18 antihistamine drugs in blood by liquid chromatography ionspray tandem mass spectrometry. *Forensic Sci Int*, 2001; 121: 108–115.

37. Decaestecker TN, Clauwaert KM, Van Bocxlaer JF, et al. Evaluation of automated single mass spectrometry to tandem mass spectrometry function switching for comprehensive drug profiling analysis using a quadrupole time-of-flight mass spectrometer. *Rapid Commun Mass Spectrom* 2000; 14: 1787–1792.

38. Decaestecker TN, Vande Casteele SR, Wallemacq PE, et al. Information-dependent acquisition-mediated LC-MS/MS screening procedure with semiquantitative potential. *Anal Chem* 2004; 76: 6365–6373.

39. Fitzgerald RL, Rivera JD, Herold DA. Broad spectrum drug identification directly from urine, using liquid chromatography-tandem mass spectrometry. *Clin Chem* 1999; 45: 1224–1234.

40. Marquet P, Saint-Marcoux F, Gamble TN, et al. Comparison of a preliminary procedure for the general unknown screening of drugs and toxic compounds using a quadrupole-linear ion-trap mass spectrometry with a liquid chromatography-mass spectrometry reference technique. *J Chromatogr B* 2003; 789: 9–18.

41. Gergov M, Weinmann W, Meriluoto J, et al. Comparison of product ion spectra obtained by liquid chromatography/triple quadrupole mass spectrometry for library searching. *Rapid Commun Mass Spectrom* 2004; 18: 1039–1046.

42. Dawson PH, Sun W-F. A round robin on the reproducibility of standard operating conditions for the acquisition of library MS/MS spectra using triple quadrupoles. *Int J Mass Spectrom Ion Process* 1984; 55: 155–170.

43. Weinmann W, Gergov M, Goerner M. MS/MS-libraries with triple quadrupole-tandem mass spectrometers for drug identification and drug screening. *Analusis* 2000; 28: 934–941.

44. Bristow AWT, Webb KS, Lubben AT, *et al*. Reproducible product-ion tandem mass spectra on various liquid chromatography/mass spectrometry instruments for the development of spectral libraries. *Rapid Commun Mass Spectrom* 2004; 18: 1447–1454.

45. Jansen R, Lachâtre G, Marquet P. LC-MS/MS systematic toxicological analysis: comparison of MS/MS spectra obtained with different instruments and settings. *Clin Biochem* 2005; 38: 362–372.

46. Weinmann W, Svoboda M. Fast screening for drugs of abuse by solid-phase extraction combined with flow-injection ion spray-tandem mass spectrometry. *J Anal Toxicol* 1998; 22: 319–328.

47. Thieme D, Sachs H. Improved screening capabilities in forensic toxicology by application of liquid chromatography-tandem mass spectrometry. *Anal Chim Acta*, 2003; 492: 171–186.

48. Niessen WMA. Advances in instrumentation in liquid chromatography-mass spectrometry and related liquid-introduction techniques. *J Chromatogr A* 1998; 794: 407–435.

49. Nordgren HK, Beck O. Multicomponent screening for drugs of abuse. Direct analysis of urine by LC-MS-MS. *Ther Drug Monit* 2004; 26: 90–97.

50. Decaestecker TN, Coopman EM, Van Peteghem CH, *et al*. Suitability testing of commercial solid-phase extraction sorbents for sample clean-up in systematic toxicological analysis using liquid chromatography-(tandem) mass spectrometry. *J Chromatogr B* 2003; 789: 19–25.

51. Decaestecker TN, Lambert WE, Van Peteghem CH, *et al*. Optimization of solid-phase extraction for a liquid chromatographic-tandem mass spectrometric general unknown screening procedure by means of computational techniques. *J Chromatogr A* 2004; 1056: 57–65.

7

Analysis of therapeutic drugs with LC-MS

Hans H. Maurer and Frank T. Peters

Introduction

Apart from alcohol and drugs of abuse, therapeutic drugs play a major role in clinical and forensic toxicology as well as in doping control. Some of them are also misused; some are rather toxic in overdose. Taken therapeutically, they may influence the pharmacodynamics and/or pharmacokinetics of alcohol and/or drugs of abuse, and they may also influence the analysis of such compounds. Therefore, efficient toxicological analysis as the basis of a competent toxicological judgement and consultation in clinical and forensic toxicology must cover therapeutic drugs.

Immunoassays (IAs) are commercially available only for the common drugs of abuse, including benzodiazepines, barbiturates and opiates, and additionally for very few opioid and non-opioid analgesics, tricyclic antidepressants, anti-epileptics, anti-arrhythmics, antibiotics and immunosuppressants. However, many relevant drug classes such as neuroleptics, selective serotonin re-uptake inhibitors, most of the opioid and non-opioid analgesics, cardiovascular drugs and anti-diabetics cannot be analysed by commercially available IAs. Therefore, chromatography-based procedures play an important role in the analysis of therapeutic drugs in biological matrices. In the last decade, liquid chromatography (LC) coupled with single-stage and, especially, with tandem mass spectrometry (MS) (LC-MS, LC-MS-MS) has become increasingly important in the analysis of therapeutic drugs in biological matrices. In the area of pharmaceutical research and development, LC-MS(-MS) has even become the standard for high-throughput determinations of drug concentrations in biological samples obtained from bioavailability, bioequivalence or pharmacokinetic studies because of its versatility [1–4]. This is reflected by a large and ever-increasing number of publications reporting LC-MS(-MS)-based methods for the analysis of new drug entities in biological matrices, especially blood, plasma or serum. While in such studies the analytes are known and only have to be quantified, the compounds to be determined are mostly unknown in clinical and forensic toxicology. Therefore, the first step, before quantification, is to screen for and identify any compounds of interest. In analytical toxicology, a high throughput of samples requires procedures that allow simultaneous screening for numerous relevant toxicants using a single method (so-called systematic toxicological analysis [STA]). For such high-throughput screening in routine analysis, thin-layer chromatography, gas chromatography (GC) with common detectors, electrokinetic techniques and high-performance LC coupled with a diode-array detector (DAD) are principally applicable, but GC-MS is by far the most widely used method in this context. The position of chromatographic techniques in screening for detection of drugs or poisons in clinical and forensic toxicology and/or doping control has been reviewed recently [5]. LC-MS or LC-MS-MS still plays only a limited role for comprehensive screening, but some applications look promising [6–19]. Most of these are target screening procedures rather than STA procedures. This means that they allow screening for a certain number of analytes in the selected-ion monitoring (SIM) or multiple-reaction monitoring (MRM) mode, whereas analytes not included *a priori* cannot be detected. Nevertheless, such methods have helped to extend the spectrum of sensitive and specific MS-based methods to analytes that are not amenable to GC-MS analysis because of their hydrophilic or thermolabile properties. In many cases, LC-MS(MS)-based target screening procedures are also applicable for quantification of the respective drugs or, at least, the same prepared sample can be used for this purpose [7, 10, 11, 13, 15, 18, 20, 21], thus saving time and resources. Such multi-analyte procedures for screening and/or quantification of therapeutic drugs in blood, plasma or serum are powerful tools in clinical and forensic toxicology, and should be used whenever possible. Therefore, the main focus of this chapter is on such multi-analyte procedures. Papers concerning LC-MS quantification of single substances [22] as well as urinalysis [5] have already been reviewed elsewhere.

Sample preparation for LC-MS(-MS) analysis of therapeutic drugs

The procedures reported in the literature for work-up of biological samples to be analysed by LC-MS(-MS) range from very simple to fully automated on-line extraction methods [2]. As in any other area of analytical toxicology, the choice of the optimum procedure for sample preparation depends on the analytical problem to be solved. However, in LC-MS(-MS) analysis, in particular, the choice of an appropriate sample preparation procedure must also take into consideration possible matrix effects (MEs), i.e. suppression or enhancement of analyte ionisation by co-eluting compounds. In case of matrix suppression, a relevant intoxicant might be overlooked [23] or the treatment outcome might be compromised. This well-known phenomenon in LC-MS(-MS) analysis depends, among other factors, on the sample matrix and the sample preparation procedure [1, 24–32].

Systematic comparison studies

In a number of recently published papers, different sample preparation procedures for LC-MS(-MS) analysis were systematically compared. Decaestecker *et al.* [33] tested the suitability of an entire series of Isolute (Argonaut Technologies, CA, USA) and Oasis (Waters, MA, USA) solid-phase extraction (SPE) sorbents in order to select the best sorbent with regard to extraction yields of 18 neutral and basic compounds (including several benzodiazepines and opioids as well as the neuroleptic haloperidol and the antidepressant trazodone). For each sorbent type, they derived a parameter from the extraction yields of all 18 compounds from aqueous solutions. Based on this parameter, the authors found the apolar Isolute C_8 sorbent to demonstrate the best overall extraction yield, while that of the mixed-mode Oasis MCX sorbent was only marginally worse. However, there were no large differences between these two sorbents and Isolute C_4, C_{18}, $C_{18}MF$, HCX and HCX_3, or the Oasis HLB sorbents, so that the practical relevance of these differences remains unclear. A further evaluation of the Isolute C_8 and Oasis MCX sorbents using spiked blood samples showed that the clean-up potential of the sorbents was comparable, but that the coefficients of variation (CVs) were considerably lower for the C_8 sorbent.

Naidong *et al.* [34] studied extraction yields and matrix-suppression effects of six therapeutic drugs (clonidine, albuterol, naltrexone, loratadine, fentanyl and ritonavir) using electrospray ionisation (ESI) LC-MS-MS after isolation from plasma by different procedures: simple protein precipitation using acetonitrile (ACN), liquid–liquid extraction (LLE) using hexane or methyl-tertiary butyl ether (MTBE) and SPE using Bond Elut Certify (mixed-mode) (Varian, CA, USA) and C_{18} or Oasis HLB columns. For protein precipitation, they reported acceptable extraction yields, but also severe matrix-suppression effects. The more selective LLE with hexane and SPE with a C_{18} sorbent resulted in little ME, but also in lower extraction yields. LLE with MTBE and SPE with the mixed-mode sorbent were reportedly a good compromise between extraction yield and MEs. However, the authors cautioned against considering any of the methods superior, and stated that, as long as extraction yields and matrix suppression are consistent among various lots of sample matrices and an acceptable sensitivity is achieved, any of the studied sample preparations may be appropriate for a certain application.

Mueller et al. [29] compared ion-suppression effects in ESI LC-MS-MS after LLE using a mixture of ethyl acetate/diethyl ether, mixed-mode SPE with Chromabond Drug (International Specialty Products, NJ, USA) and Oasis MCX columns, protein precipitation using ACN or a combination of protein precipitation and Oasis MCX. For this purpose, they performed a post-column infusion experiment in which a solution of the model compounds codeine and glafenine was added to eluents from the LC column after injection of blank matrix extracts. The authors reported ion suppression for all studied sample preparations with the exception of the basic fraction of the Chromabond Drug extract. The most severe ion suppression occurred with the methods involving protein precipitation. However, in all cases, the ion-suppression effects were limited to the LC front and not observed during the rest of the gradient elution.

In a similar experiment, Dams et al. [28] used post-column infusion of a morphine solution after injection of prepared blank urine, oral fluid and plasma samples to evaluate MEs. Apart from the influence of the ionisation modes ESI and atmospheric pressure chemical ionisation (APCI), they studied the influence of four different sample preparations, i.e. direct injection, dilution, and protein precipitation with ACN and SPE. Generally speaking, considerable to severe matrix-suppression effects were observed for all matrices and sample preparations when operating the MS in the ESI mode. However, they were less pronounced after the more selective SPE-based work-up. In the APCI mode, considerable/severe ion-suppression effects were observed only after simple work-up of urine and plasma samples. For oral fluid samples as well as plasma and urine after SPE, ion suppression was reported to be moderate or not detectable at all. Furthermore, the ion-suppression effects after SPE were limited to the LC front, confirming the findings of Mueller et al. [29].

Mallet et al. [25] studied the influence of sample preparation on ion-suppression effects by introducing a mixture of mobile phase and prepared plasma samples directly into an ESI ion source. Under such worst-case conditions without any chromatographic separation, the maximum potential for MEs after different sample preparations could be studied. The authors of this study used eight acidic and eight basic therapeutic drugs as model compounds and compared protein precipitation using methanol with SPE using Oasis HLB [reversed-phase (RP)], MCX and MAX (mixed-mode) columns. The observed matrix suppression after protein precipitation was a massive reduction of the signal by more than 90% for most studied compounds and even after RP SPE it was still considerable. Mixed-mode SPE almost completely eliminated the MEs with respect to the basic model drugs, while moderate MEs were observed for the acidic model drugs.

Another comprehensive study on the influence of different sample preparations for plasma samples with respect to MEs has been published by Souverain et al. [24]. These authors studied four off-line procedures (LLE with a mixture of hexane and amyl alcohol, mixed-mode SPE using Bond Elut Certify columns, and protein precipitation with perchloric acid or ACN) and on-line procedures using three extraction supports, namely Oasis HLB, Cyclone (UDY, CO, USA) and LiChrospher RP-4 ADS (Merck, Germany) compared with direct injection of diluted samples and injection of samples after protein precipitation with ACN. Furthermore, these experiments were carried out by operating the LC-MS system in the ESI and APCI mode. The MEs were evaluated using post-column infusion of a methadone solution while analysing blank samples. In the ESI mode, ion suppression was observed during the first minute of the chromatographic run for all off-line methods with the exception of LLE. For the on-line ESI methods, signal suppression was observed only after injection of diluted samples on the Cyclone and LiChrospher extraction supports. In the APCI mode, signal suppression was not observed for any of the off-line procedures, but for protein precipitation with perchloric acid (again only during the first minute). The on-line APCI methods were all free of signal suppression. On the contrary, a considerable enhancement was observed with the Cyclone column after injection of the diluted sample.

The studies described above clearly show that the choice of an appropriate sample preparation procedure is a key issue in LC-MS(-MS) analysis of therapeutic drugs. Due to the high risk of matrix suppression, direct injection of (diluted) samples or protein precipitation cannot be recommended, especially if ESI is used for ionisation. All other sample preparation techniques might be used successfully depending on the analytical problem. If high sensitivity is required, mixed-mode SPE is certainly a good choice, because it yields very clean extracts and thus a favourable signal-to-noise (S/N) ratio.

Applications

An enormous number of LC-MS(-MS) methods for analysis of therapeutic drugs have been published in the last decade. In the vast majority of these methods, blood, plasma or serum was used as the sample matrix. The reason is probably the fact that therapeutic effects usually correlate much better with drug concentrations in these samples than in any other routine biological sample matrix. LC-MS methods for urinalysis are comparatively rare and mostly applied in areas where analysis of urine is mandatory, as in doping analysis. LLE and SPE are by far the most common methods for sample preparation in LC-MS(-MS) analysis of therapeutic drugs, but others, such as direct injection after dilution, protein precipitation, on-line extraction or solid-phase microextraction (SPME) have also been reported. The basic information on selected multi-analyte procedures for screening, identification and/or quantification of therapeutic drugs in blood, plasma, serum or urine is summarised in Table 7.1 to simplify the rapid selection of a method suitable for an actual analytical problem.

LLE was used by many authors for extraction of a wide variety of analytes. Zhang et al. [35] used the very apolar solvent hexane for extraction of tricyclic antidepressants from human plasma at basic pH. Due to the high lipophilicity of these drugs in the unionised form, the recoveries were 69–105% and clean extracts were obtained. Thevis et al. [36] used a more polar solvent mixture of tertiary butanol and MTBE for extraction of β-blockers from urine samples after enzymatic hydrolysis and basification. No recovery data were reported, but the authors stated that the method was sufficiently sensitive for β-blocker screening in doping analysis. Gergov et al. used a two-step LLE extraction in a procedure for simultaneous screening and quantification of 18 antihistamine drugs [7], which they also later applied in a method for simultaneous screening of 238 drugs in blood [12]. Blood samples were first made basic (pH 11), extracted with butyl acetate, and subsequently acidified (pH 3) and extracted with dichloromethane/isopropanol. The two extracts were finally combined and analysed. Such two-step extractions at two pH values with intermediate polarity solvents are well suited for extraction of a wide spectrum of compounds with different physicochemical properties and are therefore ideal for screening procedures. A similar approach is used in the laboratory of the authors of this chapter. There, the routine two-step LLE procedure used for GC-MS-based screening analysis of plasma samples [37] is also used for LC-MS methods whenever possible. Thus, the number of sample preparations for routine analysis is kept at a minimum, and different GC-MS and LC-MS methods can be performed without the need for further sample preparation. Ultimately, this helps to save time and resources, which is of special importance in emergency toxicology. In the first step of this LLE procedure, plasma samples are extracted with a mixture of ethyl acetate and diethyl ether. Thereafter, the aqueous phase is made basic and extracted again with the same solvent mixture. The two extracts are finally combined and analysed together. Because of the two different pH values of extraction and because of the intermediate polarity of the solvent mixture, many neutral and some acidic compounds are extracted in the first step, whereas basic compounds are extracted in the second step. This extraction method was successfully applied to LC-MS-based screening, library-assisted identification and validated quantification of sulphonylurea-type anti-diabetics [11], the anti-epileptic oxcarbazepine and its metabolite 10-hydroxycarbazepine [38], benzodiazepines and benzodiazepine-like drugs [18], and anaesthetics, low-dose hypnotics and opioids [39].

Sample preparation using SPE with different sorbents is also common in LC-MS(-MS). An automated SPE procedure using RP Bond Elut C_{18} columns was described by Lachatre et al. [40] for simultaneous analysis of four anthracyclines and three of their metabolites in human serum – a group of anti-cancer drugs that are often therapeutically monitored because of their cardiotoxic potential. An alternative to silica-based RP sorbents are polymer resins, e.g. XAD-2, a divinylbenzene polymer with apolar properties. This sorbent was used by Thieme et al. [41] in a procedure for screening, confirmation and quantification of diuretics in urine for doping analysis. The physicochemical properties of the analytes of this method ranged from acidic drugs, such as furosemide or hydrochlorothiazide, over neutral drugs, such as canrenone, to basic drugs, such as triamterene. In these cases, SPE with an apolar sorbent and a sample pH around neutral, as used by the authors of this study, is often a good choice, and it proved to be useful for doping analysis of diuretics here. Another polymer-based SPE column, Oasis HLB, was evaluated by Venisse et al. [17] for its suitability in a general unknown screening (GUS) procedure in serum. HLB stands for 'hydrophilic–lipophilic balance', pointing to its partly polar character. The authors reported that after extraction of a serum sample spiked with 61 (mainly therapeutic) drugs using Oasis HLB columns, 47 compounds could be detected with LC-MS. However, the authors conceded that the drugs had been spiked at a concentration of 10 mg/L, which was unrealistically high for most drugs. In the same study [17], another Oasis column, i.e. Oasis MCX, was tested. The latter performed better than the Oasis HLB in the previously mentioned experiments, leading to the detection of 53 out of 61 drugs. The authors proposed that the better performance resulted from the additional cation-exchange mechanism of the mixed-mode Oasis MCX sorbent. However, the additionally detected drugs included phenobarbital and phenylbutazone, so that other effects like, for example, a more favourable S/N ratio might also play a role. The Oasis MCX was also used successfully in two other publications on GUS in serum samples [16, 17]. Mixed-mode SPE was also used in the toxicological screening method for urine samples published by Pelander et al. [19]. These authors reported extraction of enzymatically hydrolysed urine samples using Isolute HCX-5 columns. The silica-based sorbent of this product was a mixture of an RP C_4 and strong cation-exchange sorbent. After separate elution, the acidic/neutral and basic fractions were combined and analysed. In the laboratory of the authors of this chapter, a similar column with an RP C_8 and strong cation-exchange sorbent, the Isolute Confirm HCX, is used in cases where their standard LLE procedure is not applicable. For the reasons already given above, this SPE procedure is also based on a routine method that already existed in the laboratory. In contrast to the previously mentioned mixed-mode SPE procedures, only the basic fraction is used for analysis. Thus, possible interferences from the acidic-neutral fraction are eliminated and very clean extracts are obtained. Originally, this SPE method was developed for GC-MS analysis of designer drugs in plasma samples [42, 43], but it also proved to be very versatile for screening, library-assisted identification and validated quantification of neuroleptics [15] and β-blockers in plasma by LC-MS [21]. A comparable SPE column, Bond Elut Certify, was used by Josefsson et al. [14] for isolation of neuroleptics from blood samples after sonication. These authors used only the basic fraction to optimise the sample clean-up.

In the same publication [14], the use of diluted urine samples after enzymatic hydrolysis was reported for screening and quantification of the neuroleptics in human body fluids and tissues. Being aware of the problem of ion suppression, they stressed that it was important to chromatographically separate the analytes from early eluting polar compounds and salts, which is in line with the findings of Dams et al. [28]. However, Josefsson et al. [14] performed enzymatic conjugate cleavage, which might increase the risk of ion suppression, because the urine matrix is considerably altered during enzymatic hydrolysis and less hydrophilic compounds with potentially longer retention times are formed.

Nordgren and Beck [44] used simple dilution of urine samples in a multi-component screening procedure for drugs of abuse, including the therapeutic drugs zolpidem, zopiclone and zaleplon, and some of their metabolites. The detection criterion was S/N ≥ 3. Although the authors

Table 7.1 LC-MS or LC-MS-MS procedures for screening and/or quantification of therapeutic drugs in urine, blood, plasma or serum

Compound	Sample	Work-up	Stationary phase	Mobile phase	Detection mode	Validation parameters	Reference
Anaesthetics, low-dose hypnotics, opioids	plasma	LLE	Superspher 60 RP Select B	gradient, ACN, aqueous ammonium formate	single stage, APCI, positive mode, scan (screening, identification), SIM (quantification)	selectivity, linearity, LOD, LOQ, recovery, accuracy, precision, stability (bench-top, processed sample, freeze/thaw, long term), applicability	39
Anthracyclines	serum	SPE	Symmetry C$_{18}$	isocratic, ACN, aqueous ammonium formate	single stage, ESI, positive mode, SIM	selectivity, linearity, LOD, LOQ, recovery, accuracy, precision	40
Antidepressants	plasma	LLE	SB-C18 Mac Mod	gradient, ACN, aqueous ammonium acetate	TOF, ESI, positive mode, scan	selectivity, linearity, LOQ, recovery, accuracy, precision	35
Antihistamines	blood	LLE	Purospher RP-18	gradient, ACN, aqueous ammonium acetate	triple stage, ESI, positive mode, MRM, confirmation using production spectra	linearity, LOD, LOQ, recovery, accuracy, precision	7
Benzodiazepines	plasma	LLE	Superspher 60 RP Select B	gradient, ACN, aqueous ammonium formate	single stage, APCI, positive mode, scan (screening, identification), SIM (quantification)	selectivity, linearity, LOD, LOQ, recovery, accuracy, precision, stability (bench-top, processed sample, freeze/thaw, long term), applicability	18
Benzodiazepines and metabolites	blood	SPME	Supelcosil C$_{18}$	isocratic, methanol, water	single stage, ESI, positive mode	linearity, LOD, precision	47
β-Blockers	urine	LLE	Purospher STAR RP-18e	gradient, ACN, aqueous ammonium acetate	triple stage, APCI, positive mode, MRM	LOD	36
β-Blockers	plasma	SPE	Superspher 60 RP Select B	gradient, ACN, aqueous ammonium formate	single stage, APCI, positive mode, scan (screening, identification), SIM (quantification)	selectivity, linearity, LOD, LOQ, recovery, accuracy, precision, stability (bench-top, processed sample, freeze/thaw, long term), applicability	21
Diuretics	urine	SPE	XDB C8 Zorbax	gradient, ACN, aqueous ammonium acetate	triple stage, ESI, positive and negative mode, SIM, MRM	linearity	41
Neuroleptics	plasma	SPE	Superspher 60 RP Select B	gradient, ACN, aqueous ammonium formate	single stage, APCI, positive mode, scan (screening, identification), SIM (quantification)	selectivity, linearity, LOD, LOQ, recovery, accuracy, precision, stability (bench-top, processed sample, freeze/thaw, long-term), applicability	15

Table 7.1 Continued

Compound	Sample	Work-up	Stationary phase	Mobile phase	Detection mode	Validation parameters	Reference
Neuroleptics	blood, urine, hair	SPE (blood, hair)	Zorbax Stable Bond Cyano	gradient, methanol, and ACN aqueous ammonium formate	triple stage, ESI, positive mode, MRM	linearity, LOQ, recovery, accuracy, precision, stability (bench-top, processed samples, freeze/thaw, long term, HIV inactivation), applicability	14
Phenothiazines	blood, urine	SPME	Capcell Pak C$_{18}$	gradient, ACN, aqueous ammonium acetate	triple stage, ESI, positive mode, MRM	linearity, precision, accuracy, recovery, applicability	20
Protease inhibitors; non-nucleoside reverse transcriptase inhibitors	plasma	on-line SPE	Hypersil MOS	gradient, methanol, aqueous ammonium acetate	single stage, ESI, positive mode, SIM	selectivity, linearity	46
Sulphonylurea-type anti-diabetics	plasma	LLE	Superspher 60 RP Select B	gradient, ACN, aqueous ammonium formate	single stage, APCI, positive mode, scan (screening, identification), SIM (quantification)	selectivity, linearity, LOD, LOQ, recovery, accuracy, precision, stability (long term), applicability	11
9 drugs/poisons	serum	SPE	X-Terra MS C$_{18}$	gradient, ACN, aqueous ammonium formate	triple stage, ESI, positive and negative mode, IDA: scan, MRM	LOD	16
23 drugs	urine	dilution (screen), SPE (confirmation)	HyPURITY Advance	gradient, ethanol, aqueous ammonium acetate	triple stage, APCI, positive mode, MRM	linearity, precision, cut-off (screening), recovery, precision, cut-off (confirmation)	44
47 drugs	serum	SPE	Nucleosil C$_{18}$	gradient, ACN, aqueous ammonium formate	single stage, ESI, positive and negative mode, scan	–	48
53 drugs	serum	SPE	Nucleosil C$_{18}$	gradient, ACN, aqueous ammonium formate	single stage, ESI, positive and negative mode, scan	–	17
238 drugs	blood	LLE	Purospher RP-18	gradient, ACN, aqueous ammonium acetate	triple stage, ESI, positive mode, MRM	LOD	12
637 drugs	urine	SPE	Luna C$_{18}$	gradient with ACN and aqueous formic acid	TOF, ESI, positive mode, scan	LOD	19

reported two- to five-fold variations of analytical sensitivity near the cut-off levels, neither MEs nor the rate of false-negative results was studied. The authors stated only that they preferred APCI over ESI because of its lower susceptibility to MEs. This seems questionable after the findings of Dams et al. [28], who reported considerable matrix suppression for diluted urine in the first 2 min of the chromatographic run – a retention time at which many of the analytes in the method of Nordgren and Beck [44] eluted.

Sayer et al. [45] reported a procedure for the identification and quantification of six non-depolarising neuromuscular blocking agents in biological fluids, in which protein precipitation with ACN was used for sample preparation. Regarding the fact that the analytes are quaternary ammonium compounds, protein precipitation was a reasonable choice, because, in contrast to most standard LLE and SPE procedures (except weak cation-exchange SPE), it does not require the analytes to be present in un-ionised form. However, protein precipitation is associated with a high risk of MEs [24, 25, 28, 28, 29, 34], especially in the ESI mode, which was used here [45]. Unfortunately, the authors did not discuss this issue, so it remains unclear if they checked for MEs, which certainly would have been desirable.

A combination of protein precipitation and on-line SPE was used by Egge-Jacobsen et al. [46] for isolation of six protease inhibitors and three non-nucleoside reverse transcriptase inhibitors from plasma samples. These drugs are used for treatment of AIDS and are often therapeutically monitored to ensure effective treatment. After protein precipitation with methanol/aqueous $ZnSO_4$, samples were further cleaned up using a Hypersil MOS (Thermo Electron, MA, USA) extraction column. The recoveries of all analytes were higher than 76%, demonstrating the effectiveness of this approach.

SPME is becoming an alternative to SPE and LLE, mainly for GC-MS analysis. SPME is a solvent-free and concentrating extraction technique especially for rather volatile analytes. It is based on the adsorption of the analyte on a stationary phase coating a fine rod of fused silica. SPME procedures for LC-MS(-MS) determinations in body fluids have been described for phenothiazines [20] as well as for diazepam and its metabolites [47], but such procedures require special interfaces.

Separation systems for LC-MS(-MS) analysis of therapeutic drugs

In contrast to conventional LC, the parameters that may be varied to optimise separation in LC-MS(-MS) analysis are rather limited, because only volatile buffer systems can be used for routine LC-MS(-MS) analysis. Furthermore, Mallet et al. [25] reported that mobile-phase additives common in conventional LC, most notably the ion-pairing agent trifluoroacetic acid, can dramatically reduce analyte ionisation. Therefore, it is not surprising that in almost all of the procedures summarised in Table 7.1 simple gradient systems with either ammonium formate [11, 14–18, 21, 39, 40, 45, 48] or ammonium acetate [7, 12, 20, 35, 36, 41, 44, 46] buffers combined with ACN [7, 11, 12, 15–18, 20, 21, 35, 36, 39–41, 45, 48] and/or methanol [14, 44, 46] were employed. In addition, all RP stationary phases used in these procedures belonged to only a few principal types, i.e. C_8 [11, 15, 18, 21, 39, 41, 46], C_{18} [7, 12, 16, 17, 19, 20, 35, 36, 40, 45, 48] or cyano phases [14]. Despite these similarities, the chromatographic run times varied considerably. Zhang et al. [35] achieved separation of five tricyclic antidepressants within 18 s using an isocratic system, whereas the gradient systems of the general unknown procedures published by Marquet's group were 25.5 [16, 48] and 50 min [17] long. Of course, ultra-short run times of 18 s would be desirable with respect to sample throughput, but one must always be aware that shorter run times increase the risk of ion-suppression effects due to insufficient separation of the analytes from each other or from the matrix. For example, Egge-Jacobsen et al. [46] reported mutual ion suppression of saquinavir and nelfinavir co-eluting in the 5-min gradient used by these authors. Long run times can minimise this problem because separation can be optimised. However, analysis times of 50 min for a single sample (not counting sample preparation) are certainly too long for routine application. A good compromise between the need for sufficient separation and acceptable throughput can be achieved with intermediate run times, which is indicated by many publications reporting run times of about 10 min [7, 11, 14, 15, 18, 21, 36, 39, 41, 45]. The gradient system used by the working group of the authors of this chapter requires 10 min until the next injection and proved to be almost universally applicable. At least, it allowed sufficient separation of such different drug classes as sulphonylurea-type anti-diabetics, neuroleptics, benzodiazepines, β-blockers as well as anaesthetics, low-dosed hypnotics and opioids [11, 15, 18, 21, 39] with only very minor modifications.

Ionisation and detection of therapeutic drugs in LC-MS(-MS) analysis

Of the various ion sources developed for removal of the mobile phase and ionisation of analytes, ESI and APCI are the only two with practical relevance in LC-MS(-MS) analysis of therapeutic drugs. ESI has the potential to ionise a wide variety of analytes including compounds with high polarity and/or high molecular mass, while APCI is limited to moderate polarity and moderate molecular mass compounds [6]. However, as already mentioned above, APCI is much less susceptible to ion-suppression effects [24, 26, 28]. In published applications for analysis of therapeutic drugs by LC-MS(-MS), ESI [7, 12, 14, 16, 17, 20, 35, 40, 41, 45–48] was employed more often than APCI [11, 15, 18, 21, 36, 39, 44], but this might be expected to change in the future for two reasons – being comparatively small molecules with intermediate polarity, most therapeutic drugs are well suited for APCI and the risk of ion suppression is much easier to control with this ionisation technique. In the authors' working group, APCI was found to be more sensitive than ESI when analysing sulphonylurea-type anti-diabetics, neuroleptics, benzodiazepines, β-blockers as well as anaesthetics, low-dosed hypnotics and opioids in human plasma samples [11, 15, 18, 21, 39]. In most multi-analyte methods the ion sources were operated in the positive-ion mode. However, in the GUS methods of Marquet's group [16, 17, 48] and in the target screening method for diuretics by Thieme et al. [41], positive- and negative-ion mode were used to account for the acidic properties of part of the analytes.

Apart from different interfaces, different mass analysers have been used in LC-MS(-MS) analysis of therapeutic drugs. They range from single-quadrupole [11, 15, 17, 18, 21, 38–40, 45–48] over triple-quadrupole [7, 12, 14, 16, 20, 36, 41, 44] to time-of-flight (TOF) MS [19, 35] instruments. Their use and operating modes depend on the purpose of the analytical methods, and will be described together with the applications below.

Methods for screening of therapeutic drugs by LC-MS(-MS)

One of the most important tasks in analytical toxicology is the screening for and unambiguous identification of xenobiotics in body fluids and tissues. LC-MS(-MS) is increasingly employed for this purpose in routine work, especially for blood and plasma/serum analysis [5, 6, 8, 49]. However, when establishing LC-MS screening procedures in routine work, several limitations should be kept in mind [5, 8, 10, 48, 50, 51]. The spectral information of ESI and/or APCI spectra is limited compared with electron ionisation mass spectra in GC-MS, even though collision-induced dissociation (CID) caused by increasing the orifice or fragmentor voltage of single-stage LC-MS instruments (in-source CID) or by increasing the collision energy in triple-stage LC-MS-MS instruments leads to formation of structurally related fragments. However, CID fragmentation can vary considerably between different instruments [50–53]. Weinmann et al. [50] could show that different types of apparatus may lead to reproducible ESI spectra formed by in-source CID if the apparatus had been tuned using certain test compounds such as haloperidol, paracetamol, metronidazole or metamizol. They concluded that mass spectral library searching of an ESI-CID-MS library set-up with one of the two instruments should be possible with the other instrument after adjusting the CID energies by means of at least two tuning compounds. Criteria for compound identification by single- or multiple-stage LC-MS were recently reviewed by Rivier [51] and are described by the same author in Chapter 5. In his review [51], Rivier came to the conclusion that the responsibility lies with the toxicologist to decide, depending on the case, how and when the minimum requirement for identity confirmation has been reached, and not to rely exclusively on match quality parameters.

Several concepts have been developed for LC-MS screening and detection of therapeutic drugs. The groups of Marquet [17, 48] and Maurer [10, 11, 15, 18, 21, 39] have developed screening procedures for blood analysis based on different single-stage quadrupole LC-MS set-ups operated in the full-scan mode. Pelander *et al.* [19] published a screening method for urine samples based on monoisotopic masses of 637 compounds, including many therapeutic drugs, as determined by full-scan TOFMS. Such full-scan-based methods are not limited to a certain number of analytes and, in principle, are able to detect any compound amenable to LC and ionisation. However, they are generally less sensitive than methods targeting a limited number of analytes. Gergov *et al.* [12] and Nordgren *et al.* [44] reported multi-analyte screening methods for blood and urine samples, respectively, which were based on classic triple/quadrupole LC-MS-MS in the MRM mode. As this required choosing precursor ions *a priori*, such methods are limited to the target analytes. In a preliminary study, Marquet *et al.* [16] compared a new and more sensitive procedure using a quadrupole-linear ion trap mass spectrometer with the above-mentioned single-stage procedure. In this method, the MS-MS system was operated in the information-dependent acquisition (IDA) mode, in which the system is sent to full-scan mode, but switches to more sensitive product-ion-scanning mode if a certain ion current reaches a previously set threshold value. In addition to these general procedures, some LC-MS(-MS) procedures have been described for screening for single drug classes such as antihistamines [7], phenothiazines [20] or neuroleptics [14]. Marquet gives a detailed account on LC-MS-based screening procedures in Chapter 6.

As already discussed above, the working group of the authors is developing universal single-stage LC-MS procedures for screening, library-assisted identification and, in contrast to other screening procedures, additional fully validated quantification of many drug classes in blood plasma. So far, these drug classes include anaesthetics, low-dosed hypnotics and opioids [39], benzodiazepines [18], sulphonylurea-type antidiabetics [11, 54, 55], neuroleptics [15] and β-blockers [21]. In these methods, the mass spectra were recorded at two different fragmentor voltages (100 and 200 V) with a very short cycle time. However, as already discussed above, it should be kept in mind that the same fragmentor voltages selected in different apparatus set-ups may result in different abundances of the formed fragments [50–53]. Therefore, each user has to select the fragmentor voltage most suitable for their specific apparatus, which produces mass spectra comparable with those noted in the corresponding papers [11, 15, 18, 21, 38]. In the authors' experience with three different pieces of apparatus of the same type, this has allowed the successful use of the screening procedure. As an example of the screening and identification procedure, smoothed and merged mass chromatograms (scan mode, 100 V) of the ions of mass-to-charge (m/z) ratios 242, 238, 284, 417, 326, 431, 300, 342, 337, 316, 276, 271, 321, 314, 245, 387, 290 and 285 of an extract of an authentic plasma from a brain death diagnosis case indicating anaesthetics, hypnotics and opioids [39] are shown in Figure 7.1a. The mass spectrum underlying the marked peak (lower spectrum), the reference spectrum (upper spectrum), the structure and the hit-list found by computer library search (HH Maurer and AA Weber, unpublished data) are shown in Figure 7.1b.

Methods for quantification of therapeutic drugs by LC-MS(-MS)

In analytical toxicology, particularly in analysis of therapeutic drugs in blood, plasma or serum, relevant analytes must be quantified. In quantitative assays, single-stage MS set-ups were generally operated in SIM mode [11, 15, 18, 21, 38, 40, 45, 46] and LC-MS(-MS) set-ups in the

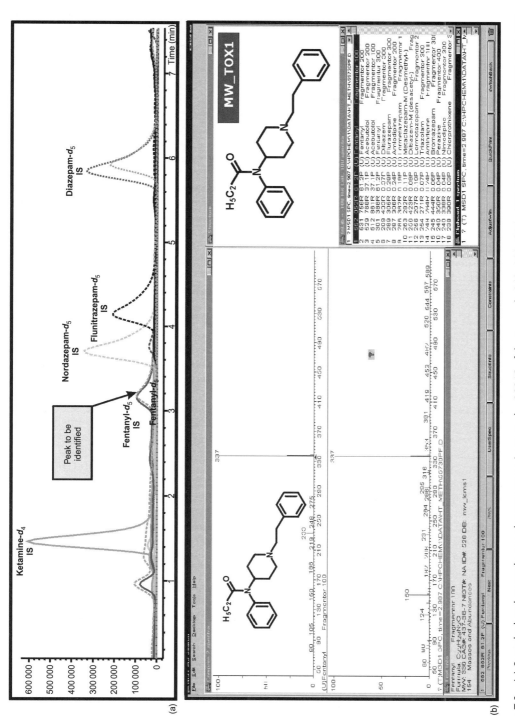

Figure 7.1 (a) Smoothed and merged mass chromatograms (scan mode, 100 V) of the ions m/z 242, 238, 284, 417, 326, 431, 300, 342, 337, 316, 276, 271, 321, 314, 245, 387, 290 and 285 of an extract of an authentic plasma from a brain death diagnosis case indicating anaesthetics, hypnotics and opioids. (b) Mass spectrum underlying the marked peak (lower spectrum), the reference spectrum (upper spectrum), the structure and the hit-list found by computer library search.

MRM mode [7, 14, 20, 36, 41] in order to achieve sufficient sensitivity, precision and accuracy. The LC-TOFMS method for the determination of tricyclic antidepressants described by Zhang et al. [35] did not require *a priori* selection of ions to be monitored, because in TOFMS analysis full mass spectra are recorded simultaneously at very high pulse rates. The number of ions or transitions to be monitored in SIM and MRM, respectively, depends on the analytical strategy. If identification and quantification are both performed in SIM or MRM mode, a sufficient number of qualifier ions for unambiguous identification should be monitored (for details, see Chapter 5). If the analytes have already been identified in the same sample extract using full-scan LC-MS under the same chromatographic conditions, monitoring only a single ion per compound for quantification is sufficient [11, 15, 18, 21, 38, 39]. This strategy has proved to be very versatile for screening, identification and quantification of different drug classes such as anaesthetics, low-dosed hypnotics and opioids [39], benzodiazepines [18], sulphonylurea-type anti-diabetics [11, 54, 55], neuroleptics [15] and β-blockers [21]. All these procedures were fully validated and the acceptance criteria were fulfilled for the majority of analytes tested. As an example of the quantification procedure, smoothed and merged mass fragmentograms [39] of an authentic plasma extract (same extract as used in Figure 7.1) indicating 0.005 mg/L of fentanyl and 0.1 mg/L of diazepam (nordazepam concentration below the limit of quantification [LOQ]) are shown in Figure 7.2. Other multi-analyte procedures allowing screening with subsequent quantification were published by Gergov et al. [7] for antihistamines, by Kumazawa et al. [20] for phenothiazines and by Josefsson et al. for neuroleptics [14].

Validation of LC-MS(-MS) procedures for the analysis of therapeutic drugs

All quantification assays should be fully validated according to the international recommendations that were critically reviewed by Peters and Maurer [56]. A detailed account on the topic of method validation, including an overview over various validation guidelines as well as recommendations for experimental designs, is given by Peters in Chapter 4. Generally, it is recommended to evaluate the following parameters: selectivity, linearity, limit of detection (LOD), LOQ, recovery, accuracy and precision, as well as analyte stability in samples on the bench-top and during work-up, in processed samples, during freeze/thaw cycles, and under the conditions of long-term storage. In the particular case of LC-MS(-MS) methods, ME studies should be considered an essential part of method validation, especially if the ESI mode is used. Finally, the applicability should be studied using authentic samples. Such validation data are the only objective basis for assessment of a procedure's performance and quality [56]. Most of the described multi-analyte procedures were more or less thoroughly validated (the evaluated validation parameters for each of these methods are listed in Table 7.1).

Unfortunately, in the papers by Kumazawa et al. [20] for the determination of phenothiazines and by Josefsson et al. [14] for the determination of neuroleptics, no validation data were presented with the exception of the LOQ. Therefore, it is not clear whether these procedures fulfil forensic or clinical toxicological standards. With the exception of the methods described by the group of the authors of this chapter [11, 15, 18, 21, 38, 39] and the method for determination of anti-AIDS drugs by Egge-Jacobsen [46], the different forms of stability have not been tested. In the authors' opinion, such stability testing should be mandatory, especially in forensic toxicology, because in most cases the specimens are not analysed directly after sampling. Repeated freezing and thawing is the rule rather than the exception.

Another important point that needs to be discussed in the context of validation is the choice of appropriate internal standards (ISs). They can compensate for variability due to sample preparation (e.g. because of differences between batches of SPE columns [57]), chromatography or even ion suppression/enhancement, and thus improve accuracy and precision data. As in any MS-based analytical methods, stable-isotope-labelled analogues of the analytes are ideal ISs. However, mutual ion suppression/enhancement

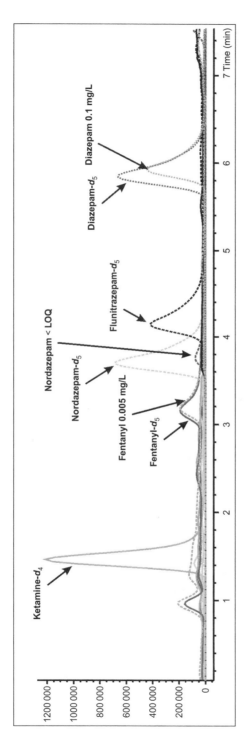

Figure 7.2 Smoothed and merged mass fragmentograms (SIM mode, 100 V, ions m/z time window 0–3.2 min, at gain 2.2: 242, 238, 417; time window 3.21–7.5 min, at gain 2.2: 276, 271, 245, 290, 285; time window 0–3.5 min, at gain 9.0: 284, 326, 431, 300, 342, 337, 316; time window 3.51–5.2 min, at gain 9.0: 342, 337, 316, 321, 314, 387; time window 5.21–7.5 min, at gain 9.0: 321, 314, 387) of an authentic plasma extract (same extract as used in Figure 7.1) indicating 0.005 mg/L of fentanyl and 0.1 mg/L of diazepam (nordazepam concentration below the LOQ).

of analytes and their stable-isotope-labelled ISs has been reported [27, 58]. Even though this phenomenon did not affect quantification, at least as long as certain conditions were fulfilled, additional MEs may lead to a change of response ratios (analyte versus IS) [59] and thus negatively affect quantification (for a more detailed discussion on this topic, *see* Chapter 4). Worse problems might have to be expected when no stable-isotope-labelled of the analyte is available as an IS, because in such cases ion-suppression effects on analytes and alternative ISs might differ considerably. At this point it must be stressed that, if no stable-isotope-labelled IS of the analyte is available and an alternative IS must be chosen, one should always avoid choosing a therapeutic drug for this purpose, at least in analytical toxicology. In this field, it can never be excluded that the patient or defendant to be monitored has taken this drug. In such cases, the peak area of the IS would be overestimated, leading to an underestimation of the analyte concentration. Therefore, if no deuterated analogue of a specific drug is available, the authors of this chapter recommend choosing a suitable IS from the pool of available deuterated compounds. Nowadays there are such deuterated substances from a large variety of structures with different physicochemical properties. This strategy was successfully used for a number of assays for various classes of drugs, all of which employed [D_3]trimipramine as IS [11, 15, 21, 38].

Conclusions and perspectives

LC-MS(-MS) has been shown to be an ideal supplement to GC-MS, especially for the detection and quantification of more polar, thermolabile and/or low-dosed therapeutic drugs, especially in blood, plasma or serum. Although urine is still the sample of choice for non-target comprehensive screening for and identification of unknown (therapeutic) drugs or poisons [5], blood, plasma or serum can, and sometimes must, be used for at least a limited screening, especially for target analytes within multi-analyte procedures. Some of these procedures also allow validated multi-analyte quantification. In the future, such procedures will certainly help to solve many analytical and time-consuming problems in clinical and forensic toxicology.

References

1. Ackermann BL, Berna MJ, Murphy AT. Recent advances in use of LC/MS/MS for quantitative high-throughput bioanalytical support of drug discovery. *Curr Top Med Chem* 2002; 2: 53–66.

2. Hopfgartner G, Bourgogne E. Quantitative high-throughput analysis of drugs in biological matrices by mass spectrometry. *Mass Spectrom Rev* 2003; 22: 195–214.

3. Hopfgartner G, Husser C, Zell M. High-throughput quantification of drugs and their metabolites in biosamples by LC-MS/MS and CE-MS/MS: possibilities and limitations. *Ther Drug Monit* 2002; 24: 134–143.

4. Niessen WM. Progress in liquid chromatography-mass spectrometry instrumentation and its impact on high-throughput screening. *J Chromatogr A* 2003; 1000: 413–436.

5. Maurer HH. Position of chromatographic techniques in screening for detection of drugs or poisons in clinical and forensic toxicology and/or doping control [review]. *Clin Chem Lab Med* 2004; 42: 1310–1324.

6. Van Bocxlaer JF, Clauwaert KM, Lambert WE, *et al.* Liquid chromatography-mass spectrometry in forensic toxicology [review]. *Mass Spectrom Rev* 2000; 19: 165–214.

7. Gergov M, Robson J N, Ojanpera I, *et al.* Simultaneous screening and quantitation of 18 antihistamine drugs in blood by liquid chromatography ionspray tandem mass spectrometry. *Forensic Sci Int* 2001; 121: 108–115.

8. Marquet P. Progress of LC-MS in clinical and forensic toxicology [review]. *Ther Drug Monit* 2002; 24: 255–276.

9. Marquet P. Is LC-MS suitable for a comprehensive screening of drugs and poisons in clinical toxicology? *Ther Drug Monit* 2002; 24: 125–133.

10. Maurer HH, Kraemer T, Kratzsch C, *et al.* Negative ion chemical ionization gas chromatography-mass spectrometry (NICI-GC-MS) and atmospheric pressure chemical ionization liquid chromatography-mass spectrometry (APCI-LC-MS) of low-dosed

and/or polar drugs in plasma. *Ther Drug Monit* 2002; 24: 117–124.

11. Maurer HH, Kratzsch C, Kraemer T, *et al.* Screening, library-assisted identification and validated quantification of oral antidiabetics of the sulfonylurea-type in plasma by atmospheric pressure chemical ionization liquid chromatography-mass spectrometry (APCI-LC-MS). *J Chromatogr B* 2002; 773: 63–73.

12. Gergov M, Ojanpera I, Vuori E. Simultaneous screening for 238 drugs in blood by liquid chromatography-ion spray tandem mass spectrometry with multiple-reaction monitoring. *J Chromatogr B* 2003; 795: 41–53.

13. Goeringer KE, McIntyre M, Drummer OH. LC-MS analysis of serotonergic drugs. *J Anal Toxicol* 2003; 27: 30–35.

14. Josefsson M, Kronstrand R, Andersson J, *et al.* Evaluation of electrospray ionisation liquid chromatography-tandem mass spectrometry for rational determination of a number of neuroleptics and their major metabolites in human body fluids and tissues. *J Chromatogr B* 2003; 789: 151–167.

15. Kratzsch C, Weber AA, Peters FT, *et al.* Screening, library-assisted identification and validated quantification of fifteen neuroleptics and three of their metabolites in plasma by liquid chromatography/mass spectrometry with atmospheric pressure chemical ionization. *J Mass Spectrom* 2003; 38: 283–295.

16. Marquet P, Saint-Marcoux F, Gamble TN, *et al.* Comparison of a preliminary procedure for the general unknown screening of drugs and toxic compounds using a quadrupole-linear ion-trap mass spectrometer with a liquid chromatography-mass spectrometry reference technique. *J Chromatogr B* 2003; 789: 9–18.

17. Venisse N, Marquet P, Duchoslav E, *et al.* A general unknown screening procedure for drugs and toxic compounds in serum using liquid chromatography-electrospray-single quadrupole mass spectrometry. *J Anal Toxicol* 2003; 27: 7–14.

18. Kratzsch C, Tenberken O, Peters FT, *et al.* Screening, library-assisted identification and validated quantification of twenty three benzodiazepines, flumazenil, zaleplone, zolpidem and zopiclone in plasma by liquid chromatography/mass spectrometry with atmospheric pressure chemical ionization. *J Mass Spectrom* 2004; 39: 856–872.

19. Pelander A, Ojanpera I, Laks S, *et al.* Toxicological screening with formula-based metabolite identification by liquid chromatography/time-of-flight mass spectrometry. *Anal Chem* 2003; 75: 5710–5718.

20. Kumazawa T, Seno H, Watanabe-Suzuki K, *et al.* Determination of phenothiazines in human body fluids by solid-phase microextraction and liquid chromatography/tandem mass spectrometry. *J Mass Spectrom* 2000; 35: 1091–1099.

21. Maurer HH, Tenberken O, Kratzsch C, *et al.* Screening for, library-assisted identification and fully validated quantification of twenty-two beta-blockers in blood plasma by liquid chromatography-mass spectrometry with atmospheric pressure chemical ionization. *J Chromatogr A* 2004; 1058: 169–181.

22. Maurer HH. Advances in analytical toxicology: current role of liquid chromatography-mass spectrometry for drug quantification in blood and oral fluid [review]. *Anal Bioanal Chem* 2005; 381: 110–118.

23. Maurer HH, Schmitt CJ, Weber AA, *et al.* Validated electrospray LC-MS assay for determination of the mushroom toxins alpha- and beta-amanitin in urine after immunoaffinity extraction. *J Chromatogr B* 2000; 748: 125–135.

24. Souverain S, Rudaz S, Veuthey J-L. Matrix effect in LC-ESI-MS and LC-APCI-MS with off-line and on-line extraction procedures. *J Chromatogr A* 2004; 1058: 61–66.

25. Mallet CR, Lu Z, Mazzeo JR. A study of ion suppression effects in electrospray ionization from mobile phase additives and solid-phase extracts. *Rapid Commun Mass Spectrom* 2004; 18: 49–58.

26. Annesley TM. Ion suppression in mass spectrometry. *Clin Chem* 2003; 49: 1041–1044.

27. Liang HR, Foltz RL, Meng M, *et al.* Ionization enhancement in atmospheric pressure chemical ionization and suppression in electrospray ionization between target drugs and stable-isotope-labeled internal standards in quantitative liquid chromatography/tandem mass spectrometry. *Rapid Commun Mass Spectrom* 2003; 17: 2815–2821.

28. Dams R, Huestis MA, Lambert WE, *et al.* Matrix effect in bioanalysis of illicit drugs with LC-MS/MS: influence of ionization type, sample preparation, and biofluid. *J Am Soc Mass Spectrom* 2003; 14: 1290–1294.

29. Muller C, Schafer P, Stortzel M, *et al.* Ion suppression effects in liquid chromatography-electrospray-ionisation transport-region collision induced

dissociation mass spectrometry with different serum extraction methods for systematic toxicological analysis with mass spectra libraries. *J Chromatogr B* 2002; 773: 47–52.

30. King R, Bonfiglio R, Fernandez-Metzler C, *et al.* Mechanistic investigation of ionization suppression in electrospray ionization. *J Am Soc Mass Spectrom* 2000; 11: 942–950.

31. Matuszewski BK, Constanzer ML, Chavez-Eng CM. Strategies for the assessment of matrix effect in quantitative bioanalytical methods based on HPLC-MS/MS. *Anal Chem* 2003; 75: 3019–3030.

32. Sun H, Naidong W. Narrowing the gap between validation of bioanalytical LC-MS-MS and the analysis of incurred samples. *Pharm Technol* 2003; 27: 74–86.

33. Decaestecker TN, Coopman EM, Van Peteghem CH, *et al.* Suitability testing of commercial solid-phase extraction sorbents for sample clean-up in systematic toxicological analysis using liquid chromatography-(tandem) mass spectrometry. *J Chromatogr B* 2003; 789: 19–25.

34. Naidong W, Bu H, Chen YL, *et al.* Simultaneous development of six LC-MS-MS methods for the determination of multiple analytes in human plasma. *J Pharm Biomed Anal* 2002; 28: 1115–1126.

35. Zhang H, Heinig K, Henion J. Atmospheric pressure ionization time-of-flight mass spectrometry coupled with fast liquid chromatography for quantitation and accurate mass measurement of five pharmaceutical drugs in human plasma. *J Mass Spectrom* 2000; 35: 423–431.

36. Thevis M, Opfermann G, Schanzer W. High speed determination of beta-receptor blocking agents in human urine by liquid chromatography/tandem mass spectrometry. *Biomed Chromatogr* 2001; 15: 393–402.

37. Maurer HH. Methods for GC-MS. In: Pfleger K, Maurer HH, Weber A, eds, *Mass Spectral and GC Data of Drugs, Poisons, Pesticides, Pollutants and their Metabolites*: Part 4, 2nd edn. Weinheim: Wiley-VCH, 2000: 3–241.

38. Maurer HH, Kratzsch C, Weber AA, *et al.* Validated assay for quantification of oxcarbazepine and its active dihydro metabolite 10-hydroxy carbazepine in plasma by atmospheric pressure chemical ionization liquid chromatography/mass spectrometry. *J Mass Spectrom* 2002; 37: 687–692.

39. Kratzsch C, Peters F T, Kraemer T, *et al.* Simple APCI-LC-MS method for screening, library-assisted identification and validated quantification of anesthetics, benzodiazepines and low dosed opioids in plasma often asked for in the context of the diagnosis of brain death. In: *Proceedings of the XIIIth GTFCh Symposium*, Mosbach, 2003: 299–309.

40. Lachatre F, Marquet P, Ragot S, *et al.* Simultaneous determination of four anthracyclines and three metabolites in human serum by liquid chromatography-electrospray mass spectrometry. *J Chromatogr B* 2000; 738: 281–291.

41. Thieme D, Grosse J, Lang R, *et al.* Screening, confirmation and quantification of diuretics in urine for doping control analysis by high-performance liquid chromatography-atmospheric pressure ionisation tandem mass spectrometry. *J Chromatogr B* 2001; 757: 49–57.

42. Peters FT, Kraemer T, Maurer HH. Drug testing in blood: validated negative-ion chemical ionization gas chromatographic-mass spectrometric assay for determination of amphetamine and methamphetamine enantiomers and its application to toxicology cases. *Clin Chem* 2002; 48: 1472–1485.

43. Peters FT, Schaefer S, Staack RF, *et al.* Screening for and validated quantification of amphetamines and of amphetamine- and piperazine-derived designer drugs in human blood plasma by gas chromatography/mass spectrometry. *J Mass Spectrom* 2003; 38: 659–676.

44. Nordgren HK, Beck O. Multicomponent screening for drugs of abuse: direct analysis of urine by LC-MS-MS. *Ther Drug Monit* 2004; 26: 90–97.

45. Sayer H, Quintela O, Marquet P, *et al.* Identification and quantitation of six non-depolarizing neuromuscular blocking agents by LC-MS in biological fluids. *J Anal Toxicol* 2004; 28: 105–110.

46. Egge-Jacobsen W, Unger M, Niemann CU, *et al.* Automated, fast, and sensitive quantification of drugs in human plasma by LC/LC-MS: quantification of 6 protease inhibitors and 3 nonnucleoside transcriptase inhibitors. *Ther Drug Monit* 2004; 26: 546–562.

47. Walles M, Mullett WM, Pawliszyn J. Monitoring of drugs and metabolites in whole blood by restricted-access solid-phase microextraction coupled to liquid chromatography-mass spectrometry. *J Chromatogr A* 2004; 1025: 85–92.

48. Saint-Marcoux F, Lachatre G, Marquet P. Evaluation of an improved general unknown screening

procedure using liquid chromatography-electrospray-mass spectrometry by comparison with gas chromatography and high-performance liquid-chromatography-diode array detection. *J Am Soc Mass Spectrom* 2003; 14: 14–22.

49. Maurer HH. Screening procedures for simultaneous detection of several drug classes used in the high throughput toxicological analysis and doping control [review]. *Comb Chem High Throughput Screen* 2000; 3: 461–474.

50. Weinmann W, Stoertzel M, Vogt S, et al. Tune compounds for electrospray ionisation/in-source collision-induced dissociation with mass spectral library searching. *J Chromatogr A* 2001; 926: 199–209.

51. Rivier L. Criteria for the identification of compounds by liquid chromatography-mass spectrometry and liquid chromatography-multiple mass spectrometry in forensic toxicology and doping analysis. *Anal Chim Acta* 2003; 492: 69–82.

52. Lips AGAM, Lameijer W, Fokkens RH, *et al.* Methodology for the development of a drug library based upon collision-induced fragmentation for the identification of toxicologically relevant drugs in plasma samples. *J Chromatogr B* 2001; 759: 191–207.

53. Hough JM, Haney CA, Voyksner RD, *et al.* Evaluation of electrospray transport CID for the generation of searchable libraries. *Anal Chem* 2000; 72: 2265–2270.

54. Holstein A, Plaschke A, Hammer C, *et al.* Hormonal counterregulation and consecutive glimepiride serum concentrations during severe hypoglycemia associated with glimepiride therapy. *Eur J Clin Pharmacol* 2003; 59: 747–754.

55. Holstein A, Plaschke A, Ptak M, *et al.* The diagnostic value of determining the hydroxy metabolite of glimepiride (M1) in blood serum in cases of severe hypoglycaemia associated with glimepiride therapy. *Diabetes Obes Metab* 2004; 6: 391–393.

56. Peters FT, Maurer HH. Bioanalytical method validation and its implications for forensic and clinical toxicology – a review [review]. *Accred Qual Assur* 2002; 7: 441–449.

57. Bogusz MJ, Maier RD, Schiwy BK, *et al.* Applicability of various brands of mixed-phase extraction columns for opiate extraction from blood and serum. *J Chromatogr B* 1996; 683: 177–188.

58. Sojo LE, Lum G, Chee P. Internal standard signal suppression by co-eluting analyte in isotope dilution LC-ESI-MS. *Analyst* 2003; 128: 51–54.

59. Jemal M, Schuster A, Whigan DB. Liquid chromatography/tandem mass spectrometry methods for quantitation of mevalonic acid in human plasma and urine: method validation, demonstration of using a surrogate analyte, and demonstration of unacceptable matrix effect in spite of use of a stable isotope analog internal standard. *Rapid Commun Mass Spectrom* 2003; 17: 1723–1734.

8

LC-MS of drugs of abuse and related compounds

Maciej J. Bogusz

Introduction

Substances of abuse form a large group consisting of various compounds, widely differing in their origin, availability and chemical nature. From the legal point of view, these substances may be divided into legal and illicit drugs. It must be stressed that tobacco products, which are legal throughout the whole world, are responsible for about 6% of all deaths worldwide. Alcohol, which is legal in most countries, contributes to 1.5%, whereas illicit drugs cause about 0.2% of deaths yearly. This chapter is focused on illicit drugs as well as legal pharmaceutical compounds that may be abused.

The analysis of drugs of abuse, particularly illicit drugs, is one of the most important challenges of forensic and clinical toxicology. As with any other branch of forensic sciences, forensic toxicological analysis must also apply and keep particularly high standards of quality. This is because the analytical result may have a direct and permanent impact on the fate of the person involved. As a consequence, forensic toxicological examinations are subjected to very tight scrutiny. This requirement should concern not only primarily forensic cases, but also extend to all clinical analysis for illicit drugs. It must be kept in mind that each result of drug analysis has potential forensic relevance, irrespective of the primary purpose of the examination.

It has often been said that the interpretation of results is the most important and most difficult part of forensic toxicology. Notwithstanding the value of correct interpretation, it must also be said that the correct result enables any further action. This is particularly true when the analytical result itself may serve as evidence of illegal action. The introduction of '*per se* laws' in traffic legislation may serve as an example. Such legislation forbids driving a vehicle if a given substance above a given concentration is present in the body fluid of a driver. '*Per se* laws' have existed for a long time with regard to ethyl alcohol. In recent years, such laws have also been introduced in several European countries for other drugs of abuse like amphetamines, cannabis, cocaine and opiates [1–3].

The enforcement of '*per se* law' is possible only when the legal limit of the substance in question is exactly defined and reliably measured. Since in the case of drugs of abuse the law forbids any use of the drug by drivers, the legal limits are defined by the quality of analytical methods and practically reflect the limits of detection (LODs). As a consequence, only methods having the highest possible selectivity and sensitivity should be applied.

The analysis of drugs of abuse not only is important in the enforcement of road traffic safety, but also enables the differentiation between the chronic and occasional drug user or makes possible the identification of the source of origin of a particular batch of illegal drugs. Such an analysis would not be possible without the application of chromatography in various forms. Only chromatographic methods successfully combine efficient separation of relevant compound(s) from a biological matrix with specific detection. Among the chromatographic techniques, liquid chromatography–mass spectrometry (LC-MS) optimally fulfils the requirements of forensic toxicological analysis due to its high selectivity, the possibility of detection of active metabolites and (emerging in recent years) efficient screening in cases of unclear death. For these reasons, in addition to the gradual introduction of relatively low-cost LC-MS instruments, this technique is now finding a more and more important place in forensic toxicological laboratories.

LC-MS underwent major evolution in the last decade. An expensive, difficult and not always reliable hyphenated technique turned into a robust analytical tool, applicable in almost all analytical situations. This was the consequence of the introduction of atmospheric pressure ionisation (API) sources. The advent of API-LC-MS made all earlier LC-MS interfaces (e.g. thermospray, particle beam ionisation or fast atom bombardment) obsolete techniques.

The following statement of Krull and Cohen [4] may illustrate the scale of this change: 'There are at least two fundamental views on the future role of MS in biotechnology HPLC. Is the mass spectrometer an expensive sophisticated LC detector? Or, is the chromatograph an expensive sample-preparation device for a mass spectrometer? One of the newer developments in LC-MS is that this distinction will cease to be important in the future.'

The change of status of LC-MS from a hyphenated method to a routine analytical tool has attracted a new generation of users who are taking the hyphenation for granted and now treat an LC-MS instrument as a single unit. Similar evolution was observed for GC-MS. Technical and methodological progress in LC-MS in recent years has been reviewed in several books and publications [5–7]. Special reviews were devoted to the application of LC-MS in forensic toxicology [8–11]. This trend of broad acceptance is also reflected in some legal acts. For example, the Substance Abuse and Mental Health Services Administration (SAMHSA) – the US federal agency responsible for the organisation of workplace drug testing – mentioned LC-MS and LC-MS-MS along with GC-MS as recommended methods for confirmatory analysis of drugs of abuse in urine [12].

General methodological considerations

It is common knowledge that poor sample preparation, even when the most modern and sophisticated techniques are applied, may negatively affect detection and quantification. LC-MS for drugs of abuse makes no exception. In earlier phases of application attempts to perform LC-MS-MS analysis in the flow-injection mode (FIA) were made, i.e. urine/serum extracts were directly injected into the MS with no chromatographic separation. FIA methods have been proposed for a few illicit drugs (morphine, codeine, amphetamine and benzoylecgonine) and potentially interfering compounds, although matrix effects (MEs) were not recognized [13]. Several authors studied the influence of the matrix on the detectability of drugs of abuse with LC-MS and this issue is also addressed in other chapters of this book (see Chapters 2–4). Müller et al. [14] evaluated the influence of the co-extracted serum matrix, as obtained with different sample preparation procedures, on the signal of codeine (positive-ion mode) and glafenine (negative-ion mode) as test substances. Extracts from protein precipitation (PP) and solid-phase extraction (SPE) produced severe ion suppression for both test substances, whereas less suppression was observed with liquid–liquid extraction (LLE), and the combination of PP with SPE fully eliminated the ME. According to the authors, ion suppression was caused by polar, non-retained matrix components eluted at the beginning of the chromatographic run. Dams et al. [15] compared signal intensities of morphine isolated from plasma, urine and oral fluid with four procedures, and analysed by MS-MS with either electrospray (ESI) or atmospheric pressure chemical ionisation (APCI). For urine, simple dilution was sufficient for the analysis, while acetonitrile (ACN) precipitation was successful for oral fluid specimens. For plasma specimens, SPE was necessary. ESI was more susceptible to MEs than APCI. Souverain et al. [16], who studied the influence of the plasma matrix on the MS signal of methadone, also found that the ESI source was more liable to MEs than APCI. The signal was altered by extraction mode and LLE was more efficient than SPE or PP precipitation.

All the above-mentioned studies indicate that the analysed extracts should be of sufficient quality and the chromatographic separation should not be neglected or sacrificed. Therefore, LC-MS is not a magic tool able to replace optimal sample preparation and separation.

The type of the column used for LC-MS analysis of drugs of abuse drugs may also influence the results. According to Naidong et al. [17], normal-phase chromatography is much better suited than reversed-phase (RP) chromatography to MS detection, even though the latter is much more widely used. This is particularly true for highly polar compounds, which are hardly retained on RP columns even with a low percentage of organic modifier in the mobile phase. In addition, a high percentage of water accounts for non-optimal spraying conditions and poor ionisation efficiency. Instead, normal-phase chromatography with an aqueous/organic mobile phase enables retention of polar species with a high percentage of organic solvent, thus assuring better dispersion and evaporation of electrospray droplets. For a number of compounds (e.g. nicotine and cotinine, salbutamol), a distinctly higher signal was measured by LC-ESI-MS-MS after separation on a silica column (70% ACN) compared with C_{18} separation with 10% ACN. A similar approach was adopted for the analysis of morphine and glucuronides in serum [18, 19], fentanyl [20] and hydrocodone/hydromorphone [21]. Normal-phase short-column (50 × 3 mm) separation combined with a high flow rate (4 mL/min) resulted in ultrafast (below 1 min) LC-MS-MS analysis of both non-polar and polar analytes, e.g. morphine and glucuronides or midazolam and its hydroxylated metabolites [22]. Another advantage of normal-phase chromatography is the possibility of directly injecting an SPE extract (100% ACN or methanol [MeOH]) with no evaporation/reconstitution in the mobile phase. Using this approach, Naidong et al. [23] developed a 96-well SPE LC-MS-MS procedure able to process of a batch of 96 samples (e.g. for the analysis of fentanyl, omeprazole or pseudoephedrine) in 1 h.

Some studies were specifically devoted to evaluating the usefulness of various ion sources or mass analysers in toxicological analysis. Dams et al. [24] compared the performance of three ion sources, i.e. pneumatically assisted ESI, sonic spray ionisation (SSI) and APCI, coupled to an ion trap (IT) mass spectrometer. The influence of solvent modifier, buffer, pH and volatile acids was studied using morphine as a model compound. Ionisation efficiency was strongly affected by the composition of the mobile phase with both ESI and SSI, and strong similarities were observed between these ion sources. The APCI source showed higher robustness, applicability to higher flow rates and positive response to acids or buffers.

LC-ESI-MS-MS with a quadrupole/time-of-flight (QTOF) mass spectrometer was evaluated by Clauwaert et al. [25] for the quantification of 3,4-methylenedioxymethamphetamine (MDMA) and 3,4-methylenedioxymphetamine (MDA) in body fluids. A limit of quantification (LOQ) of 1 µg/L was reached and the linear dynamic range extended over four decades, leading the authors to conclude that QTOF achieves a sufficient linear dynamic range for analytical toxicology applications.

Illicit drugs

Opioid agonists

Here, the use of LC-MS for confirmation and quantitative analysis of natural and synthetic opioids is reviewed. Strictly speaking, the term 'opiate' refers specifically to the products derived from the opium poppy. The chapter focuses on morphine derivatives and synthetic or semi-synthetic opioids showing agonistic action at opioid receptors OP_1 (δ), OP_2 (κ) or OP_3 (µ).

All natural and semi-synthetic opiates (including heroin) originate from opium – a dried, brown juice obtained from green, scratched poppy heads of Papaver somniferum. The composition of alkaloids in the plant is variable and depends on multiple factors, like climatic conditions, harvesting time, soil composition and plant breeding [27–35]. As a consequence, the composition of opium and, subsequently, of heroin reflects the primary variability of plant material. There are two main methods of clandestine heroin production – the lime method used in Southeast Asia and the ammonia method used in Southwest Asia. Both methods give a similar yield of morphine, codeine and thebaine, but the content of noscapine and papaverine in the end product is much higher with the ammonia method [36]. Heroin from Southeast Asia

('Golden Triangle'), known as 'China white', predominates on drug markets in the US. Southwest-Asian-type heroin, originating from Turkey, Lebanon, Afghanistan or Iran ('Golden Crescent'), is mainly present on the European market [37–40]. The latter heroin contains over 10% of noscapine and over 2% of papaverine.

From the forensic point of view, heroin is the most important opiate agonist. Heroin is usually self-administered intravenously. Over the last decade, however, a growing preference for other routes of administration has been observed, like smoking or intranasal administration ('snorting'). This has been caused by several factors, like the fear of HIV, the possibility of administering heroin without leaving external marks on the body and the decrease of the price of street heroin. Irrespective of the administration route, heroin is rapidly deacetylated to 6-acetylmorphine (6-AM). The half-life of heroin in blood after intravenous injection was estimated at 2–8 min [41, 42], after smoking at 3–5 min [43] and after intranasal or intramuscular administration at 5–6 min [44, 45]. 6-AM is deacetylated at a somewhat slower rate to morphine. The half-life after intravenous administration was 6–38 min, after smoking 5 min, and after intranasal and intramuscular administration 11 and 12 min, respectively. The half-life of morphine was estimated at around 30 min after heroin smoking and at 60–180 min after administration by other routes [46]. Figure 8.1 shows the main steps of heroin biotransformation.

Analysis of street drugs

The profiling of street heroin samples is of forensic importance since it may bring clues concerning the origin and distribution routes of the drug. This was done by Dams et al. [47] who analysed seven constituents of street heroin (morphine, codeine, 6-AM, heroin, acetylcodeine, papaverine and noscapine) with LC-SSI-MS using an IT mass spectrometer. The compounds were separated on monolithic silica column (Chromolith Performance 100 × 4.6 mm) in gradient elution using ACN:water at a flow of 5 mL/min with post-column split (1/20); the analysis time was 5 min. The protonated molecular ions were monitored. The LOD ranged from 0.25 to 1 ng on-column.

Analysis of biological samples

Natural and semi-synthetic opiates

LC-MS allows the specific detection of parent opiates and all polar metabolites without derivatisation and without acidic/enzymatic cleavage of the conjugates. Zuccaro et al. [48] developed a LC-ESI-MS method for simultaneous determination of heroin, 6-AM, morphine, morphine-3-glucuronide (M3G) and morphine-6-glucuronide (M6G) in serum of heroin-treated mice using C_2 SPE followed by normal-phase isocratic separation with a MeOH:ACN:formic acid mobile phase. The LOD was in the range 0.5–4 µg/L. Heroin metabolites (6-AM, morphine, M3G and M6G) were quantified in blood, cerebrospinal fluid, vitreous humour and urine of heroin victims by Bogusz et al. [49] using LC-APCI-MS. C_{18} SPE was used for isolation, and the drugs were separated on a C_{18} column with isocratic ACN:ammonium formate buffer conditions. The LOD for 6-AM was 0.5 µg/L. The significance of molar ratios of M3G/morphine and M6G/morphine in blood was discussed – low ratios indicated short survival time after drug intake. The method was later extended to codeine and codeine-6-glucuronide (C6G) using deuterated internal standards (ISs) for each compound (LOD 0.5–10 µg/L) [50]. This method was successfully applied to the analysis of blood samples from suspected heroin abusers [10].

LC-MS methods for the determination of morphine, codeine and the corresponding glucuronides published in recent years are summarised in Table 8.1. These methods are generally based on SPE and ESI-MS or ESI-MS-MS detection. The relative concentration of free and conjugated morphine metabolites in body fluids may provide useful information for the forensic/clinical interpretation of a given case – a high free morphine fraction generally indicating acute poisoning at a very early stage after drug intake. In addition, LC-MS allows us to discriminate between the pharmacologically active M6G and inactive M3G. Among 25 patients receiving morphine the mean ratio (free morphine + M6G)/M3G was 0.26 ± 0.09; in 20 acute heroin fatalities, it was 1.52 × 1.06 and in one case of suicidal heroin infusion it was 9.29 [56].

Unequivocal analytical demonstration of

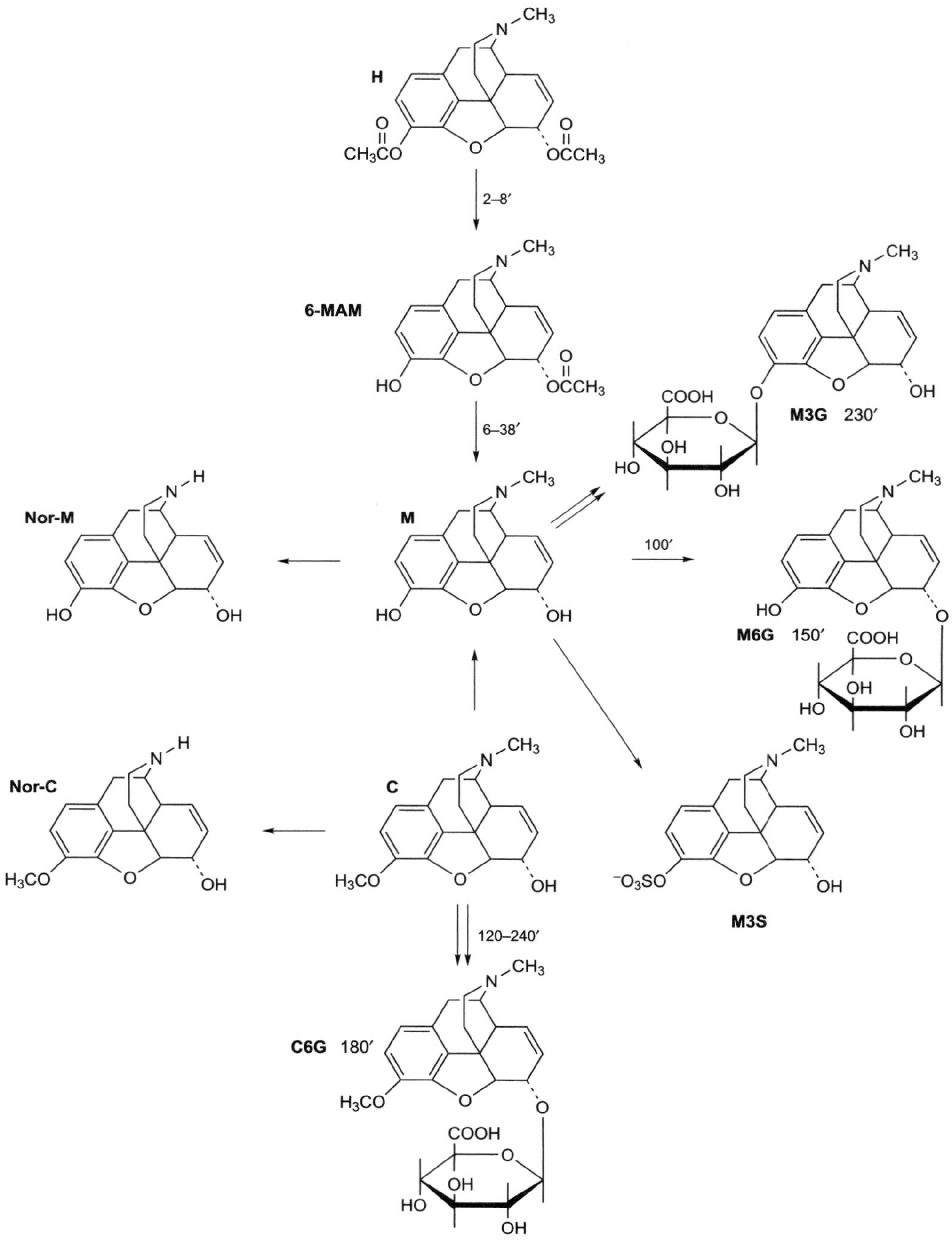

Figure 8.1 Main metabolic pathways of heroin (H), morphine (M) and codeine (C). See text for abbreviations. The half-lives in plasma (in minutes) are indicated for particular substances. (Reproduced from Bogusz [46] with permission from Elsevier Science.)

Table 8.1 LC-MS methods for opioids

Analytes	Sample	Isolation	Separation (column, elution conditions)	Detection	LOD (µg/L)	Reference
Morphine, M3G, M6G	plasma	SPE	silica, ACN:HCOOH isocratic	ESI-QqQ (MRM)	0.5–1.0	18
Morphine, M3G, M6G	plasma	96-well SPE	silica, ACN:TFA isocratic	ESI-QqQ (MRM)	0.5–10	19
Morphine, 6-AM, codeine, norcodeine, pholcodine	plasma, urine	LLE	C_8, ACN:HCOONH$_4$ isocratic	ESI-Q (SIM)	10	51
Morphine, M3G, M6G, normorphine	serum, urine	SPE	C_{18}, ACN:HCOOH gradient	ESI-QqQ (MRM)	0.3–2.5	52
Morphine, M3G, M6G, 6-AM, codeine, C6G	serum, urine	SPE	C_{18}, ACN:HCOONH$_4$ isocratic	APCI-Q (SIM)	0.1–10	50
Morphine, M3G, M6G, 6-AM, codeine, C6G	serum	SPE	C_{18}, ACN:HCOONH$_4$ gradient	ESI-Q (SIM)	0.5–5.0	53
Morphine, M3G, M6G	serum	SPE	C_{18}, ACN:HCOONH$_4$ isocratic	ESI-QqQ (MRM)	1.0–5.0	54
Morphine, M3G, M6G	plasma	SPE	C_{18}, ACN:HCOOH isocratic	ESI-QqQ, (MRM)	0.25–0.5	55
Fentanyl	plasma	96-well SPE	silica, ACN:TFA isocratic	ESI-QqQ (MRM)	0.05	20
HYM, HYC	plasma	96-well SPE	silica, ACN:TFA isocratic	ESI-QqQ (MRM)	0.1	21
BP, NBP, BPG, NBPG	plasma	SPE	C_{18}, ACN:HCOONH$_4$ gradient	ESI-QqQ (MRM)	0.1	73
Ketobemidone, Nor-K	urine	SPE	C_8, ACN:HCOOH gradient	ESI-Q (SIM)	25	88

Abbreviations: 6-AM = 6-acetylmorphine, BP = buprenorphine, BPG = buprenorphine glucuronide, C6G = codeine-6-glucuronide, DHM = dihydromorphine, H3G = hydromorphone-3-glucuronide, HYC = hydrocodone, HYM = hydromorphone, M3G = morphine-3-glucuronide, M6G = morphine-6-glucuronide, NBP = norbuprenorphine, NBPG = norbuprenorphine glucuronide, Nor-K = norketobemidone, Q = single-stage quadrupole, QqQ = triple-stage quadrupole, TFA = trifluoroacetic acid.

heroin intake as well as differentiation between administration of pure diacetylmorphine (DAM) and intake of street heroin is of great forensic importance. This differentiation became relevant since heroin prescription programmes have been introduced in some countries, e.g. Switzerland, the UK, Germany and the Netherlands. These programmes require that participants abstain from using any illicit drugs, particularly illicit heroin. After deacetylation, heroin follows common metabolic routes with morphine and (to some extent) with codeine. Therefore, 6-AM may be regarded as the only specific marker of heroin use, although it is detectable in blood and urine only for a relatively short time after intake [10, 46]. The detection of 6-AM does not allow discrimination between DAM and illicit heroin. However, in addition to DAM, several other opiates (6-AM, acetylcodeine [AC], codeine, papaverine and noscapine) as well as various adulterants are contained in illicit heroin. AC, originating from codeine during the acetylation of opium, might be regarded as specific marker of illicit heroin – its content ranging from 2 to 7% [36]. Poklis et al. [57, 58] detected AC in over 30% of morphine-positive urine specimens at concentrations from 1 to 4600 µg/L. 6-AM was found in over 70% of the samples, whereas all samples were positive for codeine. In another study, AC was detected in over 85% and 6-AM in over 94% of urine samples ($n = 71$) obtained from street heroin users [59]. Bogusz et al. [60] determined heroin markers in 25 morphine-positive urine samples with LC-APCI-MS in selected-ion monitoring (SIM) mode: C6G was found in all samples, codeine in 24, noscapine in 22, 6-AM in 16, papaverine in 14, DAM in 12 and AC in 4 samples. Additionally, morphine, M3G and M6G were found in all samples (Figure 8.2).

Katagi et al. [51] developed an automated column-switching ESI-MS method for the determination of heroin, 6-AM, morphine, AC, codeine and dihydrocodeine in urine. The method involved trapping on a cation-exchange column and separation on an analytical cation-exchange column with ACN:ammonium acetate (70:30). LODs ranging from 2 to 30 µg/L in full-scan mode and from 0.1 to 3 µg/L in SIM (monitoring of protonated quasi-molecular ions or ACN adducts) were achieved. Musshoff et al. [61] applied LC-ESI-MS-MS in selected-reaction monitoring (MRM) mode to the determination of illicit heroin markers: AC, papaverine, noscapine, as well as codeine, C6G, morphine, M3G, M6G and 6-AM in urine of heroin addicts participating in a heroin prescription programme. The LODs ranged from 0.1 (for papaverine) to 7.4 µg/L (for M6G). An LC method for the simultaneous determination of 17 opium alkaloids in urine and blood was published by Dams et al. [62]. The drugs were isolated with cation-exchange SPE, separated on a 'high-speed' phenyl column (53 × 7 mm) within 12 min, and detected with diode array (DAD) and fluorescence detectors. LODs in the range 2.5–9.7 µg/L were found.

Buprenorphine
Buprenorphine (BP), an oripavine derivative, is obtained from thebaine, and displays partial agonist and antagonist opioid activity [63]. The drug was initially used as a potent analgesic (marketed under the commercial names Temgesic and Buprenex). Further studies demonstrated the applicability of BP for the treatment of heroin addiction [64]. Unfortunately, sublingual BP tablets prescribed for addiction therapy are crushed, dissolved and taken intravenously by heroin addicts [65]. A comprehensive monograph on the application of BP in the therapy of opiate addiction was edited recently by Kintz and Marquet [66].

A number of LC-MS methods for the determination of BP and its active metabolite norbuprenorphine (NBP) are available in the literature. The main advantages in comparison with GC-MS are a simpler sample processing due to the omission of the derivatisation step, and the simultaneous determination of free and conjugated metabolites. Hoja et al. [67] determined BP and NBP in whole blood by LC-ESI-MS (single quadrupole) after β-glucuronidase hydrolysis, acetone precipitation and Extrelut (toluene/ether) extraction. The LOQ was 0.1 µg/L for both analytes. Tracqui et al. [68] applied LC-ESI-MS to the determination of BP and NBP in blood, urine and hair samples. A simple LLE with a chloroform:isopropanol:heptane mixture at pH 8.4 was applied. The mass spectra of BP, NBP and IS ([D_4]-labelled BP) exhibited only protonated molecular ions. The sensitivity was comparable with other ESI-MS methods. Moody et al. [69] developed a LC-ESI-MS-MS method for BP determination in plasma and compared it with an existing GC-MS (positive chemical ionisation mode) method. LC-MS-MS appeared more sensitive (LOQ 0.1 ng/mL) than GC-MS (LOQ 0.5 ng/mL) and allowed demonstration of the presence of drug up to 96 h after administration. The $[M + H]^+$ ion of BP remained stable up to a collision energy of 20 eV, whereas at higher energies it was shattered to

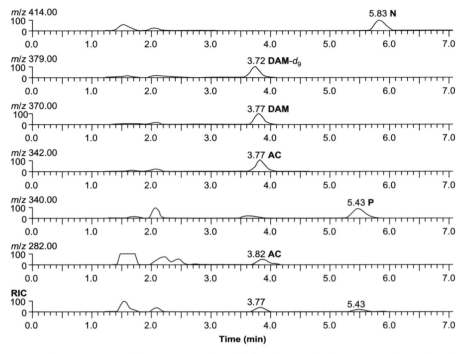

Figure 8.2 Mass chromatograms of a urine sample spiked with acetylcodeine (AC), diacetylmorphine (DAM), papaverine (P), noscapine (N) and IS at the concentration of 50 mg/L each. RIC, reconstructed ion current. (Reproduced from Bogusz et al. [60] with permission from Preston Publications, a Division of Preston Industries.)

many low-intensity product ions. Bogusz et al. [70] observed profound fragmentation of BP with LC-APCI-MS already at a collision energy of 10 eV, with a base ion at a mass-to-charge (m/z) ratio of 450 and smaller ions at m/z 468 [M + H]$^+$ and 418, respectively. In a later study, Moody et al. [71] applied a LC-ESI-MS-MS procedure for the determination of BP and NBP in human plasma. The transitions m/z 468 → 396 for B and m/z 414 → 101 for NB were monitored, and a LOQ of 0.1 µg/L was achieved for both compounds.

Gaulier et al. [72] reported a suicidal BP poisoning of a 25-year-old male heroin addict. BP and NBP were determined in body fluids and organs by sample deproteinisation, SPE and LC-ESI-MS. The following concentrations were found for BP and NBP, respectively: 3.3 and 0.4 mg/L (blood); 2035 and 536 mg/L (bile); 6.4 and 3.9 mg/L (brain). Only BP was detected in the gastric contents (899 mg/L). Apart from BP and NBP, high concentrations of 7-aminoflunitrazepam (7-amino-FLN) were also measured in blood (1.2 mg/L), urine (4.9 mg/L) and gastric contents (28.6 mg/L).

Polettini and Huestis [73] developed an LC-ESI-MS-MS method for determination of BP, NBP and BP-glucuronide (BPG) in human plasma after SPE (C_{18} cartridges) and gradient elution. [M + H]$^+$ ions were monitored for BP and NBP, as well as for their deuterated analogs. In the case of BPG, the transition from the protonated glucuronide to BP aglycone was monitored. The LOQ was 0.1 µg/L for all compounds. NBP-glucuronide (NBPG) was also tentatively detected by monitoring the transition m/z 590 → 414 (Figure 8.3), although the reference NBPG standard was not available. The authors stated that it was not possible to achieve useful fragmentation of BP as by increasing collision energy this compound dissipated to very small fragments. Therefore, they monitored the surviving quasi-molecular ions. This

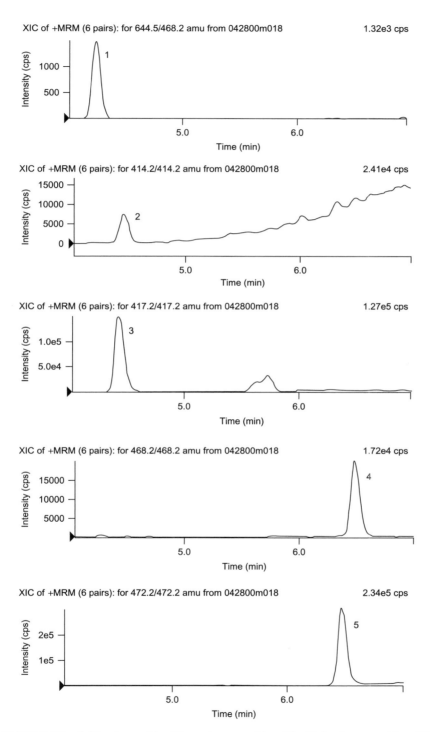

Figure 8.3 LC-ESI-MS-MS of BP glucuronide (1), norbuprenorphine (2), [D_3]norbuprenorphine (3), BP (4) and [D_3]buprenorphine (5) in plasma. (Reproduced from Polettini and Huestis [73] with permission from Elsevier Science.)

observation was later confirmed by Kronstrand et al. [74] and other authors.

Ceccato et al. [75] applied automated SPE (C_8) for isolation of BP and NBP from plasma. Reconstituted extracts were analysed with LC-APCI-MS-MS, using clonazepam as IS. The transitions m/z 468 → 468, 414 → 414 and 316 → 270 were monitored for BP, NBP and clonazepam, respectively. The LOQ was 10 ng/L for BP and 50 ng/L for NBP. BP, NBP and their glucuronides were determined in urine by Kronstrand et al. [74]. Urine samples taken from heroin addicts receiving BP were injected directly into LC-ESI-MS-MS both before and after enzymatic hydrolysis. The following transitions were monitored: m/z 468 → 468, 414 (for BP), 411 → 411, 101 (for NBP), 644 → 468, 590 (for BPG) and 590 → 414 (for NBPG). Deuterated analogs of BP and NBP were used as ISs. MEs were not studied. McAleer et al. [76] studied the pharmacokinetics of BP after sublingual administration under a naltrexone block. The drug was determined with LC-MS-MS. The mean terminal half-life was 26 h (range 9–69 h) and the C_{max} values ranged from 1.6 to 6.4 µg/L plasma after administration of 2–16 mg.

Methadone
Methadone, a morphine substitute synthesised in Germany during World War II, initially found limited application as an analgesic drug due to its very long elimination half-life and subsequent accumulation. These properties drew the attention of Dole, who first applied methadone as a heroin substitute in the therapy of addicts [77]. In the last 20 years, due to the international proliferation of methadone maintenance programmes, this drug became the most widely used opioid agonist [78]. This dictated the need for methadone monitoring in body fluids in order to control compliance and to prevent toxicity. It must be stressed that the wide availability of methadone is associated with its illicit use and with a growing number of drug-associated death cases, particularly among treated heroin addicts [79, 80].

Methadone is metabolised to inactive 2-ethylidene-1,5-dimethyl-3,3-diphenylpyrrolidine (EDDP) and, to a lesser extent, to 2-ethyl-5-methyl-3,3-diphenyl-1-pyrroline (EMDP). All these compounds contain a chiral centre. The differentiation of the enantiomers is pharmacologically relevant since R-(–)-methadone (levomethadone) shows a 25–50 times higher potency than the S-(+)-methadone. Commercial methadone preparations may contain either the racemic form or levomethadone. An LC-ESI-MS method involving enantioselective separation of methadone and EDDP was applied to hair samples by Kintz et al. [81] using deuterated analogues of all compounds for quantification. The results showed that R-(–)-methadone is preferably deposited in human hair. Enantioselective LC-MS determination of methadone was applied to saliva and serum collected from heroin addicts participating in a methadone maintenance programme by Ortelli et al. [82]. Methadone and EDDP disposition after nasal, intravenous and oral administration to volunteers was studied by LC-MS [83]. Kelly et al. [84] separated enantiomers of methadone, EDDP and EMDP using an α-glycoprotein column and MS-MS detection. The method was used for the analysis of hair samples of treated heroin addicts; the run time was around 1 h. Recently, several automated, high-throughput LC-MS-MS procedures for fast monitoring of methadone and EDDP in serum or plasma have been developed [85–87]. All these use an α-glycoprotein column for separation of enantiomers.

Other opioids
Chen et al. [21] published a LC-ESI-MS-MS procedure for the determination of hydrocodone and hydromorphone in plasma. The drugs and deuterated analogs were extracted with solvent and separated from glucuronides using a 50 × 2 mm² silica column and a mobile phase consisting of ACN:water:formic acid (80:20:1 by vol.). The LOQ was 0.1 µg/L. The same group determined fentanyl in plasma using automated 96-well SPE, straight-phase chromatography and ESI-MS-MS [20]. The LOQ was 0.05 µg/L based on a 0.25-mL sample volume.

Ketobemidone is a synthetic opioid agonist and narcotic analgesic frequently abused, particularly in Scandinavian countries. This drug and the demethylated metabolite were quantified in urine (LOD 25 µg/L) by mixed-mode SPE (recovery over 90%) and monitoring of [M + H]+ ions together with three ketobemidone fragments [88]. Sunström et al. [89] carried out the determination of ketobemidone and five phase I metabolites, and ketobemidone- and nor-ketobemidone-glucuronide in human urine using LC-ESI-MS-MS. The same group used LC-ESI-QTOF-MS for the determination of glucuronides of ketobemidone, norketobemidone and hydroxymethoxyketobemidone in urine (mass accuracy better than 2 ppm) [90].

Ceccato et al. [91] published a method for enantioselective determination of the synthetic opioid receptor agonist tramadol and active metabolite O-desmethyltramadol in plasma with LC-APCI-MS-MS. After isolation with automated SPE, analytes were separated on a Chiralpak AD column using a isohexane:ethanol:diethylamine (97:3:0.1, v/v/v) mobile phase. The $[M + H]^+$ (m/z 58 transitions for both substances and IS (ethyltramadol) were monitored. Other tramadol metabolites could be detected within this method. Juzwin et al. [92] determined tramadol N-oxide and its major metabolites in plasma of laboratory animals in pre-clinical studies. The LC-ESI-MS-MS procedure enabled an LOQ of 6 µg/L.

Naltrexone, an opioid antagonist, as well as the active metabolite 6β-naltrexol, were determined in guinea-pig plasma with LC-ESI-MS (SIM) [93]. The drugs were isolated from 0.1 mL plasma by ACN:ethyl acetate extraction; naloxone was used as IS. A LOD of 0.75 µg/L was reported (Figure 8.4).

Prenatal exposure to opiates

Pichini et al. [94] developed a LC-ESI-MS (SIM) method for the determination of morphine, 6-AM, M3G, M6G, codeine, as well as cocaine and metabolites, in meconium. The drugs were extracted with Bond Elut Certify cartridges. The procedure was applied to monitor prenatal intrauterine exposure to heroin and cocaine.

Amphetamine and Ecstasy

Amphetamine (AMP) and methamphetamine (MAMP) are synthetic compounds developed in 1895 and 1919, respectively. They have been broadly used as stimulants and anorectic drugs since 1935. Worldwide spread of AMP and MAMP use was associated with a wave of abuse and addiction caused by the high addiction potential of these compounds. Some methylenedioxy analogs of amphetamines, like 3,4-methylenedioxymethamphetamine (MDMA), 3,4-methylenedioxyethylamphetamine (MDEA) or 3,4-methylenedioxyamphetamine (MDA), were synthesised in 1914–19 and tested in the treatment of psychiatric disorders. In contrast to AMP or MAMP, these analogues show lower addiction potential and their abuse did not spread until the late 1970s. These drugs are known under various street names, e.g. 'Ecstasy', 'XTC', 'Adam' or 'Eve'. The book of the eminent pharmacologist Alexander Shulgin, *Pihkal: A Chemical Love Story* [95], contributed to the propagation of psychoactive phenethylamine abuse. In this book, procedures for synthesising 179 various drugs of this group are available together with information about recommended dosages and expected effects/symptoms. Amphetamine derivatives are usually manufactured as tablets and often distributed illegally in discotheques. MDMA and related drugs found broad acceptance in the drug scene due to their entactogenic and euphorising properties, causing the feeling of closeness to others and empathy, and facilitating social contacts. However, apart from the expected stimulating action, these compounds may alter thermoregulation and caused a growing number of death cases, mainly due to hyperactivity and heat stroke at rave parties [96–98]. The recreational use of 'Ecstasy' is very often associated with that of other psychotropic drugs and alcohol. The analysis of urine samples collected from attendees at rave parties, as well as a review of 81 Ecstasy-related deaths in England and Wales, revealed that in the majority of cases MDMA was present in combination with AMP, MAMP, other designer amphetamines or opiates [98, 99].

A general review of forensic aspects of abuse, chemistry, isolation and detection of amphetamines was presented by Cody [100]. Special attention in this review was paid to enantioselective separation of amphetamines. Human pharmacology of designer amphetamines has recently been reviewed by de la Torre et al. [101] and Maurer et al. [102]. Farre et al. [103] observed that MDMA presents non-linear

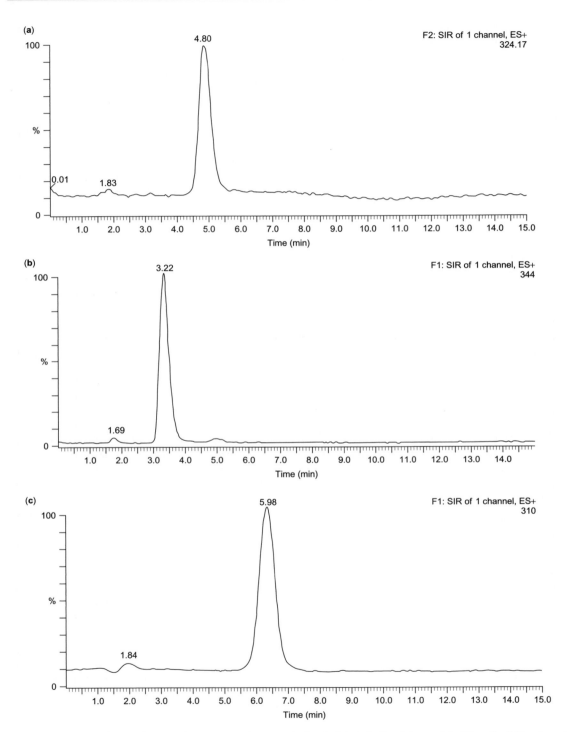

Figure 8.4 LC-MS chromatograms of guinea-pig plasma spiked with 25 μg/L of naltrexone (a), 25 μg/L of 6β-naltrexol (b) and 100 μg/L of naloxone (c). (Reproduced from Valiveti et al. [93] with permission from Elsevier Science.)

pharmacokinetics, and repeated administration caused a disproportionate increase of MDMA concentration in plasma, most probably due to metabolism inhibition, with the consecutive onset of poisoning.

LC-MS methods for the analysis of amphetamines and their methylenedioxy derivatives in biological tissues and fluids are reviewed below, and summarised in Table 8.2.

Analysis of blood and urine

Bogusz *et al*. [104] applied LC-DAD and LC-APCI-MS to the determination of psychoactive phenethylamines in serum and urine. After LLE extraction with diethyl ether, AMP, MAMP, MDMA, MDA, MDEA and other phenethylamines were derivatised with phenylisothiocyanate before analysis. LC-APCI-MS appeared about 10 times more sensitive than LC-DAD (LOD 1–5 µg/L). The same authors later proposed a different method based on SPE and LC-APCI-MS in SIM mode able to achieve the same LODs (1–5 µg/L), but this time without derivatisation [105]. Kataoka *et al*. [106] isolated AMP, MAMP, MDMA, MDEA and MDA from urine using solid-phase microextraction (SPME). Analytes were desorbed from the fibre by mobile-phase flow.

Table 8.2 LC-MS methods for amphetamines and cocaine

Drug	Sample	Isolation	Separation (column, elution conditions)	Detection	LOD (µg/L)	Reference
14 amphetamines and related compounds	plasma	SPE	C_{18}, ACN:$HCOONH_4$ isocratic	APCI-Q (SIM)	1–5	105
AMP, MAMP, MDMA, MDEA, MDA	urine	SPME	CN, ACN:CH_3COONH_4 isocratic	ESI-Q (SIM)	0.4–0.8	106
MDMA, MDEA, MDA	plasma	LLE	C_{18}, ACN:CH_3COONH_4 isocratic	ESI-QTOF-MS/SIM	1	25
AMP, MAMP, MDMA, MDEA, MDA, E	saliva	SPE	C_{18}, ACN:CH_3COONH_4 isocratic	ESI-QqQ (MRM)	1	113
AMP, MAMP, MDMA, MDA, MBDB	meconium	SPE	C_{18}, MeOH:NH_4HCO_3 gradient	ESI-Q (SIM)	1	114
Cocaine, BZE, EME	urine	SPE	C_{18}, ACN:H_2O gradient	ESI-QqQ (MRM)	0.5–2	124
Cocaine, BZE, EME, ECG	plasma	SPE	C_8, MeOH/ACN: CH_3COONH_4 isocratic	ESI-QqQ (MRM)	2.8–4.4	127
Cocaine, BZE, cocaethylene	hair	SPE	C_{18}, ACN:CH_3COONH_4 gradient	ESI-QqQ (MRM)	25 pg/mg	129
Cocaine, BZE, cocaethylene	meconium	SPE	C_8, ACN:CH_3COOH gradient	ESI-Q (SIM)	1–5	94

Abbreviations: AMP = amphetamine, BZE = benzoylecgonine, E = ephedrine, ECG = ecgonine, EME = ecgonine methyl ester, MAMP = methamphetamine, MDA = methylenedioxyamphetamine, MDEA = methylenedioxyethylamphetamine, MDMA = methylenedioxymethamphetamine, Q = single-stage quadrupole, QqQ = triple-stage quadrupole.

LC-ESI-MS analysis was carried out in SIM mode (LOD below 1 µg/L). Clauwaert et al. [25] used LC-ESI-QTOF-MS to the quantification of MDMA and its metabolite MDA in whole blood, serum, vitreous humor and urine (LOQ 1 µg/L). The linear dynamic range extended over four decades. The same group determined MDMA, MDEA and MDA in body fluids collected from rabbits in a postmortem distribution study using LC with fluorometric detection (FLD) (LOQ: blood, 2.0 µg/L; urine, 0.1 µg/L) [107]. Results were validated by LC-ESI-QTOF-MS that was found to correlate well with LC-FLD analysis. Paramethoxyamphetamine (PMA), a new designer amphetamine exhibiting relatively high toxicity, was analysed in biological fluids together with ephedrine, AMP, MDMA, MDEA and MDA by Mortier et al. [108], using LLE and LC-SSI-IT-MS. The isolation method assured negligible MEs for human urine, liver or kidney (Figure 8.5).

Several therapeutic drugs release AMP or MAMP in the course of their metabolic processes. Since illicit drug users often claim that they used a prescribed medication, it is of forensic importance to differentiate between the administration of prescribed AMP/MAMP precursors and illicit preparations. Benzphetamine (BMA) and its metabolites benzylamphetamine, hydroxybenzphetamine, hydroxybenzylamphetamine, MAMP and AMP were determined in urine by SPE and ESI-MS (SIM) detection after RP separation using an alkaline mobile phase (LOD 0.3–10 µg/L) [109]. This method allows the discrimination of BMA, AMP and MAMP intake. Dimethylamphetamine and metabolites (dimethylamine-N-oxide, MAMP and AMP) were determined in urine by SPE followed by ESI-LC-MS (SIM) (LOD 5–50 µg/L) [110]. Another prescription drug that is metabolised to MAMP and AMP is selegiline – an inhibitor of monoamine

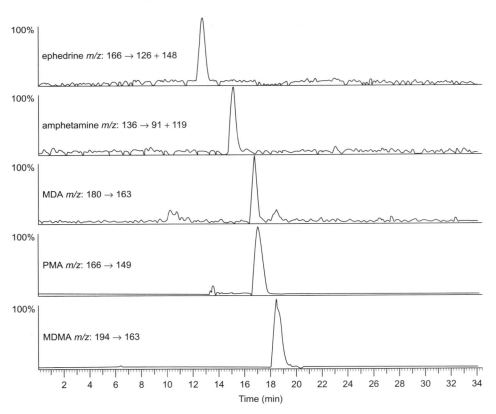

Figure 8.5 LC-MS-MS chromatogram of a urine extract containing ephedrine, amphetamine, MDA, PMA and MDMA. (Reproduced from Mortier et al. [108] with permission from John Wiley & Sons).

oxidase B used in the treatment of Parkinson's disease. Katagi et al. [111] determined selegiline as well as the specific metabolite selegiline-N-oxide, together with MAMP and AMP, in human urine. The method involved C_{18} SPE, LC on a cation-exchange column and ESI-MS detection. Urine of selegiline-treated patients was found to be positive for metabolites, although the parent compound was below the LOD. This method may allow the differentiation of selegiline and MAMP intake. The authors studied the excretion of selegiline-N-oxide after selegiline administration in volunteers [112].

Analysis of alternative matrices

Various amphetamines (AMP, MAMP, MDMA, MDA, MDEA and ephedrine) were determined in saliva by Wood et al. [113]. After dilution with 200 µL methanolic solution of the IS mixture (deuterated analogues) and centrifugation, saliva (50 µL) was directly injected into an LC-ESI-MS-MS system (RP separation with ACN:ammonium acetate mobile phase). The drugs were detected by monitoring transitions of $[M + H]^+$ ions (LOQ 1–5 µg/L). This method was later extended to the analysis of sweat wipes, plasma and oral fluid collected after controlled administration of MDMA as well as in real-life conditions (i.e. after ingestion of Ecstasy tablets at a party). Very high intra- and interindividual variability of saliva results was observed. Very low concentrations of MDMA were detected in sweat within the first 5 h after drug intake [114]. LC-APCI-MS was applied to the quantification of amphetamines and other drugs of abuse in hair [115]. Pulverised hair specimens were subjected to alkaline hydrolysis and solvent extraction. The LODs were 50 pg/mg for MAMP, MDA, MDMA and MDEA, 100 pg/mg for AMP and ephedrine, and 200 pg/mg for methcathinone and para-methoxyamphetamine. The method was applied to the analysis of 93 authentic hair samples taken from participants in a methadone programme. Cairns et al. [116] compared the levels of AMP and MAMP in hair of participants of drug rehabilitation programmes with those from workplace testing. The samples were taken from subjects who had produced MS-confirmed positive urine results. For hair analysis, LC-MS-MS was applied.

The findings were similar in both groups for target drugs; MDMA and MDA were additionally present in workplace specimens. Prenatal exposure of newborns to amphetamines was studied by Pichini et al. [117], who analysed drugs in meconium with LC-ESI-MS in SIM mode after C_{18} SPE. Only one out of 600 analysed samples contained MDMA. Other amphetamines were not detected. The method was also applied to MAMP-positive meconium samples analysed 1 year earlier with GC-MS and the results showed very good agreement.

Cocaine

Cocaine is one of the oldest drugs abused by humankind and also one of the most important 'hard drugs' of modern society. According to the National Institute of Drug Abuse, about 1.5 million Americans were identified as current cocaine users in 1997 [118]. A survey performed in 2003 indicated that American cocaine addicts accounted for 25.6% of all dependent illicit drugs users and formed the second greatest group after heroin addicts (57.4%) [119]. Although the synthesis of cocaine is possible, essentially all cocaine in illicit traffic is produced by extraction from the leaves of *Erythroxylum coca* – a shrub that is domestic in the Andes Mountains in Peru and Bolivia, and is being cultivated in various South American states. The coca leaves are mixed with water and lime, and extracted with kerosene or gasoline. The kerosene extract is then mixed with diluted sulphuric acid, and the aqueous layer is collected and made basic with ammonia or lime. The precipitate, known as coca paste, contains 40–70% cocaine as a mixture of base and salt, as well as kerosene, sulphuric acid and other impurities. The coca paste, a very dangerous intermediate product, is smoked in South America [120]. Further purification steps are required in order to obtain pure cocaine base, which is then dissolved in ether or acetone and converted to cocaine hydrochloride by the addition of hydrochloric acid. The resulting white, microcrystalline precipitate ('snow') is dried and packaged for sale. Street cocaine preparations are usually adulterated with some other local anaesthetics and diluted with neutral compounds. Cocaine hydrochloride is water soluble and may be applied intranasally ('snorting') or intravenously. In order to obtain a smokable form of drug, cocaine salt must be converted again to volatile base. The free base, known as 'crack' (the term refers to the crackling sound when the drug is smoked), is nowadays the most prevalent and most dangerous form of street cocaine.

In the human body, cocaine is converted to a multitude of active and non-active metabolites, mostly of high polarity (Figure 8.6). The metabolism, pharmacokinetics and general analytical aspects of forensic cocaine determination were reviewed by Jufer and Cone [121]. Cocaine metabolites require derivatisation before GC-MS analysis. Therefore, as for opiate metabolites, LC-MS methods enable direct analysis of cocaine and metabolites. LC-MS methods for the analysis of cocaine and metabolites in biological tissues and fluids are reviewed below, and summarised in Table 8.2.

Analysis of biological fluids

Singh et al. [122] determined cocaine and metabolites in rat plasma samples collected in a pharmacokinetic study. After simple ACN precipitation, LC-APCI-MS-MS analysis achieved an LOQ of 2 µg/L for cocaine, benzoylecgonine (BZE) and norcocaine, and an LOQ of 5 µg/L for ecgonine methyl ester (EME). Jeanville et al. [123] used fast gradient LC-ESI-MS-MS for the automated 96-well direct analysis of filtered urine samples, achieving a total analysis time of 2 min (MEs not investigated). LC-MS-MS equipped with an on-line extraction unit was used in a later work by Jeanville et al. [124] for the analysis of centrifuged urine with a total run time lower than 4 min (LOD 0.5, 2.0 and 0.5 µg/L for EME, BZE and cocaine, respectively). The same authors demonstrated that LC-ESI-QTOF-MS showed similar performance as LC-ESI-QqQ (triple quadrupole) for the direct quantification of cocaine and EME in rat plasma [125]. Skopp et al. [126, 127] studied the stability of cocaine, BZE, EME and ecgonine in human plasma at different storage temperatures. After SPE and RP separation on a narrow-bore C_8 column, MS-MS analysis was carried out in MRM mode (one transition from the $[M + H]^+$ per analyte). At room temperature, a stoichiometric conversion of cocaine to the ecgonine was observed and only ecgonine appeared to be stable. Lin et al. [128] published a LC-APCI-MS-MS procedure for cocaine and BZE in human plasma. By monitoring the transitions m/z 304 → 182 (cocaine) and 290 → 168 (BZE), an LOQ of 2.5 µg/L was obtained.

Analysis of alternative samples

Clauwaert et al. [129] determined cocaine, BZE and cocaethylene (a cocaine metabolite formed in the case of concomitant alcohol use) in hair samples. After acidic hydrolysis and SPE on Confirm HCX cartridges, the extracts were analysed with LC-ESI-MS and LC-FLD. Close agreement of results of both methods was observed in 29 forensic drug overdose cases.

Prenatal exposure to cocaine

Sosnoff et al. [130] detected BZE in dried blood spots with LC-MS-MS. An LOD of 2 µg/L was achieved based on a 12-µL sample size. This technique was used for epidemiological screening in a study involving newborns. The analysis of cocaine and metabolites in meconium, placenta or amniotic fluid may give a valuable indication of prenatal exposure to the drug. Cocaine and 12 metabolites were determined in meconium using LC-ESI-MS-MS by Xia et al. [131]. The presence of at least 8 metabolites was demonstrated in 21 out of 22 drug-positive meconium samples and ecgonine was found to be the most sensitive marker of cocaine exposure. The same authors studied the distribution of cocaine and metabolites in blood, amniotic fluid, placental and fetal tissues of pregnant rats treated intravenously with cocaine [132]. Pichini et al. [94] developed a LC-ESI-MS (SIM) method for the determination of cocaine, BZE and cocaethylene, together with opiates, in meconium. The drugs were extracted with Bond Elut Certify cartridges.

Cannabis

Cannabis products (e.g. marijuana and hashish) are surpassed only by ethyl alcohol as the most

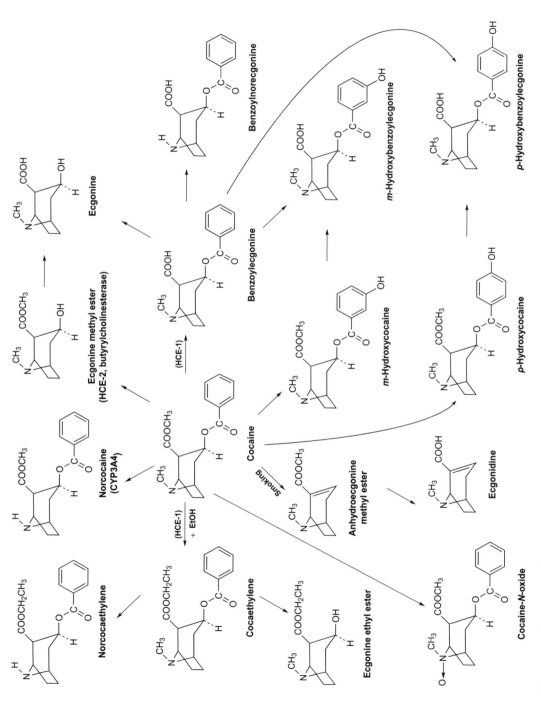

Figure 8.6 Biotransformation and thermal degradation products of cocaine. (Reproduced from Jufen et al. [121] with permission from Elsevier Science.)

commonly abused drugs. A national survey on drug use, performed in 2003, showed that about 10.6 million Americans used cannabis products alone and a further 5 million consumed cannabis together with other drugs. Generally, 75% of all illicit drug users in the US consumed marijuana or other cannabis products [119]. Known history of cannabis abuse dates as early as about 3000 BC in India. Later, the cultivation and use of *Cannabis sativa* gradually spread worldwide. Among known cannabis preparations in the illicit drug market (herbal cannabis [marijuana], cannabis resin [hashish] and liquid cannabis [cannabis oil]), marijuana is the most popular. It is the most widely used illicit drug preparation used in the world. Apart from the illicit use of cannabis as a recreational drug, the use of cannabis preparations or its active compound Δ^9-tetrahydrocannabinol (THC) in medicine has an equally long tradition [133]. The discovery of cannabinoid receptors as well as endogenous receptor agonists stimulated research on the role and possible therapeutic use of cannabis. THC as well as its synthetic analogues (nabilone, levonantradol and marinol) were thus tested for the treatment of several disorders. It was demonstrated that cannabinoid receptor agonists were effective as anti-emetics and appetite stimulants, and were applied in HIV patients [134, 135]. Among other possible uses, the suppression of muscle spasticity associated with multiple sclerosis, relief of chronic pain and therapy of behaviour disorders in Gilles de la Tourette's syndrome have been mentioned [136–138]. Martin *et al.* [139] synthesised several analogues of THC by modifying its alkyl side-chain. This manipulation produced high-affinity ligands with antagonist, partial agonist or full agonist effects on the cannabinoid receptor. It seems that the development of THC analogues, which would be devoid of the unwanted psychotropic action, may be of great interest for clinical medicine.

Analysis of street drugs

The determination of cannabinoids in plant material has several purposes. The first and most important one is to identify and quantify the main components, like active THC and other cannabinoids, e.g. cannabinol (CBN) and cannabidiol (CBD). On the basis of quantitative determination of THC, it is possible to classify the plant material as resin type (illicit drug) and fibre type (used for industrial purposes). The other goal of analysis is to identify the production site and sometimes a particular batch of drug. This may be done using high-resolution chromatography. In fact, it must be stressed that *C. sativa* contains several hundred constituents with proportions that may differ in relation to geography, climate, cultivation mode, etc. [140]. Also, the quantification of THC in plant material allows for monitoring of the potency of the preparation.

The identification of cannabinoids (THC, CBD CBN and acidic cannabinoids) in hashish samples was performed using particle beam ionisation LC-MS [141]. Bäckström *et al.* [142] determined cannabinoids in cannabis preparations by supercritical fluid chromatography coupled to APCI-MS (on-column LOD: THC, 0.69 ng; CDB, 0.55 ng; CBN, 2.10 ng).

Since September 2003, cannabis has become available in the Netherlands for medical use on prescription. Dutch authors [143] therefore developed an LC-APCI-IT-MS method for the determination of THC, 11-nor-9-carboxy-Δ^9-tetrahydrocannabinol (THC-COOH), CBN, CBD and acidic cannabinoids, using positive-ionisation MS-MS. LOQs ranged from 0.03 for CBN to 9.9 g/kg for THC-COOH.

Analysis of biological fluids

Mireault [144] analysed THC and metabolites in blood and urine samples extracted with an SPE cartridge and separated on a C_8 column using a MeOH:ammonium acetate mobile phase. Detection was performed in positive-ion mode by APCI-MS-MS using an IT instrument. An LOD of 1 μg/L was achieved for THC, 11-hydroxy-Δ^9-tetrahydrocannabinol (11-OH-THC) and THC-COOH. The same author was able to reach 10–40 times lower LODs and an LOQ of 0.25 μg/L by LC-APCI-MS-MS on a QqQ tandem mass spectrometer [145]. THC-COOH was determined in urine after alkaline hydrolysis and SPE, RP-C_8 separation with gradient ACN:formic acid elution, and ESI-MS detection in SIM mode (positive ions) by Breindahl and Andreasen [146]. Together with $[M + H]^+$, two fragments obtained by in-source collision-induced dissociation (CID) were monitored (LOD 15 µg/L). LC-ESI-MS in negative-ion mode was used for the quantification of THC-COOH in urine by Tai and Welch [147]. The compound was extracted with SPE and isocratically separated on C_{18} column in MeOH:ammonium acetate mobile phase. Only the deprotonated quasi-molecular ions of the analyte and of the deuterated IS were monitored. Weinmann et al. [148] developed an LC-MS-MS procedure for the simultaneous quantification of THC-COOH and glucuronide (THC-COOH-G) in urine. After LLE with ethyl acetate:ether (1:1) and gradient separation on a C_8 column with ACN:ammonium formate buffer, detection was performed in MRM mode by monitoring two transitions from the $[M + H]^+$ ions per analyte. The specificity of the method for the glucuronide was evaluated by re-analysing urine-positive samples after enzymatic hydrolysis and observing the disappearance of the THC-COOH-G peak (deuterated standard not available). The same authors later developed a fast method for THC-COOH determination in urine using alkaline hydrolysis and automated SPE [149]. After gradient elution on a short octadecylsilica column, APCI-MS-MS detection was carried out in negative-ion mode. Three product ions originating from the $[M - H]^-$ were used for confirmation/quantification. LOD and LOQ were 2.0 and 5.1 µg/L, respectively. Maralikova and Weinmann [150] presented an LC-MS-MS procedure for the determination of THC, 11-OH-THC and THC-COOH using positive ionisation electrospray. Two MS-MS transitions were monitored for each compound. LODs were 0.2 µg/L for THC and 11-OH-THC, and 1.6 µg/L for THC-COOH (Figure 8.7).

The stability of THC-COOH and the glucuronide (THC-COOH-G) in authentic urine samples was investigated by Skopp and Pötsch [151] using LC-ESI-MS-MS. The stability of THC-COOH-G rapidly decreased with an increase of temperature and pH of urine. Therefore, any interpretation of the ratio between free and glucuronidated forms of THC-COOH appears to be questionable. Concheiro et al. [152] determined THC in 0.2 mL of oral fluid with LC-ESI-MS after hexane extraction at pH 6 and an LOQ of 2 µg/L was reported. Selected methods for the

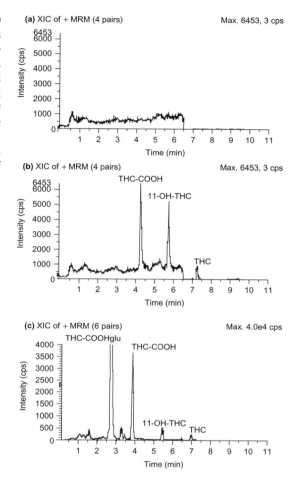

Figure 8.7 Reconstructed SRM chromatogram of (a) blank plasma extract and (b) plasma extract spiked with THC (1 µg/L), 11-OH-THC (1 g/L) and THC-COOH (5 µg/L) (middle), and (c) SRM chromatogram of plasma extract from routine analysis (THC, 4 µg/L; 11-OH-THC, 2.3 µg/L; THC-COOH, 53.6 µg/L). (Reproduced from Maralikova and Weinmann [150] with permission from John Wiley & Sons.)

determination of cannabinoids in biological samples are summarised in Table 8.3.

Hallucinogens

LSD

LSD (lysergic acid diethylamide) is a semi-synthetic compound originating from ergot alkaloids,

Table 8.3 LC-MS methods for cannabinoids and hallucinogens

Drug	Sample	Isolation	Separation (column, elution conditions)	Detection	LOD (µg/L)	Reference
THC, 11-OH-THC, THC-COOH	plasma, urine	SPE	C_8, MeOH:HCOONH$_4$ isocratic	APCI-IT (SIM)	1	144
THC, 11-OH-THC, THC-COOH	plasma, urine	SPE	C_8, MeOH:HCOONH$_4$ isocratic	APCI-QqQ (MRM)	0.25	145
THC-COOH	urine	SPE	C_{18}, MeOH: CH$_3$COONH$_4$ isocratic	ESI-Q (SIM)	0.5	146
THC-COOH, THC-COOH-G	urine	LLE	C_8, ACN:HCOONH$_4$ gradient	ESI-QqQ	10	148
THC-COOH	urine	LLE	C_{18}, ACN:HCOOH isocratic	APCI-QqQ	2	149
THC, 11-OH-THC, THC-COOH	plasma	SPE	phenylhexyl, ACN:CH$_3$COONH$_4$ gradient	ESI-QqQ	0.2–1.6	150
LSD, Nor-LSD	urine	SPE	C_{18}, ACN: CH$_3$COONH$_4$ isocratic	ESI-Q (SIM)	0.5	155
LSD, Nor-LSD	blood, urine	SPE	phenyl, ACN: CH$_3$HCOONH$_4$	ESI-QqQ (MRM)	0.025	158
LSD, O-H-LSD	urine	LLE+SPE	C_{18}, ACN: CH$_3$COONH$_4$ isocratic	APCI-IT (SIM)	0.2	161
LSD, O-H-LSD	urine	LLE + SPE	C_{18}, ACN: CH$_3$COONH$_4$ gradient	APCI-IT (SIM)	0.4	162
LSD, O-H-LSD	blood, urine	LLE + SPE	C_{18}, MeOH: HCOONH$_4$ gradient	ESI-Q (SIM)	0.1–0.4	163

Abbreviations: 11-OH-THC (11-hydroxy-Δ^9-tetrahydrocannabinol), IT = ion trap mass analyser, O-H-LSD = 2-oxo-3-hydroxy-LSD, THC = tetrahydrocannabinol, THC-COOH = 11-nor-9-carboxy-Δ^9-tetrahydrocannabinol, THC-COOH-G = THC-COOH-glucuronide, Q = single-stage quadrupole, QqQ = triple-stage quadrupole.

produced by the fungus *Claviceps purpurea*. This fungus attacks grains and has caused mass poisonings ('ergotism'). LSD was synthesised in 1938 by A. Hofmann, but its hallucinogenic properties were accidentally discovered by this scientist 5 years later [153]. LSD belongs to the most potent hallucinogenic drugs. The single hallucinogenic dose ranges from 30 to 100 µg and is usually applied on a piece of paper, known under the street name 'trip'. The drug is rapidly and extensively metabolised to oxo, hydroxy and desalkyl derivatives, a very small fraction being excreted in urine in the unmodified form [154]. Earlier LC-MS methods involved the determination of LSD and N-demethyl-LSD in urine. Webb *et al.* [155] proposed a method based on immunoaffinity extraction from urine followed by LC-ESI-MS using methysergide as IS. The [M + H]$^+$ and two fragments were monitored, and an LOD of 0.5 µg/L obtained with a 5-mL urine volume. In the following study by the same authors, D$_3$ labelled LSD was used as IS [156], thus achieving much better accuracy and precision. The LOD was not improved, though. LOQ values of 0.05

and 0.1 μg/L for LSD and N-demethyl-LSD, respectively, were obtained by Hoja et al. [157] by Extrelut extraction from urine and LC-ESI-MS. Automated mixed-phase SPE from different body fluids followed by LC-ESI-MS-MS using a phenyl column was proposed by De Kanel et al. [158] (LOD 0.025 μg/L). Bodin et al. [159] extracted LSD from urine by LLE with back extraction to acetate buffer. LC-ESI-MS analysis provided an LOD of 0.02 μg/L.

The window of detection of unmodified LSD in urine is rather limited. According to Nelson and Foltz, the drug can be detected in urine not longer than 12–22 h after use [160]. However, Poch et al. [161] demonstrated that the concentrations of the 2-oxo-3-hydroxy-LSD metabolite are distinctly higher than those of the parent drug and other LSD metabolites (Figure 8.8). Therefore, the detection time window for LSD use may be significantly widened through the determination of this metabolite. 2-Oxo-3-hydroxy-LSD, together with LSD, N-demethyl-LSD and isolysergic acid diethylamide (iso-LSD), a synthetic byproduct of LSD synthesis, were determined in urine with LC-APCI-IT-MS. These authors compared LC-APCI-MS and LC-APCI-MS-MS (IT) detection, and found very good agreement of results for LSD and metabolites [162]. Sklerov et al. [163] determined LSD and 2-oxo-3-hydroxy-LSD in blood and urine. After LLE from an alkaline environment followed by SPE (only for blood), LC-ESI-MS with in-source CID was applied (three ions per analyte).

LODs in urine were 0.1 and 0.4 μg/L for LSD and metabolite, respectively. 2-Oxo-3-hydroxy-LSD was found in high concentrations in urine, but was not detected in blood samples from authentic cases. LSD and iso-LSD were quantified in plasma by Canezin et al. [164] using LC-ESI-MS-MS (LOD 0.02 μg/L for both analytes). Different metabolites, i.e. 2-oxo-3-hydroxy-LSD, N-demethyl-LSD, N-demethyl-iso-LSD, 13- and 14-OH-LSD-glucuronides, lysergic acid ethylamide, trioxydated-LSD and lysergic acid ethyl-2-hydroxyethylamide, were detected in urine. Klette et al. [165] demonstrated that 15 compounds structurally related to LSD as well as a wide range of unrelated compounds do not interfere with the detection of the 2-oxo-3-hydroxy-LSD. This metabolite proved to be stable for 60

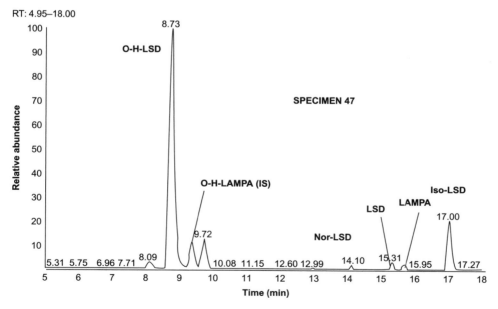

Figure 8.8 LC-APCI-MS-MS (IT) of LSD and metabolites extracted from urine sample (LAMPA: lysergic acid methylpropylamide; O-H-LAMPA: 2-oxo-3-hydroxy-LAMPA; O-H-LSD: 2-oxo-3-hydroxy-LSD). (Reproduced from Poch et al. [161] with permission from Elsevier Science.)

days in urine samples at pH 4.6–8.4 stored at –20 and 8°C. A novel extraction procedure based on disc anion-exchange SPE with subsequent LC-ESI-MS was developed for LSD and 2-oxo-3-hydroxy-LSD [166]. The LOD was 0.25 µg/L. The selectivity was tested by analysis of 93 compounds with properties similar to the target analytes, and no interference was found.

LC-MS methods for the analysis of LSD and metabolites in biological tissues and fluids are summarised in Table 8.3.

Other hallucinogens

LC-MS has been applied to the determination of some other hallucinogens. Ibogaine, an indole alkaloid contained in *Tabernanthe iboga*, a rain forest shrub, has been used for a long time by Western Africans to combat fatigue and hunger (at low dose) or as a hallucinogen in religious rituals (at higher dose). This drug and the analogue 18-methoxycoronaridine were found to promote drug abstinence from cocaine, opiates and other addictive substances, and were tested in experimental clinical trials for this purpose [167, 168]. Ibogaine has been determined, together with other drugs of abuse, by LC-APCI-MS after SPE by Bogusz et al. [70], achieving an LOD of 0.2 µg/L plasma. In a study devoted to investigating the metabolic pathways of 18-methoxycoronaridine Zhang et al. [169] determined the parent compound and three metabolites with LC-ESI-MS-MS in human liver microsomes.

Naturally occurring and synthetic indoleamines, like *N,N*-dimethyl-5-hydroxytryptamine (bufotenine), *N,N*-dimethyltryptamine (DMT), 5-methoxy-DMT, *N*-methyltryptamine (NMT) and others, show hallucinogenic properties and therefore have an abuse potential. LC-MS-MS has been used for the determination of these drugs in urine samples of psychiatric patients (with ESI ion source) [170] and in rat brain after controlled administration (APCI) [171].

Psilocybin, a hallucinogenic indole derivative, occurs in mushrooms belonging to the genera *Psilocybe, Panaeolus, Conocybe* and *Gymnopilus*. These mushrooms, known in the drug scene under the name 'magic mushrooms', were consumed by the Aztecs and are being widely abused in present times [172]. Bogusz et al. [173] analysed psilocybin and its dephosphorylated active metabolite psilocin in honey mixed with 'magic mushrooms' after MeOH extraction and LC-APCI-MS (SIM). Psilocin, but not psilocybin, was detected (Figure 8.9). Kamata et al. [174] evaluated different methods for the hydrolysis of psilocin glucuronide in urine of 'magic mushroom' consumers. The drug was determined by LC-MS and LC-MS-MS. Lechowicz et al. [175] extracted psilocin from whole blood with *n*-propyl chloride/dichloromethylene at pH 11. The drug was determined with LC-APCI-MS (SIM) against the deuterated analog with an LOD and an LOQ of 1.7 and 5.2 µg/L, respectively. The concentration of drug in casework samples ranged from above the LOD to 8 µg/L.

Therapeutic drugs with abuse potential

Incapacitating drugs

Some drugs may be fraudulently used to render a victim defenceless. These drugs, which are typically tasteless, colourless and odourless, are usually slipped into the drink of the unknowing partner and cause fast loss of control up to unconsciousness after a low dose. As their use occurs typically in the context of sexual abuse, they are known as 'date rape drugs'. Other street names are 'Mind Eraser' or 'Forget Pill' as often the victim develops amnesia and cannot give a reliable report of the attack. Drugs used for incapacitating purposes have been recently reviewed [176]. The most frequently used are FLN and other benzodiazepines, usually in combination with alcohol, γ-hydroxybutyrate (GHB), hallucinogens and opioids.

Benzodiazepines

FLN is a potent hypnotic drug most frequently misused in drug-facilitated sexual assaults by addition to alcoholic beverages. It is known under the street names 'Roofies', 'Mexican Valium', 'La Rocha' and others. FLN was developed by Roche in Switzerland and is

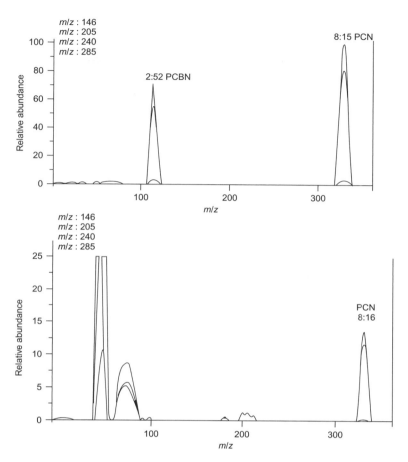

Figure 8.9 LC-APCI-MS chromatogram of psilocybin (PCBN) and psilocin (PCN), 50 ng each (upper picture), and chromatogram of the extract of 'herbal honey' (lower picture). (Reproduced from Bogusz et al. [173] with permission from Springer Verlag.)

available in Europe as a prescription drug under the name Rohypnol, but is illegal in US. FLN quickly disappears from blood (also in vitro) and is converted to active, polar metabolites, mainly to 7-amino-FLN. Contrary to GC-MS, LC-MS enables the analysis of these compounds without derivatisation. FLN, 7-amino-FLN, and the other metabolites N-desmethyl-FLN and 3-OH-FLN were determined in blood/plasma by LC-APCI-MS (SIM) of the [M + H]$^+$ ions [177]. A preliminary comparison of the ionisation efficiency with ESI and APCI showed that the latter provided a signal 7–30 times stronger for all compounds except 7-amino-FLN, which gave a similar response with both ion sources. The method involved C_{18} SPE and RP separation with ACN:ammonium formate buffer (pH 3.0). The method, with an LOD that ranged from 0.2 to 1 µg/L, has been applied in routine toxicological casework [10]. LeBeau et al. determined FLN, 7-amino-FLN and N-desmethyl-FLN in blood and urine after mixed-mode SPE [178] and LC-ESI-MS-MS (IT) detection. The method was based on the monitoring of [M + H]$^+$ ions and confirmation of identity by MS-MS product-ion-scan mode (LOD in blood 0.5–1 µg/L). Darius et al. [179] determined FLN (but not metabolites) in serum with LC-ESI-MS-MS (IT) after LLE with t-butylmethyl ether and separation on a C_{18} column in ACN:water mobile phase. Detection of FLN and IS (clonazepam) was carried out in MRM mode, achieving an LOD of 0.2 µg/L. Automated in-tube SPME and LC-ESI-MS was applied by Yuan et al. [180] for the quantification of the benzodiazepine derivatives diazepam, nordazepam, temazepam, oxazepam, 7-amino-FLN and N-desmethyl-FLN in serum and urine (FLN parent compound not included). The extraction procedure, providing fully automated solvent-free isolation of analytes in 15 min for each consecutive sample, was optimised by comparing the performance using six different extraction capillaries under various extraction conditions. LC-ESI-MS analysis was performed in full-scan (m/z 100–400) and SIM mode (LOD 0.02–2 µg/L). However, the analytical recovery from serum was below 50% and automated desorption causing peak broadening was observed. Kollroser et al. [181] applied LC-APCI-MS-MS (IT) to the determination of FLN, 7-amino-FLN and N-desmethyl-FLN in plasma after SPE on Oasis MCX cartridges. Analytical recoveries were higher than 90% and LODs in the range 0.25–2 µg/L were obtained.

Midazolam is a short-acting benzodiazepine used as a hypnotic and in preoperative medication. Due to its high potency and fast onset of symptoms, it may be abused as an incapacitating agent. Midazolam and its active hydroxylated metabolite were extracted from serum with ether:isopropanol (98:2) at alkaline pH and separated on a Nucleosil C_{18} (150 × 1 mm) column. LC-ESI-MS analysis monitored both [M + H]$^+$ ions and fragments of both compounds (LOQ 0.5 µg/L) [182]. Ware et al. [183] quantified midazolam and hydroxymidazolam in plasma using triazolam and hydroxytriazolam as ISs. After SPE (Oasis HLB cartridges), LC-APCI-MS-MS analysis in SRM mode was carried out [183]. All drugs eluted within 2 min and the LOQ was 0.1 µg/L. Cheze et al. [184] developed an LC-ESI-MS-MS method for determination of the benzodiazepines bromazepam, 3-hydroxy-bromazepam, clonazepam and 7-aminoclonazepam in urine and hair. The drugs were isolated with solvent extraction and subjected to SRM analysis. The application to two forensic cases of criminal assault was described. The same research group demonstrated the usefulness of hair analysis by LC-MS-MS in drug-facilitated crimes [185]. Among 90 forensic cases, various benzodiazepines and zolpidem were detected in 21 cases. Smink et al. [186] determined 33 benzodiazepines, their metabolites and related drugs in whole blood using LC-APCI with MS or MS-MS (IT) detection (Figure 8.10). The drugs were separated using a gradient of MeOH:0.006 mol/L formic acid. Comparison of signal intensities in ESI and APCI mode showed that the latter source assures much better sensitivity for most compounds. The LODs ranged from 0.1 to 12.6 µg/L.

Ketamine

Ketamine has been used as a general anaesthetic drug since the 1970s. This drug also produces hallucinations and shows addiction potential. For these reasons, ketamine found its place in the drug scene as 'Special-K' or 'Kit-Kat'. Ketamine abuse at rave parties was reported in France [187] and Hong Kong [188]. This drug was also used to facilitate sexual assault [189]. An LC-ESI-IT-MS procedure for the determination of ketamine and the metabolites norketamine and (presumptive) dehydronorketamine in urine was proposed by Moore et al. [190] (LOD 3 µg/L for ketamine and norketamine). The method, involving SPE with Clean Screen cartridges, was applied in 33 forensic cases. Concentration ranges of 6–7744 and 7–7986 µg/L were measured for ketamine and norketamine, respectively (Figure 8.11). Cheng and Mok [188] published a procedure for the fast determination of ketamine in urine. The drug was isolated with automated SPE and detected with LC-MS-MS. Total run time was 2.5 min and a LOD of 5 µg/L was reported.

Zolpidem

Zolpidem has been commercially available as a hypnotic drug for around 15 years. This drug is used in drug-facilitated crimes, due to the fast onset of symptoms and amnesic properties. French authors developed an LC-ESI-MS-MS method for the determination of zolpidem in urine and hair [191] as well as in oral fluid (Figure 8.12) [192] of the victims of drug-facilitated crimes. In urine, zolpidem was detectable for up to 60 h after administration of a single 10-mg dose; in oral fluid, it was detectable for up to 8 h. The drug was also detectable in hair after single-dose administration.

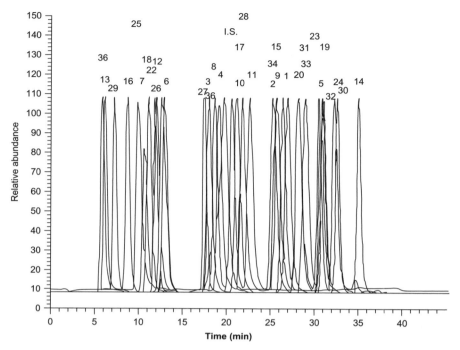

Figure 8.10 Superimposed ion chromatograms of 36 tested benzodiazepines and related substances: [(1) alprazolam, (2) α-OH-alprazolam, (3) bromazepam, (4) OH-bromazepam, (5) brotizolam, (6) chlordiazepoxide, (7) norchlordiazepoxide, (8) demoxepam, (9) clobazam, (10) desmethylclobazam, (11) clonazepam, (12) acetamidoclonazepam, (13) aminoclonazepam, (14) diazepam, (15) flunitrazepam, (16) 7-aminoflunitrazepam, (17) N-desmethylflunitrazepam, (18) flurazepam (19) desalkylflurazepam, (20) OH-ethylflurazepam, (21) IS, (22) loprazolam, (23) lorazepam, (24) lormetazepam, (25) desmethylmedazepam, (26) midazolam, (27) 1-OH-midazolam, (28) nitrazepam, (29) acetamidonitrazepam, (30) nordazepam, (31) oxazepam, (32) temazepam, (33) triazolam, (34) OH-triazolam, (35) zolpidem, (36) zopiclone. (Reproduced from Smink et al. [186] with permission from Elsevier Science.)

γ-Hydroxybutyrate

GHB is an endogenous metabolite of γ-aminobutyric acid and is therefore always present in urine. GHB has been used clinically as an intravenous anaesthetic and hypnotic agent. It has been abused as a steroid alternative at gyms and 'health food' stores until the US Food and Drug Administration banned it in 1990. GHB is widely abused as a recreational drug ('liquid Ecstasy') among disco goers. In 2001, GHB, classified as a Schedule 1 substance in the US, surpassed FLN as the most common substance used in drug-facilitated sexual assaults [193]. This drug was usually determined in body fluids by GC-MS. Wood et al. [194] published a rapid method for the determination of GHB and its precursors γ-butyrolactone (GBL) and 1,4-butanediol (1,4-BD) in urine using LC-MS-MS (Figure 8.13). Diluted urine samples were subjected to SPE clean-up on Oasis MCX cartridges and injected on an Atlantis C_{18} column. An LOQ of 1 mg/L was achieved for all compounds determined. GHB was differentiated from its isomers α- and β-hydroxybutyric acids. The method was applied in 182 routine cases in five forensic centres and appeared much simpler than GC-MS procedures.

Other therapeutic compounds with abuse potential

Methylphenidate is a phenethylamine derivative with stimulating properties used in the treatment

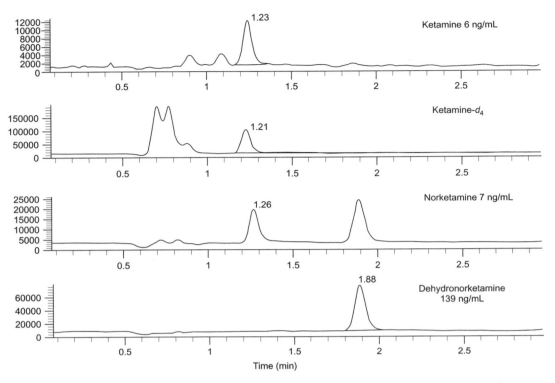

Figure 8.11 Mass chromatogram (ESI-IT) of a urine extract containing ketamine, metabolites (norketamine and dehydronorketamine) and IS. (Reproduced from Moore et al. [190] with permission from Preston Publications, a Division of Preston Industries.)

of narcolepsy. Methylphenidate has a history of abuse, mainly by intravenous injection of the crushed and dissolved tablets. The drug has two chiral centres and is marketed as a racemic mixture, although D-*threo*-methylphenidate is more potent than the L-isomer. Ramos *et al.* [195] described an LC-APCI-MS-MS procedure for the determination of methylphenidate enantiomers in plasma. A vancomycin-based Chirobiotic V column was used for separation. The LOQ was 0.087 µg/L. The same authors developed a procedure for the determination of racemic methylphenidate [196]. A 96-well plate LLE with cyclohexane from basified plasma was utilised. After RP separation (3 min total run time), the eluent was sent to APCI-MS-MS using a switching valve in order to divert the solvent front to waste. A LOQ of 0.050 µg/L was obtained.

Hyoscine is a naturally occurring alkaloid, present in *Hyoscyamus niger* and *Datura stramonium*. It may be abused due to its hallucinogenic properties. Hyoscine was quantified in serum in pharmacokinetic studies with an LOQ of 0.02 µg/L by LC-ESI-MS-MS positive-ion MRM mode. Two product ions from the $[M + H]^+$ of both the analyte and the IS (atropine) were monitored [197]. SPE with Oasis HLB cartridges using 0.2 mL of serum was followed by separation on a C_{18} column in gradient elution on a mobile phase consisting of ACN, ammonium acetate and formic acid.

Sildenafil (Viagra) has recently become a very popular drug. It has been estimated that no fewer than 100 million tablets have been prescribed worldwide. Apart from being used in the therapy of erectile disorders, sildenafil has also become a party drug, sold on college campuses, and taken at clubs and rave parties. However, severe adverse effects in the sexual sphere have been reported

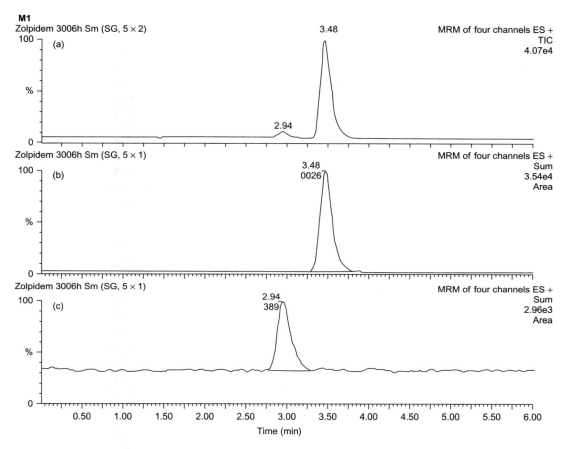

Figure 8.12 Chromatogram obtained after extraction of an oral fluid specimen containing zolpidem (0.4 ng/mL). (a) SRM chromatograms (sum of all transitions monitored); (b) sum of the two transitions for the IS; (c) sum of the two transitions for zolpidem. (Reproduced from Kintz et al. [192] with permission from Elsevier Science.)

[198] in addition to life-threatening cardiac failure. Particularly dangerous is the combination of sildenafil with Ecstasy [199]. Weinmann et al. [200] described a procedure for the detection and confirmation of sildenafil and three metabolites in postmortem samples (urine and organs from a heavily putrefied body) by LC-ESI-MS and LC-ESI-MS-MS. Dumestre-Toulet et al. [201] determined sildenafil and verapamil in blood, organs and hair by LC-ESI-MS. The method was applied in a case of 43-year-old man with a history of cardiovascular disease and erectile dysfunction who was found dead in a hotel room after a sexual relationship. Eerkes et al. [202] developed an LC-ESI-MS-MS method for the determination of sildenafil and desmethylsildenafil in human plasma, using a silica column and aqueous:organic mobile phase (composition not given) for separation (Figure 8.14). The samples were extracted using automated SPE on 96-well plates and an LOQ of 1 μg/L was reported for both compounds.

Barbituric acid derivatives, drugs with a long history of abuse, are seldom present in the drug scene nowadays. Heller [203] determined pentobarbital in dog food samples with LC-MS and found that the sensitivity was worse that that of GC-MS. In our own experience, barbiturates can be determined with negative ESI-LC-MS-MS at nanogram levels. The method was applied for the

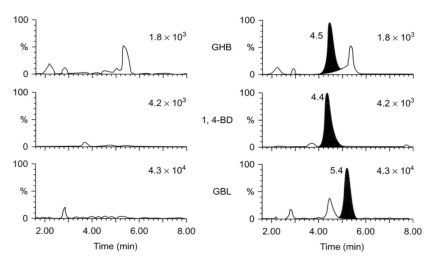

Figure 8.13 SRM chromatograms obtained with a single injection of a blank urine sample (left-hand column), and the same sample spiked with 10 mg/L of GHB, GBL and 1,4-BD (right-hand column top, middle and bottom, respectively). (Reproduced from Wood *et al.* [194] with permission from Elsevier Science.)

detection of phenobarbital in herbal remedies (Figure 8.15).

LC-MS screening for drugs of abuse

The tendency to use multiple drugs is an increasingly common feature of drug addicts all over the world [204, 205], and analytical toxicologists face the challenge to identify and determine a wide range of compounds, often in very limited amount of sample. De Zeeuw [206] recently summarised the difficulties that arise in proper substance identification by MS procedures. Various application fields in analytical toxicology, e.g. forensic toxicology, food testing and doping in sport, developed their own criteria for confirmation of the presence of a particular substance. However, the issue of identification has not been properly addressed. The criteria used for confirmation and published in various guidelines are often discrepant and sometimes not realistic. Criteria for compound identification are reviewed in Chapter 5.

The use of LC-MS for toxicological screening of drugs of abuse initially appeared to be an extremely attractive possibility, since this technique possesses a much broader detection spectrum than GC-MS and is more sensitive than LC-DAD. However, the large interlaboratory variability of LC-MS mass spectra has not allowed us to establish a generally applicable LC-MS library, similar to GC-MS EI databases. In a study performed in three laboratories, mass spectra of identical substances of abuse, analysed on the same instruments under nominally identical conditions, showed large differences in the degree of fragmentation [207]. However, controlled changes in the composition of the mobile phase, i.e. in the percentage of organic modifier or in the molarity of ammonium formate buffer, did not have any relevant influence on mass spectra. Existing LC-MS screening procedures may be divided into two groups: those comprising substances belonging to similar pharmacological classes or used for similar applications (e.g. for drugs of abuse) and those developed for undirected searches, i.e. for 'general unknown screening', which are reviewed in Chapter 6. In the first case, the task is much easier, since the number of substances is limited and sample preparation is more effective due to the similar chemical features of compounds involved. Screening procedures applied for abused substances usually comprise the most important drugs which are

LC-MS screening for drugs of abuse 177

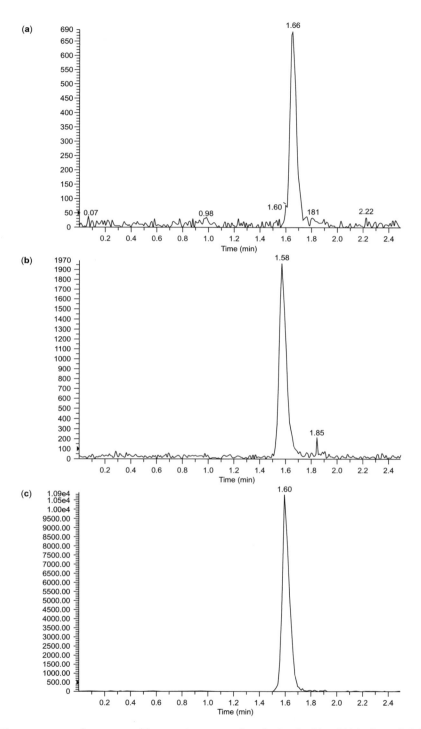

Figure 8.14 Chromatograms of an extracted human plasma sample spiked with sildenafil (a), desmethylsildenafil (b) and IS (c) at concentrations of 1 μg/L each. (Reproduced from Eerkes *et al.* [202] with permission from Elsevier Science.)

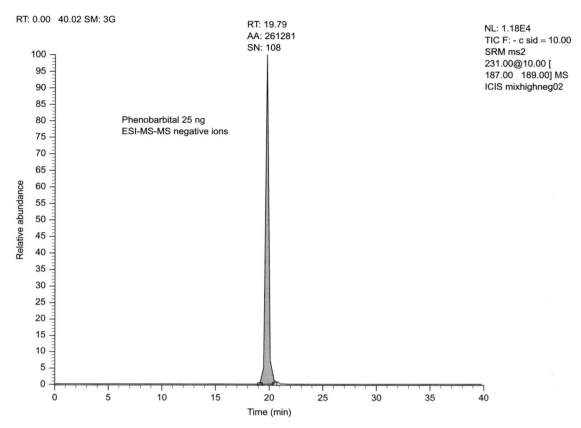

Figure 8.15 MRM chromatogram of 25 ng phenobarbital analysed with LC-ESI-MS-MS in negative-ion mode. Chromatographic separation was done on a C_{18} column in a mobile phase consisting of ACN and 10 mmol/L ammonium formate buffer, pH 3.0, in gradient elution mode. (M. J. Bogusz, unpublished data.)

difficult to detect in other ways, e.g. by LC-DAD or GC-MS. Bogusz et al. [70] developed an LC-APCI-MS procedure for the determination of opiates, cocaine and metabolites, LSD, and ibogaine, using a single SPE isolation procedure and a mobile phase consisting of the same components (ACN and ammonium formate buffer, pH 3.0). This method was later extended to 14 psychoactive phenethylamines, like AMP and designer amphetamines [105]. For each compound, the protonated molecular ion and one to three fragments were monitored. Practical experience from this procedure was reported [10]. Cailleux et al. [208] extracted the opiate agonists (morphine, 6-AM, codeine, norcodeine, pholcodine and codethyline) as well as nalorphine and cocaine and its metabolites (BZE, EME, cocaethylene and anhydromethylecgonine) from blood, plasma or urine with chloroform:isopropanol (95:5) at pH 9. The drugs were separated on a C_8 column in ACN:ammonium formate: formic acid. Protonated molecular ions and one fragment for each substance were monitored using ESI-MS-MS. The quantification was performed using deuterated ISs. The limits of quantification were 10 μg/L for opiates and 5 μg/L for cocaine/metabolites. Rittner et al. [209] developed a library of LC-ESI-MS spectra of 70 psychoactive drugs and metabolites. After optimisation of the isolation and separation procedures by comparing different SPE methods and LC columns, C_{18} SPE and C_{18} RP gradient separations using an ACN:water:MeOH:formic acid mobile phase were selected. Mass spectra of drugs (m/z 100–650, positive ions) were recorded at two fragmentation energies. Sodium, potassium and ACN adducts were observed for many drugs. This method was used to screen serum samples taken from road traffic offenders: in 9.8% of 140 cases various drugs were detected, most of them belonging to the class of benzodiazepines. Maralikova and Weinmann [210] applied EU criteria published in 2002 [211] for confirmatory LC-ESI-MS-MS analysis of 12 drugs belonging to various groups (opioids, cannabinoids, cocaines and LSD) in plasma and urine. Automated SPE was applied for isolation. For each drug, two transitions were monitored, fulfilling the criteria of confirmation. The lowest levels for validation ranged from 5 to 15 µg/L. Nineteen drugs (benzodiazepines, opioids, amphetamines and hypnotics) were analysed in urine by LC-ESI-MS-MS after SPE on an Oasis HLB column. The LODs ranged from 0.02 to 21 µg/L [212].

In addition to plasma or urine, alternative specimens have also been screened for drugs of abuse with LC-MS. Mortier *et al.* [213] stressed the importance of proper sample preparation for LC-ESI-MS-MS determination of drugs of abuse in saliva. Amphetamines (AMP, MAMP, MDMA, MDEA and MDA), opiates (morphine, codeine and 6-AM) as well as cocaine and BZE were isolated from spiked saliva with three different protein precipitation procedures. In each case, serious MEs were observed. The authors warned that with the use of a highly selective method, like LC-MS-MS in SRM mode, the influence of the matrix may go unnoticed and postulated the use of optimised sample preparation methods. The same Belgian group [214] established an LC-ESI-QTOF-MS method for the determination of various drugs of abuse: opiates (morphine and codeine), amphetamines (AMP, MAMP, MDMA, MDEA and MDA) as well as cocaine and BZE in oral fluid. A 200-µL volume of sample was submitted to mixed-mode SPE. The drugs were separated in a MeOH:ammonium formate gradient on a narrow-bore phenyl column. Recoveries varied from 52% to 99%, and the LOQ was 2 µg/L for all compounds. The method was applied to real samples obtained from suspected car drivers. Dams *et al.* [215] determined opiates, cocaine/metabolites and methadone in oral fluid with LC-APCI-MS-MS (IT). Sample pre-treatment was limited to ACN protein precipitation. The LODs ranged from 0.25 to 5 µg/L. The method was applied for monitoring illicit drug use and methadone concentrations in pregnant heroin/cocaine addicts taking methadone. Jenkins *et al.* [216] extracted acidified oral fluid samples with SPE and subjected them to LC-ESI-MS-MS analysis for cocaine, opiates, amphetamines and phencyclidine. The method was calibrated in the range of 5–250 µg/L. Scheidweiler and Huestis [217] determined morphine, 6-AM, codeine, norcodeine cocaine, norcocaine, BE, EME and cocaethylene in human hair after MeOH sonication. LC-APCI-MS-MS was applied and the quantification was performed with deuterated analogues for each compound. Only one transition was monitored for each drug. The LODs ranged from 8.5 to 41.5 ng/g hair. A screening procedure in hair for 14 drugs of abuse, belonging to various classes (nicotine and cotinine, opiates, cocaines, amphetamines and benzodiazepines) was described by Kronstrand *et al.* [218]. The drugs were extracted from 20–50 mg hair using the LC mobile phase and subjected to LC-ESI-MS-MS analysis. One transition for each drug was monitored. LODs ranged from 3 to 70 ng/g hair. The authors postulated the use of this procedure as an alternative to immunoassay screening in workplace drug testing.

Drugs of abuse and related compounds were also included in several LC-MS and LC-MS-MS screening procedures developed for broader applications [219–233].

Future perspectives

The future perspectives for the application of LC-MS in forensic determination of drugs of abuse, no different from other fields of analytical toxicology, can be deduced from the existing problems as well as from the trends observed in recent years. With regard to sample preparation, major efforts in the development of fast, automated and efficient isolation procedures are foreseeable. It has been fully recognised that co-eluting interferents from the matrix may affect detection and quantification (i.e. by ion suppression/enhancement, overlapping peaks/spectra, etc.), and there is a general agreement between scientists that current and future published procedures must take these into account during method development.

A trend towards the shortening of the overall sample throughput is evident and in this regard chromatographic separation, together with sample preparation, plays an important role. Fast separation may be achieved by using short, fine-grain or monolithic columns, by applying fast elution (rapid gradients, high percentage or organic modifier in the mobile phase, high flow rate) or by column switching.

The use of LC-MS as a tool for the general screening of drugs and toxic substances will be more common in the future as it provides substantial advantages over other techniques that have been used so far for this purpose (i.e. GC-MS, LC-DAD). LC-MS commercial mass spectral libraries (either in source-CID or product-ion spectra) will probably become increasingly available. Another promising option for the identification of unknown compounds is offered by accurate mass determination of molecular ions.

Finally, mass spectrometers will provide better performance (in terms of sensitivity, selectivity, scan speed and mass accuracy), and will become smaller and more affordable, and thus will favour the further spread of LC-MS in analytical toxicology.

References

1. Wennig R, Verstraete A. Drugs and driving. In: Bogusz MJ, ed., *Handbook of Analytical Separations.* Vol 2: *Forensic Science.* Amsterdam: Elsevier, 2000: 439–457.

2. Aderjan R, Bonte W, Daldrup T, *et al.* Änderung des § 24a des Strassenverkehrsgesetzes und Bericht der Grenzwertkommission. *Toxichem Krimtech* 1998; 65: 70–73.

3. *European Legal Database on Drugs. Drugs and Driving.* http://eldd.emcdda.eu.int (accessed January 2005).

4. Krull IS, Cohen SA. Analytical chemistry and biotechnology. *LC-GC Int* 1998; 11: 139–140.

5. Willoughby R, Sheehan E, Mitrovich S. *A Global View on LC/MS*. Pittsburgh, PA: Global View, 1998.

6. Niessen WMA. Advances in instrumentation in liquid chromatography-mass spectrometry and related liquid-introduction techniques. *J Chromatogr A* 1998; 794: 407–435.

7. Niessen WMA. State-of-the-art in liquid chromatography-mass spectrometry. *J Chromatogr A* 1999; 856: 179–197.

8. Bogusz MJ. Hyphenated liquid chromatographic techniques in forensic toxicology. *J Chromatogr B* 1999; 733: 65–91.

9. Marquet P, Lachatre G. Liquid chromatography-mass spectrometry: potential in forensic and clinical toxicology. *J Chromatogr B* 1999; 733: 93–118.

10. Bogusz MJ. Liquid chromatography-mass spectrometry as a routine method in forensic science a proof of maturity. *J Chromatogr B* 2000; 748: 3–19.

11. Marquet P. Progress of liquid chromatography-mass spectrometry in clinical and forensic toxicology. *Ther Drug Monit* 2002; 24: 255–276.

12. Department of Health and Human Services. SAMSHA. Proposed Revisions to Mandatory Guidelines for Federal Workplace Drug Testing Programs. *Federal Register* 2004; 69: 19673–19732.

13. Weinmann W, Svoboda M. Fast screening for drugs of abuse by solid phase extraction combined with flow-injection ionspray tandem mass spectrometry. *J Anal Toxicol* 1998; 22: 319–328.

14. Müller C, Schäfer P, Störtzel M, *et al.* Ion suppression effects in liquid chromatography-electrospray-ionisation transport-region collision induced dissociation mass spectrometry with different serum extraction methods for systematic toxicological analysis with mass spectra libraries. *J Chromatogr B* 2002; 773: 47–52.

15. Dams R, Huestis MA, Lambert WE, *et al.* Matrix effect in bioanalysis of illicit drugs with LC-MS/MS: influence of ionization type, sample preparation, and biofluid. *J Am Soc Mass Spectrom* 2003; 14: 1290–1294.

16. Souverain S, Rudaz S, Veuthey JL. Matrix effect in LC-ESI-MS and LC-APCI-MS with off-line and on-line extraction procedures. *J Chromatogr A* 2004; 1058: 61–66.

17. Naidong W, Shou W, Chen YL, *et al.* Novel liquid chromatographic-tandem mass spectrometric

method using silica columns and aqueous–organic mobile phases for quantitative analysis of polar analytes in biological fluids. *J Chromatogr B* 2002; 754: 387–399, 2001.

18. Naidong W, Lee JW, Jiang X, *et al*. Simultaneous assay of morphine, morphine-3-glucuronide, and morphine-6-glucuronide in human plasma using normal-phase liquid chromatography-tandem mass spectrometry with a silica column and an aqueous organic mobile phase. *J Chromatogr B* 1999; 735: 255–269.

19. Shou WZ, Pelzer M, Addison T, *et al*. An automatic 96-well solid phase extraction and liquid chromatography-tandem mass spectrometry method for the analysis of morphine, morphine-3-glucuronide and morphine-6-glucuronide in human plasma. *J Pharm Biomed Anal* 2002; 27: 143–152.

20. Shou WZ, Jiang X, Beato BD, *et al*. A highly automated 96-well solid phase extraction and liquid chromatography/tandem mass spectrometry method for the determination of fentanyl in human plasma. *Rapid Commun Mass Spectrom* 2001; 15: 466–476.

21. Chen YL, Hanson GD, Jiang X, *et al*. Simultaneous determination of hydrocodone and hydromorphone in human plasma by liquid chromatography with tandem mass spectrometric detection. *J Chromatogr B* 2002; 769: 55–64.

22. Shou WZ, Chen YL, Eerkes A, *et al*. Ultrafast liquid chromatography/tandem mass spectrometry bioanalysis of polar analytes using packed silica columns. *Rapid Commun Mass Spectrom* 2002; 16: 1613–1621.

23. Naidong W, Shou WZ, Addison T, *et al*. Liquid chromatography/tandem mass spectrometric bioanalysis using normal-phase columns with aqueous/organic mobile phases – a novel approach of eliminating evaporation and reconstitution steps in 96-well SPE. *Rapid Commun Mass Spectrom* 2002; 16: 1965–1975.

24. Dams R, Benijts T, Gunther W, *et al*. Influence of the eluent composition on the ionization efficiency for morphine of pneumatically assisted electrospray, atmospheric pressure chemical ionization and sonic spray. *Rapid Commun Mass Spectrom* 2002; 16: 1072–1077.

25. Clauwaert KM, Van Bocxlaer JF, Major HJ, *et al*. Investigation of the quantitative properties of the quadrupole orthogonal acceleration time-of-flight mass spectrometer with electrospray ionization using 3,4-methyylenedioxymethamphetamine. *Rapid Commun Mass Spectrom* 1999; 13: 1540–1545.

26. Vorce SP, Sklerov JH, Kalasinsky KS. Assessment of the ion-trap mass spectrometer for routine qualitative and quantitative analysis of drugs of abuse extracted from urine. *J Anal Toxicol* 2000; 24: 595–601.

27. Paul BD, Dreka C, Knight ES, *et al*. Gas chromatographic/mass spectrometric detection of narcotine, papaverine and thebaine in seeds of *Papaver Somniferum*. *Planta Med* 1996; 62: 544–547.

28. Fater Z, Samu S, Szatmary M, *et al*. Combinations of liquid chromatographic methods for the breeding of high alkaloid-containing poppies [in Hungarian]. *Acta Pharm Hung* 1997; 67: 211–219.

29. Pelders MG, Ros JW. Poppy seeds: differences in morphine and codeine content and variation in inter- and intraindividual excretion. *J Forensic Sci* 1996; 41: 209–212.

30. Fritschi G, Prescott WR. Morphine levels in urine subsequent to poppy seed consumption. *Forensic Sci Int* 1985; 27: 111–117.

31. Hayes LW, Krasselt WG, Mueggler PA. Concentrations of morphine and codeine in serum and urine after ingestion of poppy seeds. *Clin Chem* 1987; 33: 806–808.

32. Petitt BC, Dyszel SM, Hood LV. Opiates in poppy seed: effect on urinalysis results after consumption of poppy seed cake-filling. *Clin Chem* 1987; 33, 1251–1152.

33. Struempler RE. Excretion of codeine and morphine following ingestion of poppy seeds. *J Anal Toxicol* 1987; 11: 97–99.

34. Lo DS, Chua TH. Poppy seeds: implications of consumption. *Med Sci Law* 1992; 32: 296–302.

35. Meadway C, George S, Braithwaite R. Opiate concentrations following the ingestion of poppy seed products – evidence for the 'poppy seed defense'. *Forensic Sci Int* 1998; 96: 29–38.

36. Huizer H. Analytical studies on illicit heroin. PhD Thesis. Technical University Delft, 1988.

37. Gough TA, The examination of drugs in smuggling offences. In: Gough TA, ed., *The Analysis of Drugs of Abuse*. Chichester: Wiley, 1991: 511–566.

38. Johnston A, King LA. Heroin profiling: predicting the country of origin of seized heroin. *Forensic Sci Int* 1998; 95: 47–55.

39. Janhunen K, Cole MD. Development of a predictive model for batch membership of street samples of heroin. *Forensic Sci Int* 1999; 102: 1–11.

40. Kaa K. Impurities, adulterants and diluents of illicit heroin. Changes during a 12-year period. *Forensic Sci Int* 1994; 64: 171–179.

41. Umans JG, Chiu TSK, Lipman RA, et al. Determination of heroin and its metabolites by high performance liquid chromatography. *J Chromatogr* 1982; 233: 213–225.

42. Inturrisi CE, Max MB, Foley KM, et al. The pharmacokinetics of heroin in patients with chronic pain. *N Engl J Med* 1984; 310: 1213–1217.

43. Jenkins AJ, Keenan RM, Henningfield JR, et al. Pharmacokinetics and pharmacodynamics of smoked heroin. *J Anal Toxicol* 1994; 18: 317–330.

44. Cone EJ, Holicky BA, Grant TM, et al. Pharmacokinetics and pharmacodynamics of intranasal 'snorted' heroin. *J Anal Toxicol* 1993; 17: 327–337.

45. Skopp G, Gansmann G, Cone EJ, et al. Plasma concentrations of heroin and morphine-related metabolites after intranasal or intramuscular administration. *J Anal Toxicol* 1997; 21: 105–111.

46. Bogusz MJ. Opioid agonists In: Bogusz MJ, ed., *Handbook of Analytical Separations*. Vol 2: *Forensic Science*. Amsterdam: Elsevier, 2000: 3–66.

47. Dams R, Benijst T, Gunther W, et al. Sonic spray ionization technology: performance study and application to a LC/MS analysis on a monolithic silica column for heroin impurity profiling. *Anal Chem* 2002; 74, 3206–3212.

48. Zuccaro P, Ricciarello R, Pichini S, et al. Simultaneous determination of heroin, 6-monoacetylmorphine, morphine, and its glucuronides by liquid chromatography-atmospheric pressure ionspray-mass spectrometry. *J Anal Toxicol* 1997; 21: 268–277.

49. Bogusz MJ, Maier RD, Driessen S. Morphine, morphine-3-glucuronide, morphine-6-glucuronide and 6-monoacetylmorphine determined by means of atmospheric pressure chemical ionization-mass spectrometry-liquid chromatography in body fluids of heroin victims. *J Anal Toxicol* 1997; 21: 346–355.

50. Bogusz MJ, Maier RD, Erkens M, et al. Determination of morphine and its 3-and 6-glucuronide, codeine, codeine-6-glucuronide and 6-monoacetylmorphine in body fluids by liquid chromatography-atmospheric pressure chemical ionization mass spectrometry. *J Chromatogr B* 1997; 703: 115–127.

51. Katagi M, Nishikawa M, Tatsuno M, et al. Column switching high-performance liquid chromatography-electrospray ionization mass spectrometry for identification of heroin metabolites in human urine. *J Chromatogr B* 2001; 751: 177–185.

52. Schänzle G, Li S, Mikus G, et al. Rapid, highly sensitive method for the determination of morphine and its metabolites in body fluids by liquid chromatography-mass spectrometry. *J Chromatogr B* 1999; 721: 55–65.

53. Dienes-Nagy A, Rivier L, Giroud G, et al. Method for quantification of morphine and its 3- and 6-glucuronides, codeine, codeine glucuronide and 6-monoacetylmorphine in human blood by liquid chromatography-electrospray mass spectrometry for routine analysis in forensic toxicology. *J Chromatogr A* 1999; 854: 109–118.

54. Blanchet M, Bru G, Guerret M, et al. Routine determination of morphine, morphine 3-β-D-glucuronide and morphine 6-β-D-glucuronide in human serum by liquid chromatography coupled to electrospray mass spectrometry. *J Chromatogr A* 1999; 854: 93–108.

55. Slawson MH, Crouch DJ, Andrenyak DM, et al. Determination of morphine, morphine-3-glucuronide, and morphine-6-glucuronide in plasma after intravenous and intrathecal morphine administration using HPLC with electrospray ionization and tandem mass spectrometry. *J Anal Toxicol* 1999; 23: 468–473.

56. Bogusz MJ. LC/MS as a tool for case interpretation in forensic toxicology. Presented at 49th ASMS Conference on Mass Spectrometry and Allied Topics, Chicago, IL, 2001.

57. O'Neal CL, Poklis A. Simultaneous determination of acetylcodeine, monoacetylmorphine and other opiates in urine by GC-MS. *J Anal Toxicol* 1997; 21: 427–432.

58. O'Neal CL, Poklis A. The detection of acetylcodeine and 6-acetylmorphine in opiate-positive urines. *Forensic Sci Int* 1998; 95: 1–10.

59. Staub C, Marset M, Mino A, et al. Detection of acetylcodeine in urine as an indicator of illicit heroin use: method validation and results of a pilot study. *Clin Chem* 2001; 47: 301–307.

60. Bogusz MJ, Maier RD, Erkens M, et al. Detection of non-prescription heroin markers in urine with

liquid chromatography-atmospheric pressure chemical ionization mass spectrometry. *J Anal Toxicol* 2001; 25: 431–438.

61. Musshoff F, Trafkowski J, Madea B. Validated assay for the determination of markers of illicit heroin in urine samples for the control of patients in a heroin prescription program. *J Chromatogr B* 2004; 811: 47–52.

62. Dams R, Benijst T, Lambert WE, et al. Simultaneous determination of 17 opium alkaloids and opioids in blood and urine by fast liquid chromatography-diode array detection-fluorescence detection, after solid phase extraction. *J Chromatogr B* 2002; 773: 53–61.

63. Walsh SL, Preston KL, Bigelow GE, et al. Acute administration of buprenorphine in humans: partial agonists and blockade effects. *J Pharmacol Exp Ther* 1995; 274: 361–372.

64. Amass L, Kamien JB, Mikulich SK. Efficacy of daily and alternate dosing regimens with the combination buprenorphine–naloxone tablet. *Drug Alcohol Depend* 2000; 58: 143–152.

65. Vidal-Trecan G, Varescon I, Nabet N, et al. Intravenous use of prescribed sublingual buprenorphine tablets by drug users receiving maintenance therapy in France. *Drug Alcohol Depend* 2003; 69: 175–181.

66. Kintz P, Marquet P, eds. *Buprenorphine Therapy of Opiate Addiction*. Totowa, NJ: Humana Press, 2002.

67. Hoja H, Marquet P, Verneuil B, et al. Determination of buprenorphine and norbuprenorphine in whole blood by liquid chromatography-mass spectrometry. *J Anal Toxicol* 1997; 21: 160–165.

68. Tracqui A, Kintz P, Mangin P. HPLC/MS determination of buprenorphine and norbuprenorphine in biological fluids and hair samples. *J Forensic Sci* 1997; 42: 111–114.

69. Moody DE, Laycock JD, Spanbauer AC, et al. Determination of buprenorphine in human plasma by gas chromatography-positive ion chemical ionization mass spectrometry and liquid chromatography-tandem mass spectrometry. *J Anal Toxicol* 1997; 21: 406–414.

70. Bogusz MJ, Maier RD, Kruger KD, et al. Determination of common drugs of abuse in body fluid using one isolation procedure and liquid chromatography-atmospheric pressure chemical ionization mass spectrometry. *J Anal Toxicol* 1998; 22: 549–558.

71. Moody DE, Slawson MH, Strain EC, et al. A liquid chromatographic-electrospray ionization tandem mass spectrometric method for determination of buprenorphine, it metabolite norbuprenorphine, and a coformulant, naloxone, that is suitable for in vivo and in vitro metabolism studies. *Anal Biochem* 2002; 306: 31–39.

72. Gaulier JM, Marquet P, Lacassie E, et al. Fatal intoxication following self-administration of a massive dose of buprenorphine. *J Forensic Sci* 2000; 45: 226–228.

73. Polettini A, Huestis MA. Simultaneous determination of buprenorphine, norbuprenorphine, and buprenorphine-glucuronide in plasma by liquid chromatography-tandem mass spectrometry. *J Chromatogr B* 2001; 754: 447–459.

74. Kronstrand R, Selden TG, Josefsson M. Analysis of buprenorphine, norbuprenorphine, and their glucuronides in urine by liquid chromatography-mass spectrometry. *J Anal Toxicol* 2003; 27: 464–470.

75. Ceccato A, Klinkenberg R, Hubert P, et al. Sensitive determination of buprenorphine and its N-dealkylated metabolite norbuprenorphine in human plasma by liquid chromatography coupled to tandem mass spectrometry. *J Pharm Biomed Anal* 2003; 32: 619–631.

76. McAleer SD, Mills RJ, Polack T, et al. Pharmacokinetics of high-dose buprenorphine following single administration of sublingual formulations in opioid naïve healthy male volunteers under a naltrexone block. *Drug Alcohol Depend* 2003; 72: 75–83.

77. Dole VP, Nyswander M. A medical treatment for diacetylmorphine (heroin) addiction. *J Am Med Assoc* 1965; 193: 646–650.

78. Newman RG. Methadone: prescribing maintenance, pursuing abstinence. *Int J Addict* 1995; 30: 1303–1309.

79. Heinemann A, Iversen-Bergmann S, Stein S, et al. Methadone-related fatalities in Hamburg 1990–1999; implications for quality standards in maintenance treatment? *Forensic Sci Int* 2000; 113: 449–455.

80. Vormfelde SV, Poser W. Death attributed to methadone. *Pharmacopsychiatry* 2001; 34: 217–222.

81. Kintz P, Eser HP, Tracqui A, et al. Enantioselective separation of methadone and its main metabolite

in human hair by liquid chromatography/ion spray-mass spectrometry. *J Forensic Sci* 1997; 42: 291–295.

82. Ortelli D, Rudaz S, Chevalley AF, et al. Enantioselective analysis of methadone in saliva by liquid chromatography-mass spectrometry. *J Chromatogr A* 2000; 871: 163–172.

83. Dale O, Hoffer C, Sheffels P, et al. Disposition of nasal, intravenous, and oral methadone in healthy volunteers. *Clin Pharmacol Ther* 2002; 72: 536–545.

84. Kelly T, Doble P, Dawson M. Chiral analysis of methadone and its major metabolites (EDDP and EMDP) by liquid chromatography-mass spectrometry. *J Chromatogr B* 2005; 814: 315–323.

85. Souverain S, Eap C, Veuthey JL, et al. Automated LC-MS method for the fast stereoselective determination of methadone in plasma. *Clin Chem Lab Med* 2003; 41: 1615–1621.

86. Liang HR, Foltz RL, Meng M, et al. Method development and validation for quantitative determination of methadone enantiomers in human plasma by liquid chromatography/tandem mass spectrometry. *J Chromatogr B* 2004; 806: 191–198.

87. Whittington D, Sheffels P, Kharash ED. Stereoselective determination of methadone and the primary metabolite EDDP in human plasma by automated on-line extraction and liquid chromatography mass spectrometry. *J Chromatogr B* 2004; 809: 313–321.

88. Breindahl T, Andreasen K. Validation of urine drug-of-abuse testing for ketobemidone using thin-layer chromatography and liquid chromatography-electrospray mass spectrometry. *J Chromatogr B* 1999; 736: 103–113.

89. Sundstrom I, Bondesson U, Hedeland M. Identification of phase I and phase II metabolites of ketobemidone in patient urine using liquid chromatography-electrospray tandem mass spectrometry. *J Chromatogr B* 2001; 763, 121–131.

90. Sundstrom I, Hedeland M, Bondesson U, et al. Identification of glucuronide conjugates of ketobemidone and its phase I metabolites in human urine utilizing accurate mass and tandem time-of-flight mass spectrometry. *J Mass Spectrom* 2002; 37: 414–420.

91. Ceccato A, Vanderbist F, Pabst JY, et al. Automated determination of tramadol enantiomers in human plasma using solid-phase extraction in combination with chiral liquid chromatography. *J Chromatogr B* 2000; 748: 65–76.

92. Juzwin SJ, Wang DC, Anderson NJ, et al. The determination of RWJ-38705 (tramadol-N-oxide) and its metabolites in preclinical pharmacokinetic studies using LC-MS/MS. *J Pharm Biomed Anal* 2000; 22: 469–480.

93. Valiveti S, Nalluri BN, Hammell DC, et al. Determination and validation of a liquid chromatography-mass spectrometry method for the determination of naltrexone and 6β-naltrexol in guinea pig plasma. *J Chromatogr B* 2004; 810: 259–267.

94. Pichini S, Pacifici R, Pellegrini M, et al. Development and validation of a liquid chromatography-mass spectrometry assay for the determination of opiates and cocaine in meconium. *J Chromatogr B* 2003; 794: 281–292.

95. Shulgin A, Shulgin A. *Pihkal: A Chemical Love Story*. Berkeley, CA: 1995.

96. Henry AS, Jeffrys KJ, Dawling S. Toxicity and deaths from 3,4-methylenedioxymethamphetamine ('ecstasy'). *Lancet* 1992; 340: 384–387.

97. Lora-Tamayo A, Tena T, Rodriguez A. Amphetamine derivative related deaths. *Forensic Sci Int* 1997; 85: 149–157.

98. Schifano F, Oyefeso A, Webb L, et al. Review of deaths related to taking ecstasy, England and Wales, 1997–2000. *BMJ* 2003; 326: 80–81.

99. Zhao H, Brenneisen R, Scholer A, et al. Profiles of urine samples taken from ecstasy users at rave parties: analysis by immunoassays. *J Anal Toxicol* 2001; 25: 258–269.

100. Cody JT. Amphetamines. In: Bogusz MJ, ed., *Handbook of Analytical Separations*. Vol 2: *Forensic Science*. Amsterdam: Elsevier, 2000: 107–141.

101. de la Torre R, Farre M, Roset PN, et al. Human pharmacology of MDMA: pharmacokinetics, metabolism and disposition. *Ther Drug Monit* 2004; 26: 137–144.

102. Maurer HH, Kraemer T, Springer D, et al. Chemistry, pharmacology, toxicology, and hepatic metabolism of designer drugs of the amphetamine (ecstasy), piperazine, and pyrrolidinophenone types: a synopsis. *Ther Drug Monit* 2004; 26: 127–131.

103. Farre M, de la Torre R, Mathuna BO, et al. Repeated

doses administration of MDMA in humans: pharmacological effects and pharmacokinetics. *Psychopharmacology* 2004; 173: 364–375.

104. Bogusz MJ, Kala M, Maier RD. Determination of phenylisothiocyanate derivatives of amphetamine and its analogues in biological fluids by HPLC-APCI-MS or DAD. *J Anal Toxicol* 1997; 21: 59–69.

105. Bogusz MJ, Kruger KD, Maier RD. Analysis of underivatized amphetamines and related phenethylamines with high-performance liquid chromatography-atmospheric pressure chemical ionization mass spectrometry. *J Anal Toxicol* 2000; 24: 77–84.

106. Kataoka H, Lord HL, Pawliszyn J. Simple and rapid determination of amphetamine, methamphetamine, and their methylenedioxy derivatives in urine by automated in-tube solid-phase microextraction coupled with liquid chromatography-electrospray ionization mass spectrometry. *J Anal Toxicol* 2000; 24: 257–265.

107. Clauwaert KM, VanBocxlaer JF, De Letter EA, *et al.* Determination of the designer drugs 3,4-methylenedioxymethamphetamine, 3,4-methylenedioxyethylamphetamine, and 3,4-methylenedioxyamphetamine in whole blood and fluorescence detection in whole blood, serum, vitreous humor, and urine. *Clin Chem* 2000; 46: 1968–1977.

108. Mortier KA, Dams R, Lambert WE, *et al.* Determination of paramethoxyamphetamine and other amphetamine-related designer drugs by liquid chromatography/sonic spray mass spectrometry. *Rapid Commun Mass Spectrom* 2002; 16: 865–870.

109. Sato M, Mitsui T, Nagase H. Analysis of benzphetamine and its metabolites in rat urine by liquid chromatography-electrospray ionization mass spectrometry. *J Chromatogr B* 2001; 751: 277–289.

110. Katagi M, Tatsuno M, Miki A, *et al.* Discrimination of dimethylamphetamine and methamphetamine use: simultaneous determination of dimethylamphetamine-*N*-oxide and other metabolites in urine by high-performance liquid chromatography-electrospray ionization mass spectrometry. *J Anal Toxicol* 2000; 24: 354–358.

111. Katagi M, Tatsuno M, Miki A, *et al.* Simultaneous determination of selegiline-*N*-oxide, a new indicator for selegiline administration, and other metabolites in urine by high performance liquid chromatography-electrospray ionization mass spectrometry. *J Chromatogr B* 2001; 759: 125–133.

112. Katagi M, Tatsuno M, Tsutsumi H, *et al.* Urinary excretion of selegiline-*N*-oxide, a new indicator for selegiline administration in man. *Xenobiotica* 2002; 32: 823–831.

113. Wood M, De Boeck D, Samyn N, *et al.* Development of a rapid and sensitive method for the quantitation of amphetamines in human saliva. Presented at 49th ASMS Conference on Mass Spectrometry and Allied Topics, Chicago, IL, 2001.

114. Samyn N, De Boeck G, Wood M, *et al.* Plasma, oral fluid and sweat wipe ecstasy concentrations in controlled and real life conditions. *Forensic Sci Int* 2002; 128: 90–97.

115. Stanaszek R, Piekoszewski W. Simultaneous determination of eight underivatized amphetamines in hair by high-performance liquid chromatography-atmospheric pressure chemical ionization mass spectrometry (HPLC-APCI-MS). *J Anal Toxicol* 2004; 28: 77–85.

116. Cairns T, Hill V, Schaffer M, Thistle W. Amphetamines in washed hair of demonstrate users and workplace subjects. *Forensic Sci Int* 2004; 145: 137–142.

117. Pichini S, Pacifici R, Pellegrini M, *et al.* Development and validation of a high-performance liquid chromatography-mass spectrometry assay for determination of amphetamine, methamphetamine, and methylenedioxy derivatives in meconium. *Anal Chem* 2004; 76: 2124–2132.

118. National Institute of Drug Abuse Research Report. *Cocaine Abuse and Addiction*. NIH publ. no. 99-4342. Bethesda, MD: NIH, 1999.

119. Office of Applied Statistics. *2003 National Survey on Drug Use and Health*. Bethesda, MD: SAMHAS, 2003.

120. El Sohly MA, Brenneisen R, Jones AB. Coca paste: chemical analysis and smoking experiments. *J Forensic Sci* 1991; 36: 93–103.

121. Jufer RA, Darwin WD Cone EJ. Current methods for the separation and analysis of cocaine analytes. In: Bogusz MJ, ed., *Handbook of Analytical Separations*. Vol 2: *Forensic Science*. Amsterdam: Elsevier, 2000: 67–106.

122. Singh G, Arora V, Fenn PT, *et al.* A validated stable isotope dilution liquid chromatography tandem mass spectrometry assay for the trace analysis of

cocaine and its major metabolites in plasma. *Anal Chem* 1999; 71: 2021–2027.

123. Jeanville PM, Estape ES, Needham SR, et al. Rapid confirmation/quantitation of cocaine and benzoylecgonine in urine utilizing high performance liquid chromatography and tandem mass spectrometry. *J Am Soc Mass Spectrom* 2000; 11: 257–263.

124. Jeanville PM, Estape ES, Torres-Negron I, et al. Rapid confirmation/quantitation of ecgonine methyl ester, benzoylecgonine, and cocaine using on-line extraction coupled with fast HPLC and tandem mass spectrometry. *J Anal Toxicol* 2001; 25: 69–75.

125. Jeanville PM, Woods JH, Baird TJ, et al. Direct determination of ecgonine methyl ester and cocaine in rat plasma, utilizing on-line sample extraction coupled with rapid chromatography/ quadrupole orthogonal acceleration time-of-flight detection. *J Pharm Biomed Anal* 2000; 23: 897–907.

126. Skopp G, Klingmann A, Potsch L, et al. In vitro stability of cocaine in whole blood and plasma including ecgonine as a target analyte. *Ther Drug Monit* 2001; 23: 174–181.

127. Klingmann A, Skopp G, Aderjan R. Analysis of cocaine, benzoylecgonine, ecgonine methyl ester, and ecgonine by high pressure liquid chromatography-API mass spectrometry and application to a short-term degradation study of cocaine in plasma. *J Anal Toxicol* 25: 425–430.

128. Lin SN, Moody DE, Bigelow GE, et al. A validated liquid chromatographic-atmospheric pressure chemical ionization tandem-mass spectrometry method for quantitation of cocaine and benzoylecgonine in human plasma. *J Anal Toxicol* 2001; 25: 497–503.

129. Clauwaert KM, Van Bocxlaer JF, Lambert WE, et al. Narrow-bore HPLC in combination with fluorescence and electrospray mass spectrometric detection for the analysis of cocaine and metabolites in human hair. *Anal Chem* 1998; 70: 2336–2344.

130. Sosnoff CS, Ann Q, Bernert JT, Jr, et al. Analysis of benzoylecgonine in dried blood spots by liquid chromatography-atmospheric pressure chemical ionization mass spectrometry. *J Anal Toxicol* 1996; 20: 179–184.

131. Xia Y, Wang P, Bartlett MG, et al. An LC-MS-MS method for the comprehensive analysis of cocaine and cocaine metabolites in meconium. *Anal Chem* 2000; 15: 764–771.

132. Srinivasan K, Wang PP, Eley AT, et al. Liquid chromatography-tandem mass spectrometry analysis of cocaine and its metabolites from blood, amniotic fluid, placenta, and fetal tissues: study of the metabolism and distribution of cocaine in pregnant rats. *J Chromatogr B* 2000; 745: 287–303.

133. Hollister LE. Health aspects of cannabis. *Pharmacol Rev* 1986; 38: 2–20.

134. Tramer MR, Carroll D, Campbell FA, et al. Cannabinoids for control of chemotherapy induced nausea and vomiting: quantitative systematic review. *BMJ* 2001; 323: 16–21.

135. Darmani NA. Delta(9)-tetrahydrocannabinol and synthetic cannabinoids prevent emesis produced by the cannabinoid CB(1) receptor antagonist/ inverse agonist SR 141716A. *Neuropsychopharmacology* 2001; 24: 198–203.

136. Campbell FA, Tramer MR, Carroll D, et al. Are the cannabinoids an effective and safe option in the management of pain? A general systematic review. *BMJ* 2001; 232: 13–16.

137. Williamson EV, Evans FJ. Cannabinoids in clinical practice. *Drugs* 2000; 60: 1303–1314.

138. Müller-Vahl KR, Koblenz A, Jobges A, et al. Influence of treatment of Tourette syndrome with delta9-tetrahydrocannabinol (delta9-THC) on neuropsychological performance. *Pharmacopsychiatry* 2001; 34: 19–24.

139. Martin BR, Jefferson R, Wiley JL et al. Manipulation of the tetrahydrocannabinol side chain delineates agonists, partial agonists, and antagonists. *J Pharmacol Exp Ther* 1999; 290: 1065–1078.

140. Turner CE, ElSohly MA, Boeren EG. Constituents of *Cannabis sativa* L. A review of the natural constituents. *J Nat Prod* 1980; 43: 169–234.

141. Rustichelli C, Ferioli V, Vezzalini F, et al. Simultaneous separation and identification of hashish constituents by coupled liquid chromatography-mass spectrometry (HPLC-MS). *Chromatographia* 1996; 43: 129–136.

142. Bäckström B, Cole MD, Carrott MJ, et al. A preliminary study on the analysis of *Cannabis* by supercritical fluid chromatography with atmospheric pressure chemical ionization mass spectroscopic detection. *Science & Justice* 1997; 37: 91–97.

143. Stolker AAM, van Schoonhoven J, de Vries AJ, et al. Determination of cannabinoids in cannabis products using liquid chromatography-ion trap

mass spectrometry. *J Chromatogr A* 2004; 1058: 143–151.

144. Mireault P. Analysis of Δ⁹-tetrahydrocannabinol and its two major metabolites by APCI-LC/MS. Presented at 46th ASMS Conference on Mass Spectrometry and Allied Topics, Orlando, FL, 1998.

145. Picotte P, Mireault P, Nolin G. A rapid and sensitive LC/APCI/MS/MS method for the determination of Δ⁹-tetrahydrocannabinol and its metabolites in human matrices. Presented at 46th ASMS Conference on Mass Spectrometry and Allied Topics, Long Beach, CA, 2000.

146. Breindahl T, Andreasen K. Determination of 11-nor-delta9-tetrahydrocannabinol-9-carboxylic acid in urine using high-performance liquid chromatography and electrospray ionization mass spectrometry. *J Chromatogr B* 1999; 732: 155–164.

147. Tai SS, Welch MJ. Determination of 11-nor-delta9-tetrahydrocannabinol-9-carboxylic acid in a urine-based standard reference material by isotope-dilution liquid chromatography-mass spectrometry with electrospray ionization. *J Anal Toxicol* 2000; 24: 385–389.

148. Weinmann W, Vogt S, Goerke R, *et al.* Simultaneous determination of THC-COOH and THC-COOH-glucuronide in urine samples by LC/MS/MS. *Forensic Sci Int* 2000; 113: 381–387.

149. Weinmann W, Goerner M, Vogt S, *et al.* Fast confirmation of 11-nor-9-carboxy-Δ⁹-tetrahydrocannabinol (THC-COOH) in urine by LC/MS/MS using negative atmospheric pressure chemical ionization (APCI). *Forensic Sci Int* 2001; 121: 103–107.

150. Maralikova B, Weinmann W. Simultaneous determination of Δ⁹-tetrahydrocannabinol, 11-hydroxy-Δ⁹-tetrahydrocannabinol and 11-nor-9-carboxy-Δ⁹-tetrahydrocannabinol in human plasma by high-performance liquid chromatography/tandem mass spectrometry. *J Mass Spectrom* 2004; 39: 526–531.

151. Skopp G, Pötsch L. An investigation of the stability of free and glucuronidated 11-nor-Δ⁹-tetrahydrocannabinol-9-carboxylic acid in authentic urine samples. *J Anal Toxicol* 2004; 28: 35–40.

152. Concheiro M, de Castro A, Quintela O, *et al.* Development and validation of a method for the quantitation of Δ⁹-tetrahydrocannabinol in oral fluid by liquid chromatography-electrospray mass spectrometry. *J Chromatogr B* 2004; 810: 319–324.

153. Hofmann A. How LSD originated. *J Psychedel Drugs* 1979; 11: 53–60.

154. Baselt RC. Lysergic acid diethylamide. In: *Disposition of Toxic Chemicals and Drugs in Man*, 5th edn. Foster City, CA: Chemical Toxicology Institute, 2000: 486–489.

155. Webb KS, Baker PB, Cassells NP, *et al.* The analysis of lysergide (LSD): the development of novel enzyme immunoassay and immunoaffinity extraction procedures with an HPLC-MS confirmation procedure. *J Forensic Sci* 1996; 41: 938–946.

156. White SA, Kidd AS, Webb KS. The determination of lysergide (LSD) in urine by high-performance liquid chromatography-isotope dilution mass spectrometry (IDMS). *J Forensic Sci* 1999; 44: 375–379.

157. Hoja H, Marquet P, Verneuil B, *et al.* Determination of LSD and N-demethyl-LSD in urine by liquid chromatography coupled to electrospray ionization mass spectrometry. *J Chromatogr B* 1997; 692: 329–335.

158. de Kanel J, Vickery WE, Waldner B, *et al.* Automated extraction of lysergic acid diethylamide (LSD) and N-demethyl-LSD from blood, serum, plasma, and urine samples using the Zymark RapidTrace with LC/MS/MS confirmation. *J Forensic Sci* 1998; 43: 622–625.

159. Bodin K, Svensson JO. Determination of LSD in urine with liquid chromatography-mass spectrometry. *Ther Drug Monit* 2001; 23: 389–393.

160. Nelson CC, Foltz RL. Chromatographic and mass spectrometric methods for determination of lysergic acid diethylamide (LSD) and metabolites in body fluids. *J Chromatogr* 1992; 580: 97–109.

161. Poch GK, Klette KL, Hallare DA, *et al.* The detection of metabolites of LSD in human urine specimens: 2-oxo-3-hydroxy-LSD, a prevalent metabolite of LSD. *J Chromatogr B* 1999; 724, 23–33.

162. Poch GK, Klette KL, Anderson CJ. The quantitation of 2-oxo-3-hydroxy LSD in human urine specimens, a metabolite of LSD: comparative analysis using liquid chromatography-selected ion monitoring mass spectrometry and liquid chromatography-ion trap mass spectrometry. *J Anal Toxicol* 2000; 24: 170–179.

163. Sklerov JH, Magluilo J, Shannon KK, *et al.* Liquid chromatography-electrospray ionization mass

spectrometry for the detection of lysergide and a major metabolite, 2-oxo-3-hydroxy LSD, in urine and blood. *J Anal Toxicol* 2000; 24, 543–549.

164. Canezin J, Cailleux A, Turcant A, et al. Determination of LSD and its metabolites in human biological fluids by high-performance liquid chromatography with electrospray tandem mass spectrometry. *J Chromatogr B* 2001; 765: 15–27.

165. Klette KL, Horn CK, Stout PR, et al. LC-MS analysis of human urine specimens for 2-oxo-3-hydroxy LSD: method validation for potential interferants and stability study for 2-oxo-3-hydroxy LSD under various storage conditions. *J Anal Toxicol* 2002; 26: 193–200.

166. Horn CK, Klette KL, Stout PR. LC-MS analysis of 2-oxo-3-hydroxy LSD from urine using a Speedisk positive-pressure processor with Cerex Poly-Chrom CI II columns. *J Anal Toxicol* 2003; 27: 459–463.

167. Mash DC, Kovera CA, Pablo J, et al. Ibogaine. Complex pharmacokinetics, concerns for safety, and preliminary efficacy measures. *Ann NY Acad Sci* 2000; 914: 394–401.

168. Glick SD, Maisonneuve IM, Szumlinski KK. 18-Methoxycoronaridine (18-MC) and ibogaine: comparison of antiaddictive efficacy, toxicity, and mechanism of action. *Ann NY Acad Sci* 2000; 914: 369–386.

169. Zhang W, Ramamoorthy Y, Tyndale RF, et al. Metabolism of 18-methoxycoroaridine, an ibogaine analog, to 18-hydroxycoronaridine by genetically variable CYP2C219. *Drug Metab Disp* 2002; 30: 663–669.

170. Forstrom T, Tuominen J, Karkkainen J. Determination of potentially hallucinogenic N-dimethylated indoleamines in human urine by HPLC/ESI-MS-MS. *Scand J Clin Lab Invest* 2001; 61: 547–556.

171. Barker SA, Littlefield-Chabaud MA, David C. Distribution of the hallucinogens N,N-diethyltryptamine and 5-methoxy-N,N-dimethyltryptamine in rat brain following intraperitoneal, injection: application of a new solid-phase extraction LC-APCI-MS-MS-isotope dilution method. *J Chromatogr B* 2001; 751: 37–47.

172. Brenneisen R, Stalder AB. Psilocybe. In: Hansel R, Keller K, Romples H, et al., eds, *Hagers Handbuch der Pharmazeutischen Praxis 6*. Berlin: Springer, 1994: 287.

173. Bogusz MJ, Maier RD, Schäfer AT, et al. Honey with Psilocybe mushrooms: a revival of a very old preparation on the drug market? *Int J Legal Med* 1998; 111: 147–150.

174. Kamata T, Nishikawa M, Katagi M, et al. Optimized hydrolysis for the detection of psilocin in human urine samples. *J Chromatogr B* 2003; 796: 421–427.

175. Lechowicz W, Parczewski A, Chudzikiewicz E. Rapid determination of psilocin in whole blood by LC-MS/APCI. Abstracts of the Joint SOFT/TIAFT Meeting, Washington, DC, 2004.

176. LeBeau MA, Mozayani A, eds. *Drug-Facilitated Sexual Assault*. San Diego, CA: Academic Press, 2001.

177. Bogusz MJ, Maier RD, Krüger KD, et al. Determination of flunitrazepam in blood by high-performance liquid chromatography-atmospheric pressure chemical ionization mass spectrometry. *J Chromatogr B* 1998; 713: 361–369.

178. LeBeau MA, Montgomery MA, Wagner JR, et al. Analysis of biofluids for flunitrazepam and metabolites by electrospray liquid chromatography/mass spectrometry. *J Forensic Sci* 2000; 45: 1133–1141.

179. Darius J, Banditt P. Validated method for the therapeutic drug monitoring of flunitrazepam in human serum using liquid chromatography-atmospheric pressure chemical ionization tandem mass spectrometry with a ion trap detector. *J Chromatogr B* 2000; 738: 437–441.

180. Yuan H, Mester Z, Lord H, et al. Automated in-tube solid-phase microextraction coupled with liquid chromatography-electrospray ionization mass spectrometry for the determination of selected benzodiazepines. *J Anal Toxicol* 2000: 24: 718–725.

181. Kollroser M, Schober C. Simultaneous analysis of flunitrazepam and its major metabolites in human plasma by high performance liquid chromatography tandem mass spectrometry. *J Pharm Biomed Anal* 2002; 28: 1173–1182.

182. Marquet P, Baudin O, Gaulier JM, et al. Sensitive and specific determination of midazolam and 1-hydroxymidazolam in human serum by liquid chromatography-electrospray mass spectrometry. *J Chromatogr B* 1999; 734: 137–144.

183. Ware ML, Hidy BJ. LC/MS/MS determination of midazolam and 1-hydroxymidazolam in human plasma. Presented at 49th ASMS Conference on

Mass Spectrometry and Allied Topics, Chicago, IL, 2001.

184. Cheze M, Villain M, Pepin G. Determination of bromazepam, clonazepam and metabolites after a single intake in urine and hair by LC-MS/MS. Application to forensic cases of drug facilitated crimes. *Forensic Sci Int* 2004; 145: 123–130.

185. Duffort G, Cheze M, Pepin G. Drug-facilitated crimes: usefulness of hair analysis by liquid chromatography-tandem mass spectrometry. Poster presented at Joint SOFT/TIAFT Meeting, Washington, DC, 2004.

186. Smink BE, Brandsma JE, Dijkhuizen A, *et al*. Quantitative analysis of 33 benzodiazepines, metabolites and benzodiazepine-like substances in whole blood by liquid chromatography-(tandem) mass spectrometry. *J Chromatogr B* 2004; 811: 13–20.

187. Gaulier JM, Canal M, Pradeille JL, *et al*. New drugs at 'rave parties': ketamine and prolintane. *Acta Clin Belg Suppl* 2002; 1: 41–46.

188. Cheng JYK, Mok VKK. Rapid determination of ketamine in urine by liquid chromatography-tandem mass spectrometry for a high throughput laboratory. *Forensic Sci Int* 2004; 142: 9–15.

189. Smith KM. Drugs used in acquaintance rape. *J Am Pharm Assoc (Wash)* 1999; 39: 519–525.

190. Moore KA, Sklerov J, Levine B, *et al*. Urine concentrations of ketamine and norketamine following illegal consumption. *J Anal Toxicol* 2001; 25: 583–588.

191. Villain M, Cheze M, Tracqui A, *et al*. Windows of detection of zolpidem in urine and hair: application to two drug facilitated sexual assaults. *Forensic Sci Int* 2004; 16: 157–161.

192. Kintz P, Cirimele V, Villain M, *et al*. Testing for zolpidem in oral fluid by liquid chromatography-tandem mass spectrometry. *J Chromatogr B* 2004; 811: 59–63.

193. *Gamma Hydrohybutyrate – Factsheet*. Washington, DC: Office of the National Drug Control Policy, 2002: www.whitehousedrugpolicy.gov/publications/factsht/gamma/ (accessed October 2004).

194. Wood M, Laloup M, Samyn N, *et al*. Simultaneous analysis of gamma-hydroxybutyric acid and its precursors in urine using liquid chromatography-tandem mass spectrometry. *J Chromatogr A* 2004; 1056: 83–90.

195. Ramos L, Bakhtiar R, Majumdar T, *et al*. Liquid chromatography/atmospheric pressure chemical ionization tandem mass spectrometry enantiomeric separation of *dl-threo*-methylphenidate (Ritalin®) using a macrocyclic antibiotic as the chiral selector. *Rapid Commun Mass Spectrom* 1999; 13: 2054–2062.

196. Ramos L, Bakhtiar R, Tse F. Liquid–liquid extraction using 96-well plate format in conjunction with liquid chromatography/tandem mass spectrometry for quantitative determination of methylphenidate (Ritalin®) in human plasma. *Rapid Commun Mass Spectrom* 2000; 14: 740–745.

197. Oertel R, Richter K, Ebert U, *et al*. Determination of scopolamine in human serum and microdialysis samples by liquid chromatography-tandem mass spectrometry. *J Chromatogr B* 2002; 750: 121–128.

198. Goldmeier D, Lamba H. Prolonged erections produced by dihydrocodeine and sildenafil. *BMJ* 2002; 324: 1555.

199. Peterson KS. Young men add Viagra to their arsenal. *USA Today* 2001; 21 March.

200. Weinmann W, Lehmann N, Miller C, *et al*. Postmortem detection and identification of sildenafil (Viagra) and its metabolites by LC/MS and LC/MS/MS. *Int J Legal Med* 2001; 114: 252–258.

200. Dumestre-Toulet V, Cirimele V, Gromb S, *et al*. Last performance with Viagra: post-mortem identification of sildenafil and its metabolites in biological specimens including hair sample. *Forensic Sci Int* 2002; 126: 71–76.

202. Eerkes A, Addison T, Naidong W. Simultaneous assay of sildenafil and desmethylsildenafil in human plasma using liquid chromatography-tandem mass spectrometry on silica column with aqueous–organic mobile phase. *J Chromatogr B* 2002; 768: 277–284.

203. Heller DN. Liquid chromatography/mass spectrometry for timely response regulatory analyses: identification of pentobarbital in dog food. *Anal Chem* 2000; 72: 2711–2716.

204. Peel HW, Jeffrey WK. A report on the incidence of drugs and driving in Canada. *Can Soc Forensic Sci J* 1990; 23: 75–79.

205. Schiwy-Bochat KH, Bogusz M, Vega JA, *et al*. Trends in occurrence of drugs of abuse in blood and urine of arrested drivers and drug traffickers in the border region of Aachen. *Forensic Sci Int* 1995; 71: 33–41.

206. De Zeeuw RA. Substance identification: the weak link in analytical toxicology. *J Chromatogr B* 2004; 811: 3–12.

207. Bogusz MJ, Maier RD, Krüger KD, et al. Poor reproducibility of in-source collisional atmospheric pressure ionization mass spectra of toxicologically relevant drugs. *J Chromatogr B* 1999; 844: 409–418.

208. Cailleux A, Le Bouil A, Auger B, et al. Determination of opiates and cocaine and its metabolites in biological fluids by high performance liquid chromatography with electrospray tandem mass spectrometry. *J Anal Toxicol* 1999; 23: 620–624.

209. Rittner M, Pragst F, Bork WR, et al. Screening method for seventy psychoactive drugs or drug metabolites in serum based on high-performance liquid chromatography-electrospray ionization mass spectrometry. *J Anal Toxicol* 2001; 25: 115–124.

210. Maralikova B, Weinmann W. Confirmatory analysis for drugs of abuse in plasma and urine by high-performance liquid chromatography-tandem mass spectrometry with respect to criteria for compound identification. *J Chromatogr B* 2004; 811: 21–30.

211. Commission Decision of 12 August 2002 implementing Council Directive 96/23/EC concerning the performance of analytical methods and the interpretation of results. *Official J European Union* 2002; L 221/8.

212. Gao VCX, Zhang Y, Liang C, et al. Simultaneous determination of 19 illicit drugs by LC/MS/MS. Presented at 52nd ASMS Conference on Mass Spectrometry and Allied Topics, Nashville, TN, 2004.

213. Mortier KA, Clauwaert KM, Lambert WE, et al. Pitfalls associated with liquid chromatography/electrospray tandem mass spectrometry in quantitative bioanalysis of drugs of abuse in saliva. *Rapid Commun Mass Spectrom* 2001; 15: 1773–1775.

214. Mortier K, Maudens K, Lambert W, et al. Simultaneous, quantitative determination of opiates, amphetamines, cocaine and benzoylecgonine in oral fluid by liquid chromatography quadrupole time-of-flight mass spectrometry. *J Chromatogr B* 2002; 779: 321–330.

215. Dams R, Murphy CM, Choo RE, et al. LC-atmospheric pressure chemical ionization-MS/MS analysis of multiple illicit drugs, methadone, and their metabolites in oral fluid following protein precipitation. *Anal Chem* 2003; 75: 798–804.

216. Jenkins KM, Young MS, Wood M. Solid phase extraction and tandem LC/MS determination of drugs of abuse in preserved saliva. Presented at 52nd ASMS Conference on Mass Spectrometry and Allied Topics, Nashville, TN, 2004.

217. Scheidweiler KB, Huestis M. Simultaneous quantification of opiates, cocaine, and metabolites in hair by LC-APCI-MS/MS. *Anal Chem* 2004; 76: 4358–4363.

218. Kronstrand R, Nyström I, Strandberg J, et al. Screening for drugs of abuse in hair with ion spray LC-MS-MS. *Forensic Sci Int* 2004; 145: 183–190.

219. Marquet P, Dupuy JL, Lachaitre G. Development of a general unknown screening procedure using liquid chromatography-ionspray-mass spectrometry. Presented at 46th ASMS Conference on Mass Spectrometry and Allied Topics, Orlando, FL, 1999.

220. Marquet P, Venisse N, Lacassie E, et al. In-source CID mass spectral libraries for the 'general unknown' screening of drugs and toxicants. *Analusis* 2000; 28; 41–54.

221. Venisse N, Marquet P, Duchoslav E, et al. A general unknown procedure for drugs and toxic compounds in serum using liquid chromatography-electrospray-single quadrupole mass spectrometry. *J Anal Toxicol* 2003; 27: 7–14.

222. Saint-Marcoux F, Lachatre G, Marquet P. Evaluation of an improved general unknown screening procedure using liquid chromatography-electrospray mass spectrometry by comparison with as chromatography and high performance liquid chromatography-diode array detection. *J Am Soc Mass Spectrom* 2003; 14: 14–22.

223. Marquet P, Saint-Marcoux F, Gambie TN, et al. Comparison of a preliminary procedure for the general unknown screening of drugs and toxic compounds using a quadrupole-linear ion-trap mass spectrometer with a liquid chromatography-mass spectrometry reference technique. *J Chromatogr B* 2003; 789: 9–18.

224. Weinmann W, Wiedemann A, Eppinger B, et al. Screening for drugs in serum by electrospray ionization/collision-induced dissociation and library searching. *J Am Soc Mass Spectrom* 1999; 10: 1028–1037.

225. Weinmann W, Stoertzel M, Vogt S, et al. Tune compounds for electrospray ionization/in source collision-induced dissociation with mass spectra library searching. *J Chromatogr A* 2001; 926: 199–209.

226. Weinmann W, Stoertzel M, Vogt S, *et al*. Tuning compounds for electrospray ionization/in source collision-induced dissociation and mass spectra library searching. *J Mass Spectrom* 2001; 36: 1013–1023.

227. Mueller C, Schaefer P, Vogt S, *et al*. Combining an ESI-CID mass spectra and a UV-spectra library of drugs with an Access database for clinical and forensic-toxicological analysis. Presented at 50th ASMS Conference on Mass Spectrometry and Allied Topics, Orlando, FL, 2002.

228. Lips AG, Lameijer W, Fokkens RH, *et al*. Methodology for the development of a drug library based upon collision-induced fragmentation for the identification of toxicologically relevant drugs in plasma samples. *J Chromatogr B* 2001; 759: 191–207.

229. Gergov M, Boucher B, Ojanpera I, *et al*. Toxicological screening of urine for drugs by liquid chromatography/time-of-flight mass spectrometry with automated target library search based on elemental formulas. *Rapid Commun Mass Spectrom* 2001; 15: 521–526.

230. Pelander A, Ojanpera I, Gergov M, *et al*. Qualitative screening analysis of autopsy urine samples by improved LC/TOFMS method. Poster at 40th TIAFT Meeting, Paris, 2002.

231. Weinmann W, Gergov M, Goerner M. MS/MS-libraries with triple quadrupole-tandem mass spectrometers for drug identification and drug screening. *Analusis* 2000; 28; 934–948.

232. Gergov M, Ojanpera I, Vuori E, Qualitative screening for blood for 238 therapeutic and illegal drugs using liquid chromatography/tandem mass spectrometry. *J Chromatogr B* 2003; 795: 41–53.

233. Gergov M, Weinmann W, Meriluoto J, *et al*. Comparison of product ion spectra obtained by liquid chromatography/triple quadrupole mass spectrometry for library search. *Rapid Commun Mass Spectrom* 2004; 18: 1039–1046.

9

LC-MS in doping control

Detlef Thieme

Introduction

Definition of doping

Doping analysis comprises a diversity of substance classes with different pharmaceutical and chemical properties. Therefore, the discussion of the suitability of liquid chromatography-mass spectrometry (LC-MS) in doping analysis needs to distinguish various categories.

According to its formal definition, a doping violation in sports can be caused by various events, e.g.:

- the detection of a prohibited substance or metabolites or markers of that substance (as defined by the recent document [1] of the World Anti-Doping Agency [WADA]) in the athlete's specimen
- the use of prohibited substances or methods
- possession or trafficking prohibited substances
- refusing without compelling justification to submit a sample.

This definition is clearly legally motivated and does not support the discussion of technical issues.

The number and classes of prohibited substances is very complex in human sports, where selected stimulants, narcotics, hormones, β_2-agonists, anti-oestrogenic agents and diuretics are covered. The situation becomes even more confusing if the term 'doping' is extended to animal (e.g. equestrian) sports, where any application of pharmaceutical drugs is totally prohibited and even substance classes like muscle relaxants or mild stimulants are included.

Moreover, doping analysis is closely related to adjacent fields with similar analytical prospects, like veterinary residue control (predominantly dealing with identification of growth promoters in various matrices), forensic sciences (the majority of doping-relevant substances are scheduled as controlled substances in most countries), environmental analysis (e.g. steroids in waste water) or clinical chemistry (e.g. due to the increasing relevance of steroid hormone replacement therapy).

This chapter describes the key fields of application of LC-MS in routine doping control (i.e. screening analysis, confirmation and quantification of positive results) extra to particular research activities. The latter are focused on the intended technical improvements (e.g. extension of detection time windows, reduction of turn-around times and costs) of conventional analytical procedures and, in particular, on the detection of prohibited substances that cannot be adequately identified so far (e.g. growth hormone).

The arrangement follows mainly historical and technical considerations, and does not necessarily represent the frequency or relevance of the application of LC-MS.

LC-MS in doping control – historical and technical aspects

Some peculiar legal and technical principles in doping control influence analytical strategies in athletes drug testing:

- The substance-based doping definition prioritises target analyses compared with general unknown screening procedures. Sensitive and specific selected ion monitoring (SIM) or

selected reaction monitoring (SRM) experiments are much more frequent than scanning experiments.
- Urine, which is the preferred specimen for doping control due to the ease of sample collection and relatively high concentrations of xenobiotics, requires a careful consideration of substance metabolism, including conjugation. Minor biochemical pathways leading to long-term metabolites are often more important than active parent compounds. The relevance of quantitative analyses is reduced to a few 'threshold substances'. This group comprises compounds that are accepted below certain threshold concentrations, because low amounts may be due to a permitted administration (e.g. inhalation of salbutamol) or an endogenous origin of the substance (e.g. natural levels of testosterone). In general, qualitative substance identification is a sufficient proof of a doping offence.
- The differentiation between substance prohibition 'in competition' and 'out of competition' requires modified analytical procedures with respect to numbers of included substances and threshold concentrations.
- A major analytical challenge consists in the verification of the prohibited administration of endogenous substances like testosterone, human growth hormone (hGH) or erythropoietin (EPO). In such cases, minor quantitative (e.g. amount of steroids compared with endogenous precursors or biochemical byproducts) or qualitative (e.g. glycosylation of proteins) deviations need to be identified.
- MS plays an outstanding role in doping analysis and was originally considered as a mandatory analytical technique for confirmation of substance identity. Exceptions were later acknowledged in the field of peptide hormones.

Approaches to the application of an LC-MS coupling in the framework of doping control were already reported in 1981 [2], when a combination of LC-MS equipped with a moving belt was used for MS confirmation of corticosteroids. The relevance of LC-MS application in doping control was ruled by practical demands, resulting in an early implementation of the technique in anti-doping research and routine [3] analyses. The issue of peptide hormones was already tackled in the mid-1990s, because gas chromatography (GC)-MS analysis could not solve the problem of identification of macromolecular compounds. The potential of LC-MS to differentiate intact growth hormone obtained from different manufacturers, quantify the insulin-like growth factor (IGF-1) and characterise human chorionic gonadotrophin (hCG) after tryptic digestion had already been reported in 1994 [4, 5]. However, these 'proofs of principles' demonstrate the general usefulness of LC-MS for the identification of peptide hormone doping, but are not used routinely to date, mainly due to sensitivity limitations.

The subsequent developments were mainly characterised by practical improvements. Substances (e.g. mesocarb [6]) and substance groups (e.g. diuretics [7]) causing severe analytical problems (stability) or inconvenience (time-consuming derivatisation reactions) were covered by efficient LC-MS assays, while other screenings remained unchanged, due to the availability of well-established and validated GC-MS procedures.

In contrast, there is an obvious preference to use LC-MS in cases of upcoming new substances (like the 'designer steroid' tetrahydrogestrinone [THG] and the stimulant modafinil) or substance groups (e.g. corticosteroids, included in the list of prohibited substances in 2003).

There is no preferred default LC-MS instrumentation in doping control analyses. Almost any technical variant of ionisation – electrospray ionisation (ESI), atmospheric pressure chemical ionisation (APCI) or photo-ionisation (APPI) – has been applied in combination with quadrupole (Q), ion trap (IT) or time-of-flight (TOF) mass analysers, whether as single or tandem mass spectrometers.

Certain reports of related technical developments of LC-MS seem to be just coincidentally associated with doping, e.g.:

- The introduction of isotope ratio MS linked to LC enables the identification of the origin of substances. In particular, a differentiation of endogenous production from synthetic material becomes possible in principle. However,

applications presented so far [8] are rather insensitive (requiring 400 ng substance on-column) and therefore of no practical value in routine cases.
- The introduction of Fourier-transform ion cyclotron resonance (FTICR) MS to identify corticosteroids [9] is probably technically motivated. The high expense of this technique does not permit routine applications. Nevertheless, it is clear that high-resolution (HR)-MS is essential for identification of multiply charged intact peptide hormones (hGH) [10]. Affordable routine instruments could greatly improve the detection and characterisation of proteins.

Additionally, progress in chromatographic separation (e.g. column switching [11], use of graphitised carbon [12] or chiral [13] columns) needs to be achieved, particularly in the field of peptide hormones. The improvement of ionisation efficiency appears to be a crucial aspect of steroid analysis. Derivatisation (dansylation [14]) of steroids and attempts to improve APPI ionisation (e.g. using anisol as dopant gas) are both specifically focused on a sensitivity enhancement [15].

Small molecules

The majority of doping-relevant substances belong to the category of small molecules. The most common definition of substance groups is based on pharmaceutical activity, distinguishing stimulants, narcotics, cannabinoids, anabolic agents, β-agonists, anti-oestrogens, masking agents and glucocorticoids. However, this schedule is not suitable for analytical consideration as structurally and analytically unrelated species may be arbitrarily grouped together (e.g. clenbuterol and testosterone as anabolic agents) because of their equal intended activity. However, one substance may be scheduled in different groups. The $β_2$-agonist salbutamol, for instance, may be considered as a permitted anti-asthmatic, as a stimulant or as an anabolic agent, depending on the occasion of the doping control (in/out of competition) and on its urinary concentration. The group of masking agents is rather complex too. It summarises any substance that may interfere with doping analysis by:

- diluting the urine and accelerating excretion (diuretics)
- suppression of reabsorption of xenobiotics (uricosurics, e.g. probenicid)
- manipulation of endogenous steroid profiles (administration of epitestosterone, used as an endogenous reference of urinary steroid concentrations, is able to conceal elevated levels of testosterone).

Diuretics

The main motivation to prohibit the use of diuretics in sports is the intended reduction of body mass in weight-classified sports. The second reason is a masking effect. Due to forced diuresis, the clearance of prohibited substances (e.g. anabolic steroids) may be accelerated and the urinary concentration may drop below the detection limit or threshold. Different types of diuretics (thiazide-like, loop and potassium-sparing diuretics) may be distinguished on the basis of their pharmacological properties (Table 9.1). However, these classes are not differentiated with regard to their doping relevance. The first two groups are supposed to be most efficient in doping because of their high potency. They are characterised by acidic groups (acid amides, typically sulphonamides) and are therefore suitable for negative ionisation. In contrast to these compounds, the group of potassium-sparing diuretics is characterised by steroid structures (aldosterone antagonists, e.g. canrenone) or cycloamidine structures (triamterene), and is more appropriate to protonation and subsequent detection in positive-ionisation mode.

Other compounds with diuretic (side) effects, like osmodiuretics (mannitol) or xanthine derivatives (caffeine), are no longer mentioned in the list of doping substances, because prohibited use could not be certainly discriminated from their permitted applications as food ingredients.

The identification of diuretics in doping control was originally carried out by GC-MS and LC-UV diode array detection (LC-DAD) [7, 16]. The reduced polarity of the methyl derivatives

Table 9.1 Typical classes and examples of prohibited diuretics

Class	Typical modifications	Examples	Chemical structure
Thiazide	R_1 = H, alkyl, subst. phenyl R_2 = Cl, CF_3	benzthiazide hydrochlorothiazide	
Thiazide analogues	R_1 = subst. amide R_2 = H, OH	mefruside xipamide	
Loop diuretics (furosemide type)	R_1 = Cl, phenoxy R_2 = amine	furosemide bumetanide	
Potassium-sparing diuretics, aldosterone antagonists		canrenone spironolactone	
Cycloamidine derivatives		triamterene amiloride	

permitted a GC separation and subsequent identification of the majority of diuretics in SIM mode. This approach includes analytical limitations, in addition to the requirement for a time-consuming derivatisation step. Certain substances do not form stable, reproducible and uniform methyl derivatives. Chromatographic artefacts (Figure 9.1) need to be factored into the screening [17] and there remained at least one diuretic (benzthiazide) undetectable in GC-MS. Therefore, LC-DAD was additionally applied as a complementary analytical technique. The obvious benefits of LC-ESI-MS are revocation of derivatisation, improvement of comprehensiveness, reduction of turn-around times and increase of sensitivity [3, 18]. As stated earlier, there are strongly acidic as well as basic compounds among the class of diuretics, requiring either negative- or positive-ionisation modes, respectively. Basic diuretics may well be combined with other screening procedures (e.g. anabolic steroids) to avoid a re-injection of samples in different ionisation modes. Alternatively, the option of a scan-to-scan polarity switching was utilised to detect both groups of diuretics simultaneously [19, 20].

The relatively high urinary concentrations combined with the sensitivity of the technique reduces chromatographic separation to a minimum and diminishes turn-around times to a

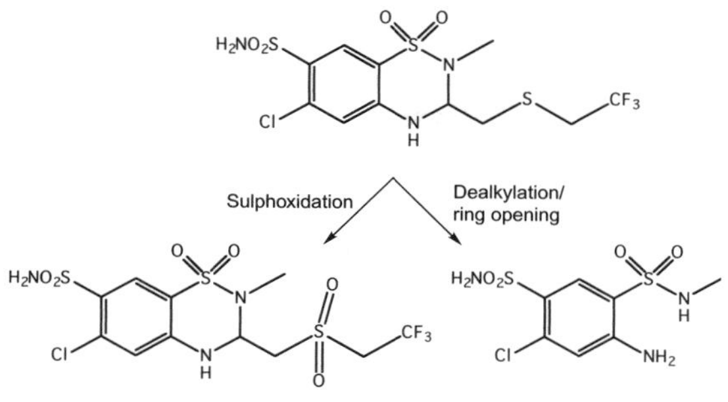

Figure 9.1 Diuretics with thiazide structure are converted by hydrolysis or sulphoxidation during sample transportation, storage or preparation, which may cause analytical problems in GC-MS. Respective artefacts need to be factored into LC-MS screening.

few minutes. A combination with automated solid-phase extraction (SPE) may be applied to establish high-throughput LC-MS screening [21]. The assay is sufficiently selective to differentiate concentrations above a minimum required performance level (MRPL [17]) of 250 µg/L from a blank matrix (Figure 9.2). The potential influence of ion suppression is not critical, because there are no threshold values and the quantification of diuretics is not required in screening analyses.

β_2-Agonists

The classification of these sympathomimetic agents in doping control has been frequently modified. At present, there is a differentiation between substances that are permitted by inhalation to treat asthma (requiring a therapeutic use exemption), while others are prohibited due to their potential stimulating or anabolic effect. Clenbuterol, which is supposed to be the most potent anabolic β-agonist, constituted a particular analytical challenge as typical urinary concentrations are of the order of magnitude of 1 µg/L. Due to its two chlorine atoms, it is well suited for HR-MS and its trimethylsilyl (TMS) derivative was preferably identified by either GC-HR-MS or GC-MS-MS. Typical LC-MS screening procedures are slightly less sensitive than GC-MS procedures, but more comprehensive, which appears to be more important in the field of horse testing [22] because of the general prohibition of all β_2-agonists, regardless of their intended action, concentration or administration route. Typically, β_2-agonists (Table 9.2) are identified in positive ESI mode using the protonated molecule as precursor ion. Characteristic fragmentation reactions are losses of the terminal isobutene group, resulting in $[M - 56]^+$ fragments, whether or not combined with losses of water [23].

Approaches to differentiate between inhalational and prohibited systemic (e.g. oral) application of salbutamol were based on quantitative examinations (values greater than 1 mg/L were, according to the WADA regulations, considered as an adverse finding) or investigations of salbutamol enantiomers. A discrimination function was derived from the higher amounts of the $S(+)$ relative to $R(-)$ isomer after oral administration. This approach requires the combination of chiral LC separation with MS detection, whether on-line or off-line [13].

β-Blockers

β-Adrenergic blocking agents (β-blockers) are prohibited due to their reduction of heart rate, blood pressure and hand tremor. Doping controls are consequently restricted to competition controls in particular sports where steadiness is important (archery, shooting, etc.).

β-Blockers are characterised by a very similar chemical structure. With a few exceptions (e.g. sotalol or carvedilol), they represent derivatives of oxypropanolamine terminated by t-butyl or isopropyl groups and aromatic substituents (Table 9.3).

GC-MS was the conventional screening technique for β-blockers in doping control. Derivatisation with N-methyl-N-trimethylsilyltrifluoroacetamide (MSTFA), if necessary combined with N-methyl-bis-trifluoroacetamide (MBTFA), was applied after hydrolysis of the conjugates and isolation from urine [24]. The application of LC-MS represents a useful alternative to avoid this time-consuming derivatisation step, and the formation of unstable derivatives and artefacts in some cases (e.g. acebutolol). All β-blockers contain a secondary amino group accounting for

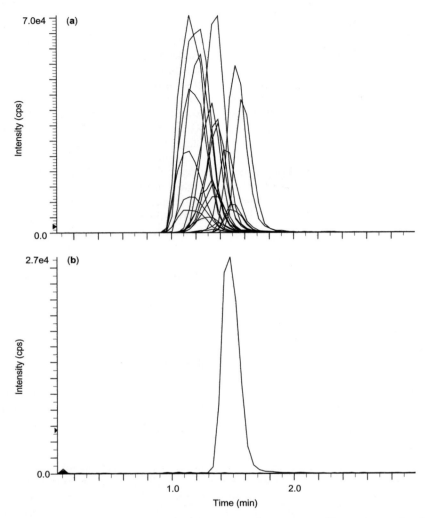

Figure 9.2 Multiple-reaction monitoring (MRM) experiments permit the screening of low amounts of prohibited diuretics in negative-ionisation mode. The turnaround time and sample preparation may be reduced to a minimum due to the outstanding selectivity. A mixture of 20 diuretics extracted from a urine matrix (a) is compared with a blank urine containing mefruside as internal standard (IS) (b).

their protonation and detection in positive mode. According to their high structural similarity, there are group-specific fragmentation reactions for both classes of β-blockers (Table 9.3). Substances with a terminal *t*-butyl group are characterised by loss of isobutene (M − 56), usually in combination with loss of water, while most of isopropyl terminated compounds undergo a loss of an isopropylamino group (M − 77) in addition to the formation of an *N*-isopropyl-propanolamine fragment (mass-to-charge [*m*/*z*] ratio 116) [25].

The proposal of a combined screening of diuretics and β-blockers agents, utilising a scan-to-scan polarity-switching technique [19], is very promising for clinical purposes, because both substance groups are frequently combined in the treatment of hypertension. However, this

Table 9.2 Chemical structures of typical β-agonists

Class	Typical modifications	Examples	Chemical structure
Aniline	R_1, R_2 = Cl, Br, CN, OH R_3 = C(CH$_3$)$_3$, subst. phenyl	clenbuterol brombuterol fenoterol	
Phenol	R_1 = OH, subst. alkyl R_2 = alkyl, subst. phenyl	salbutamol orciprenaline	
Benzazepinone		zilpaterol	

Table 9.3 Chemical structures of typical β-blocking agents prohibited in particular sports

Class	Modifications	Examples	Chemical structure
Isopropylamine	R_1 = substituted aromatic rings	atenolol acebutolol propranolol	
t-Butylamine	R_1 = substituted aromatic rings	bupranolol carteolol timolol	

approach seems to be less rational in doping control, because the intention of an abuse of both substance classes and the scope of their prohibition (concerning sports, in competition versus out of competition) are different.

Steroids

This substance class is characterised by a uniform structure consisting of a modified sterane skeleton (Table 9.4). Due to this apolar structure, the efficiency of ionisation and suitability of LC-MS mainly depend on molecular substitutions, resulting in tremendous variations of sensitivity among different steroids. Saturated steroid molecules (e.g. by saturation of rings or reduction of keto groups) may hardly be ionised by protonation or formation of ammonium adducts. The presence of conjugated double bonds (3-keto-4-ene, steroids), oxidation of the sterane moiety or other polar ring substitutions (e.g. the pyrazol ring in stanozolol) improves the ionisation rate significantly. Therefore, this apparently uniform substance class needs to be divided into the following groups accounting for their LC-MS properties.

Synthetic anabolic steroids

The group of anabolic steroids still includes a diversity of similar structures, which cannot be systematically separated. The subgroups listed in Table 9.4 (biologically active anabolic steroids, precursors and metabolites) overlap each other from legal as well as biochemical perspectives. These compounds (e.g. androstenedione) may be considered as precursors and metabolites in the biosynthesis of steroids, but they also originate from prohibited application of synthetic analogues (pro-hormones) of endogenous steroids or even from synthetic hormones. The analytical result of a urine analysis does not necessarily allow the differentiation of an approved medication, the abuse of pro-hormones (legally

Table 9.4 Chemical structures of selected endogenous and synthetic steroids

Class	Typical modifications	Example	Chemical structure
Endogenous steroids (including their precursors and metabolites)	oxidation/reduction in positions 3/17, saturation of 4 double bond, 5-α/β isomers	testosterone	
Synthetic steroids	17α alkylation, double bonds at position 1–2, 4–5 or 5–6, A-ring condensation, 4-chlorination, 19-demethylation	oxandrolone	
Esters	acetate, propionate, decanoate	testosterone esters	

tolerated in certain countries) or the application of scheduled prohibited steroids.

The attempts to develop LC-MS methods for the identification of anabolic steroids of the so-called 'free fraction' (i.e. slightly polar steroids which are excreted unconjugated in urine) demonstrated significant diversity of ionisation principles [26]. All anabolic steroids are detected in positive mode. However, the appearance of the most abundant precursor ion depends significantly on structural modifications of the sterane skeleton and is almost unpredictable. Protonated ions $[M + H]^+$, ammonium adducts $[M + NH_4]^+$ and fragments resulting from loss of up to three molecules of water $[M - nH_2O]^+$ were reported as base peaks. The balance between these ions is determined by proton affinity and therefore conjugated double bonds (3 keto-4-ene steroids, pyrazol ring condensation, aromatic rings) are the most obvious structural indicators for an improved protonation. Physical conditions in the ion source appear to represent another sensitive factor of ionisation efficiency. Comparing ESI, APCI and APPI for various anabolic steroids, it turned out that the choice of a precursor depends significantly on the ionisation technique. Oxandrolone (Table 9.4), for example, formed predominantly the $[M + H]^+$ ion in ESI, the adduct $[M + NH_4]^+$ in APCI and the $[M - H_2O]^+$ fragment in APPI [27].

Moreover, there is no general consensus about the preferences of various ionisation techniques for steroid analysis (Figure 9.3). Controversial evaluations of APPI [27, 28] suggest that technical differences between manufacturers' concepts are more significant than physical constraints. The typical acquisition mode of these methods is MRM, including two to three fragmentation reactions per analyte. LC-MS has contributed to the structural elucidation in the case of the new upcoming anabolic steroid tetrahydrogestrinone (THG) (Figure 9.3) [29] and provided an efficient complementary screening method, directed to the identification of polar steroids (e.g. trenbolone, THG, stanozolol).

However, unpredictable formation of precursor ions, relatively low ion abundances, and unspecific fragmentation reactions in combination with the existence of numerous isomeric steroids and metabolites complicate the design of comprehensive and sensitive LC-MS methods for the identification of anabolic steroids. Application in routine steroid analysis is mainly focused on selected polar steroid molecules (Figure 9.4), e.g. stanozolol [26, 30], boldenone [31], trenbolone [32, 33] or THG [29]. These polar steroids often encounter analytical difficulties in GC-MS, due to the formation of instable derivatives and artefacts after silylation. The application of an additional derivatisation (e.g. methoxime derivatives of THG) followed by an extra GC-MS procedure would be required to identify these substance groups. Therefore, LC-MS is a beneficial alternative to the extra effort of additional derivatisations or modified GC techniques.

Endogenous steroids

The problem of endogenous steroids in drug testing human athletes is mainly reduced to the quantitative balance between testosterone and its biochemical byproduct epitestosterone (see 'Steroid conjugates'). In addition, there is an upcoming interest in the quantification of endogenous steroids related to the therapeutic administration of steroids (treatment of testosterone deficiency [34]) and to their increasing relevance as lifestyle drugs. Therefore, the quantitative evaluation of endogenous steroid profiles using LC-MS-MS is of increasing importance in clinical chemistry.

The growing number of precursors of endogenous anabolic steroids (so-called prohormones) that are widely available as nutrition supplements and abused in sports and bodybuilding constitutes another analytical challenge. Biochemical precursors of testosterone (e.g. androstenedione, dehydroepiandrosterone) were the first products on the market, but they have been recently replaced by synthetic steroids, representing slight structural modifications of the endogenous compounds (e.g. 1-testosterone, 1,5α-androstenedione, where the location of the double bond is shifted from the 4 to the 1 position). Analytical problems to identify these compounds by LC-MS are comparable to those encountered for their endogenous counterparts [35]. The technique is probably not sufficient for the unambiguous identification of all potentially

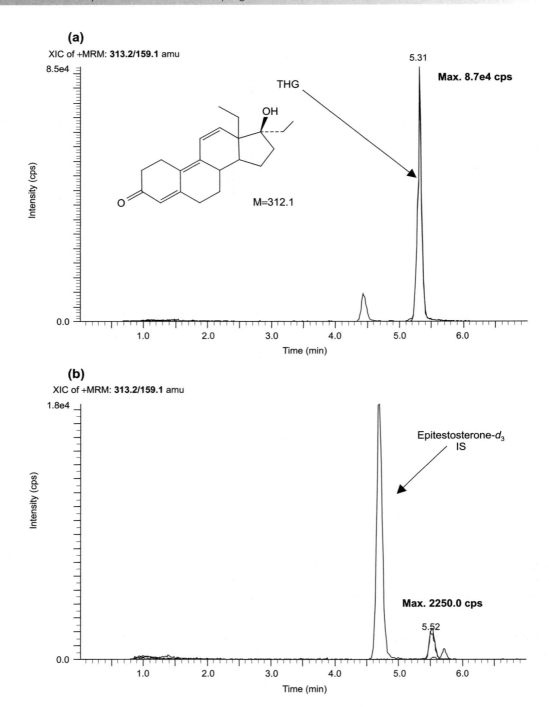

Figure 9.3 Comparison of LC-MS-MS detection of epitestosterone and THG in (a) APPI and (b) ESI. Both chromatograms were run under identical LC conditions. APPI shows an outstanding sensitivity for detection of THG, but is limited to steroids with chromophoric molecular structures, while ESI appears to be the most versatile ionisation technique.

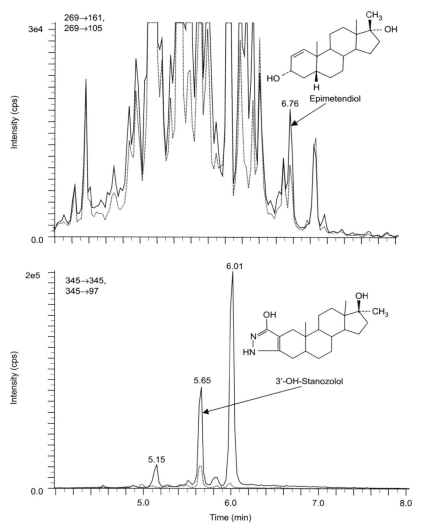

Figure 9.4 Threshold concentrations of 2 μg/L of 3'-OH-stanozolol and epimetendiol in extracts from a urinary matrix. The polar stanozolol metabolite is exceptionally well ionised and permits a sensitive identification, while detection of the medium polar epimetendiol is hampered by a limited intensity and specificity.

relevant steroids, but represents a helpful supplement to GC-MS.

Steroid esters

Steroids are typically administered as fatty acid esters by intramuscular injection. These compounds are stored in adipose tissues, rapidly hydrolysed in blood, not markedly excreted in urine as such and, therefore, not included in routine doping analysis.

The general benefit of the identification of steroid esters is an unequivocal proof of illegal administration of a synthetic compound. Any trace amount of the exogenous esters provides a clear indication for a doping offence, while the corresponding free steroids need to be distinguished from natural endogenous levels.

LC-APCI-MS-MS was reported to facilitate a sensitive identification of testosterone [36] or 19-nortestosterone esters [37] in equine plasma. The formation of dominant $[M + H]^+$ precursor ions was reported for all steroid esters.

Steroid conjugates

The detection of the abuse of endogenous steroids (testosterone or its precursors) is based on a quantitative evaluation of the urinary concentrations of testosterone and epitestosterone. The latter steroid is a byproduct of the biosynthesis of steroids and supposed to be suppressed after the administration of steroids. According to WADA criteria, further investigations to exclude physiological deviations are mandatory, if the ratio of testosterone/epitestosterone exceeds a value of 6. This definition is mainly empirical and derived from the conventional analytical GC-MS procedure, which was traditionally based on the quantification of bis-TMS derivatives of both epimers. Hydrolysis of the conjugates (mainly glucuronides and sulphates, Table 9.5) was carried out after hydrolysis with glucuronidase from *Helix pomatia*, which does not cleave the corresponding sulphates. Logically, the testosterone/epitestosterone ratios are by consensus referred to the total amount of free and glucuronidated steroids. It was suggested that elevated relative amounts of epitestosterone sulphate may cause increased testosterone/epitestosterone ratios when measured by conventional GC-MS. A potential racial bias of phase 2 biotransformation was discussed.

The quantification of intact conjugates, which is made possible by LC-MS, seems to be a conclusive option to examine the individual influence of glucurono- and sulpho-conjugation on testosterone/epitestosterone ratios directly and circumvent the uncertainty of hydrolysis recovery [38, 39]. Steroid glucuronides were reported to form different adducts (protonated, ammoniated and sodiated ions) in positive ESI. Depending on the declustering potentials, the $[M + H]^+$ was found to be most suitable due to its high signal-to-noise (S/N) ratios. Reasonably specific fragments (aglycone and its singly or doubly dehydrated fragment ions) were chosen for sensitive MRM experiments. The use of a deuterated IS for each

Table 9.5 Chemical structures of steroid conjugates

Conjugation	Examples	Chemical structure
Glucuronides	androsterone glucuronide	
Sulphates	testosterone sulphate	

conjugate was found to be mandatory due to the potential-ion suppression in the urinary matrix. The assumed influence of sulphation on elevated testosterone/epitestosterone ratios was not supported by LC-MS-MS examinations [39].

Quantitative uncertainties of conjugate hydrolysis are less crucial in the case of exogenous steroids, because anti-doping legislation does not require a quantitative threshold. Approaches to identify glucuronides of synthetic steroids by LC-MS were directed to structural investigations of steroid glucuronides and automation of the analytical process by application of on-line micro-extraction [40]. Steroid glucuronides proved to exhibit similar ionisation principles to their free analogues. The presence of conjugated double bonds increases the proton affinity, resulting in an elevated abundance of $[M + H]^+$ pseudo-molecular ions, while saturated molecules tend to form ammonium adducts rather than protonated ions. The steroid conformation (5α versus 5β linkage of the A and B rings of the sterane skeleton) was reported to influence the fragmentation of A-ring-saturated steroids; 5β isomers were found to produce considerable higher amounts of the dehydrated aglycone fragments (relative to the deconjugated steroid) than 5α isomers [41].

Corticosteroids

Glucocorticosteroids (Table 9.6) are prohibited when administered systemically (orally or by injection), whereas all other administration routes require medical notification. Due to their influence on protein and carbohydrate metabolism, they are known to be abused as growth promoters in food-producing animals and may reasonably be abused in sports. At the moment, there is no analytical solution for a reliable differentiation of the administration pathway and analytically positive cases may be rejected by the availability of a therapeutic use exemption. ESI in negative mode was consistently reported to be the most efficient ionisation for a corticosteroid screening method [42, 43]. Typical fragmentation reactions are loss of the CH_2OH moiety, water or hydrofluoric acid. The LC-MS assays are carried out either in single-MS mode using diagnostic

Table 9.6 Chemical structures of synthetic corticosteroids

Class	Typical modifications	Examples	Chemical structure
Cortisone	1–2 double bond	prednisone	
Cortisol	1–2 double bond R_1 = H, OH R_2 = H, OH, CH_3 $R_1, R_2 = C(CH_3)_2$ R_3, R_4 = H, F	prednisolone betamethasone triamcinolone	

fragments or in MS-MS mode, where [M + CH$_3$COO]$^-$ adducts serve as precursor ions due to the absence of [M – H]$^-$ pseudo-molecular ions.

Alternatively, the identification of intact corticosteroid conjugates in bovine urine was evaluated [44]. Base peaks detected in positive ESI mode are sodium adducts, while a predominant [M – H]$^-$ ion is observed in negative mode. Both precursors exhibit a low fragmentation rate. MRM experiments in negative-ionisation mode using a [M – H]$^-$ → [M – H]$^-$ pseudo-transition (monitored at elevated collision energy) were found to produce specific signals due to the high ion stability.

Other prohibited substances (stimulants, narcotics, cannabinoids)

There are numerous other prohibited substances in doping control eligible for application of LC-MS, like stimulants or narcotics. Respective analytical procedures are available in clinical, veterinary or forensic toxicology and may certainly be adopted in doping analysis. However, conventional screening procedures based on GC with nitrogen phosphorous-selective detection (NPD) or GC-MS are still state of the art, because the concentration and stability of respective compounds are sufficiently high. Moreover, the list of prohibited stimulants and narcotics has been revised and condensed recently, simplifying the analytical challenges.

Analytical development in these areas is focused on upcoming (e.g. modafinil) or critical substances (e.g. mesocarb). The latter stimulant is thermally instable, forms irreproducible artefacts and is therefore difficult to identify by conventional GC-MS. The long history of attempts to use various ionisation techniques (particle beam, thermospray, ESI, APCI) to detect mesocarb and elucidate its biotransformation [45] demonstrates the analytical difficulties when tackling this substance. Mesocarb does not fit analytically into its native pharmacological group of stimulants and is typically integrated into other LC-MS screening procedures (e.g. diuretics).

Other isolated substances causing GC problems due to their high polarity and/or low thermal stability (e.g. the anti-oestrogenic clomiphene or the cocaine metabolite benzoylecgonine) are comparable candidates for an insertion into LC-MS screening procedures.

The identification of the major urinary metabolite of Δ^9-tetrahydrocannabinol (THC) (the glucuronide of 11-nor-9-carboxy-THC) in doping control does not represent any particular exception compared with forensic urine analysis (see Chapter 8). A threshold value of 15 µg/L specifies a potential doping violation; however, there are no uniform sanctions and cannabinoids were downgraded to the group of 'specified substances, which are particularly susceptible to unintentional anti-doping rule violations . . . and which are less likely to be successfully abused as doping agents' [1]. Any eligible LC-MS assay for cannabinoids [46] may be applied in sports drug testing without modifications.

The identification of alkylated xanthine derivatives is no longer essential in human doping analysis after its removal from the list of prohibited substances in 2004, although recent publications have dealt with its identification and quantification by LC-MS in horses [47, 48]. According to the required performance specification of the Association of Official Racing Chemists (AORC), concentrations as low as 100 µg/L need to be detected in equine urine.

Among numerous pharmaceutical substances that are controlled in equine doping control, there are quaternary ammonium anti-cholinergic agents (e.g. isopropamide, glycopyrrolate) that appear to be well suited for LC-MS. According to their ionic structure, quaternary ammonium drugs require special pre-treatment (ion pair formation) to enable conventional liquid–liquid extraction (LLE) or LC separation. Examining the identification of eight quaternary ammonium drugs in horse urine, the application of capillary electrophoresis was found to be more appropriate (enhanced sensitivity and separation power) than LC [49].

Large molecules

Introduction

Several proteins and peptide hormones are included in the list of prohibited substances, due to their anabolic effect, as in the case of hGH or hCG, or due to an increase in the oxygen transportation capacity of blood, like EPO and haemoglobin-based oxygen carriers (HBOCs). These substances are synthetic or recombinant analogues of endogenous hormones. The possibility to discriminate between an abuse of prohibited substances and endogenous production depends on the structural modifications of the respective exogenous compound. Possible analytical approaches are:

- quantitative evaluations of compounds with a significant concentration difference between basal levels and exogenous administrations (hCG, IGF-1)
- identification of variations of the primary structure (insulin, HBOCs)
- discrimination of mass variants of proteins (20- and 22-kDa isomers of hGH)
- investigation of charge variants resulting from post-translational modifications (e.g. glycosylation of EPO).

In fact, the analytical approach depends on the current availability of pharmaceutical substances which are potentially abused. EPO, for instance, was originally available as recombinant protein with a primary structure identical to the endogenous hormone. The introduction of synthetic variants with modified amino acid sequences required an immediate adaptation of the analytical approach.

The practical application of LC-MS techniques to identify large molecules in routine doping control is still restricted by sensitivity limitations and the requirement of extensive and selective sample clean-up. A shotgun approach based on digestion of proteins followed by HR separation of the digests is most frequently applied. Selective clean-up procedures (e.g. immunoaffinity enrichment) of the resulting complex peptide mixtures need to be applied to reduce ion suppression and to be able to identify relevant proteins at low amounts.

The characterisation of structural particularities based on the identification of intact proteins requires relatively high amounts of sample material and HR-MS for identification of the multiply charged precursor ions. Adequate analytical techniques (e.g. FTICR MS combined with linear ITs) are used for research investigations, but are not available for routine applications.

The identification of macromolecular compounds by LC-MS in routine doping analysis is so far restricted to HBOCs and synthetic insulin.

Human chorionic gonadotrophin

This gonadotrophic hormone stimulates the endogenous production of testosterone and is therefore prohibited in men. Quantification of hCG by two different immunological assays is recognised by WADA as sufficient analytical technique. Different conventional cut-off values of 10 or 25 IU/L were suggested (but not officially adopted) to discriminate normal values and pathological situations or misuse. An LC-MS procedure using an IT MS in positive ESI mode was found to be suitable for a sensitive quantification procedure of hCG [50]. Immunoaffinity extraction was used to enrich the peptide from the urinary matrix prior to the tryptic digestion of the glycoprotein. A residue of the β-subunit containing 17 amino acids was chosen as a significant marker for hCG; in particular, a distinct structural specificity compared with the similar subunit of the luteinising hormone was confirmed. MRM experiments of the doubly charged peptide allowed a quantification down to threshold concentrations of 5 IU/L. Examination of the glycosylation of hCG revealed a considerable micro-heterogeneity [51], which does not affect the evaluation of doping cases because a reliable differentiation is possible based on a wide concentration gap between endogenous and abnormal hCG levels in men.

Human growth hormone

Human GH stimulates the production of IGF-I, leading to a promotion of protein synthesis, increase of muscle mass, reduction of the amount of stored fat and an induction of the growth of long bones. The secretion from pituitary glands occurs in three to five daily pulses, during which the basal serum concentrations (around 3 μg/L) are temporarily greatly exceeded and return rapidly (half-life around 15 min) to normal.

Therefore, hGH serum concentrations are not suitable for the detection of its abuse. Instead, a combined evaluation of concentrations of IGF-I, its binding protein (IGF-BP3) and bone markers was proposed to reveal abuse of growth hormone in human athletes [52].

In horses, a cut-off serum concentration of 700 µg/L of IGF-I was anticipated as the criterion for growth hormone administration [53]. A sensitive LC-ESI-MS quantification procedure (limit of quantification = 30 µg/L IGF-I) was reported to be eligible for routine screening. The use of Arg^3-IGF-I, which is derived from IGF-I by replacement of Glu^3, as IS, appears to be essential, due to its similar properties in the affinity chromatography clean-up. Full-scan MS was applied to detect the intact pseudo-molecular ion ($[M + 7H]^{7+}$, m/z 1093.4) of IGF-I, which is used for quantification in equine serum samples [53].

Another analytical approach is based on the evaluation of qualitative variations of growth hormone. There are several mass variants of the predominant protein containing 191 amino acids corresponding to a molecular mass of 22 kDa. The most interesting one is a 20-kDa molecule, which is derived by deletion of residues 32–46 and exhibits comparable physiological activity. The concentration ratio of both variants does not depend significantly on gender, age, body height and weight [54]. Administration of any hGH isoform suppresses the endogenous secretion of both variants. Accordingly, abuse of commercially available hGH may be detected by an elevated 22/20 kDa ratio, because it contains exclusively the 22-kDa variant. Studies to prove the appropriateness of this approach based on enzyme-linked immunoassays are documented [55]. Current alternative quantification procedures of low-level proteins by LC-MS, based on sophisticated analytical equipment, e.g. FTICR MS, require extensive sample preparation and do not yet achieve the required sensitivity [10, 56]. Identification of two tryptic peptides of hGH (22 kDa), which was obtained from digestion of LC fractionated plasma, provided limits of detection 10-fold higher than relevant clinical plasma concentrations [57]. Another approach was capable of identifying clinically relevant levels (5 µg/L) from unfractionated plasma, applying tagged peptides, isolated by affinity chromatography [10]. The detection was carried out by HR-LC-MS (FTICR), which was coupled to a linear IT.

In another study, qualitative investigations on hGH were carried out using LC-MS-MS (FTICR). Based on the isolation and accurate mass measuring of the $[M + 17H]^{17+}$, a deviation of the molecular mass of 4 Da (compared with non-post-translationally modified hGH) could be detected and attributed to the formation of two disulphide bonds [56].

The identification of structural modifications like disulphide linkage or variations of glycosylation, influencing proper folding of proteins, may be of diagnostic value in doping analysis. Present technical constraints (relatively high cut-off values and/or high amounts of sample material, requirement of high mass ranges and MS resolution, and high MS accumulation times that are incompatible with typical LC peak widths and prohibit on-line coupling) seem to impede the application of LC-MS to the identification of hGH abuse in the near future.

Erythropoietin

EPO is a 34-kDa glycoprotein hormone controlling red blood cell production and is therefore a potent doping agent to enhance the oxygen transport capacity of blood. Recombinant human EPO (rhEPO) has been available since 1989 and cannot be directly distinguished from endogenous EPO due to its identical amino acid sequence. Indirect methods (e.g. haematological parameters) served as an indicator for a potential abuse. The variable glycosylation is determined by post-translational modification, which depends on the availability of enzymes in the respective cells, resulting in a wide diversity of carbohydrates, typically terminated by one or two sialic acid residues.

The number and location of sialic acid residues per molecule (Figure 9.5) determine the formation of quaternary isoforms and influence the plasma half-life of EPO. Removal of sialic acid groups leads to a total inactivity, due to a rapid clearance from blood circulation, while new EPO variants (i.e. darbepoetin) provide a prolonged activity, triggered by the inclusion of additional oligosaccharides, which are terminated by sialic acid.

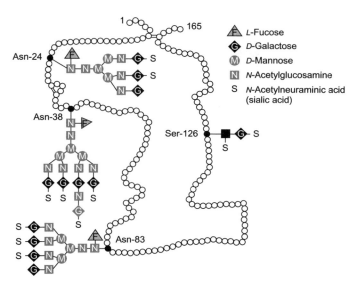

Figure 9.5 Chemical structure of EPO.

This alteration is based on a modification of five amino acids of the polypeptide and respective substances may be distinguished from natural EPO by analysis of the primary structure. Another variant of EPO (SEPO) consists of a synthetic peptide that is conjugated with a polymer, aiming for a reduction of the biological diversity.

The identification of EPO variants in doping control urine samples is to date based on a chemiluminescence detection of its isoforms after immunological enrichment and electrophoretic separation [58]. The logical MS approaches to an identification in combination with LC or capillary electrophoresis are hampered by the limited sensitivity.

Application of capillary electrophoresis- and LC-IT MS enabled the quantitative characterisation of EPO reference material after cleavage and derivatisation. The relative amount of sialic acid per molecule of rhEPO was found to be 17.6 mol/mol. This was supposed to provide a potential parameter to differentiate recombinant from endogenous EPO [59]. Alternatively, a characteristic sulphation of N-linked oligosaccharides in EPO molecules from different cell lines was detected, based on MS examinations (negative-ionisation mode) after LC separation of the protein digest using graphitised carbon columns [60]. An LC-ESI-MS investigation of various commercially available EPO forms (epoetin α, epoetin β, darbepoetin) resulted in astonishingly good LC separation. The full-scan MS of the three variants shows characteristic particularities in ESI (on-line micro LC coupling) as well as in matrix-assisted laser desorption ionisation (MALDI) mode (examination of the LC fractions) [61]. The injected quantity of protein was about 100 ng and, hence, far beyond the amount available in routine analyses. The limited MS sensitivity is due to the heterogeneity of glycosylation, because the ion abundance is spread over a large number of individual molecules. A deglycosylation of the EPO molecule leads to a reduction of diversity and increase of MS sensitivity. Analyses of deglycosylated EPO (rhEPO and darbepoetin) by MALDI showed a good match of its molecular mass with the assumed structure [62]. LC-ESI-MS experiments allowed a sensitive identification of $[M + nH]^{n+}$ pseudo-molecular ions (n = 5–8) of rhEPO after application of 250 fmol on-column. The additional identification of peptide, obtained from endoprotease Glu-C digestion, provides complementary structural information and is supposed to permit a sufficient identification of rhEPO and darbepoetin in race horses and greyhounds [62].

Insulin

The presumptive abuse of insulin among athletes and body-builders is due to its stimulation of endogenous protein synthesis. Several cases of hypoglycaemia (i.e. significantly reduced levels of blood sugar) were observed in fatal incidents in body-building, indicating its high popularity in this field. Insulin (Figure 9.6) belongs to the group of prohibited peptide hormones according to the recent WADA definition [1]. There are various medications available, based on structural alterations of insulin. The main purpose of these structural variations consists in the regulation of its bioavailability. Insulin tends to self-association to biologically inactive hexamers, which can be suppressed by structural modifications like switching of the positions of lysine and proline (B28 versus B29, Humalog) or replacement of proline with an aspartic acid residue (B29, Novolog). Alternatively, long-acting insulin was synthesised by elevation of its isoelectric point (modification in A and B chains, Lantus).

The identification of intact insulin derivatives, carried out on a QTOF instrument equipped with a nano-ESI ion source, revealed the predominant formation of corresponding multiply charged molecules ($z = 5$–7). This enables a mass-specific differentiation of the synthetic derivatives from human insulin [63] (Figure 9.7) after LC separation. Alternatively, the identification of the B chains can be applied after reduction of the disulphide bonds.

Haemoglobin-based oxygen carriers

Similar to the administration of EPO or synthetic perfluorocarbons, the intention of an abuse of HBOCs in sports is an elevation of the oxygen transportation capacity of blood. Various cross-linked bovine or human haemoglobin preparations (e.g. Hemopure, Hemolink, Oxyglobin) are approved for the treatment of animals or humans. The intact oxyglobin was reported to be inadequate for an efficient confirmation of haemoglobin by LC-MS in blood samples, because of the formation of one single dominant fragment that was interfered with by impurities from haemolysed plasma. Sufficient variations between the primary structures of human and bovine haemoglobin permit the differentiation of both proteins, based on the identification of diagnostic tryptic peptides. Specific peptides resulting from α and β chains of haemoglobin were found to be suitable for a differentiation of both species [64, 65]. Subunits of the α (residue 69–90, 2367 Da) and the β chain (residue 40–58, 2090 Da) were commonly used as markers for bovine haemoglobin. Human haemoglobin was identified by recording of a diagnostic peptide

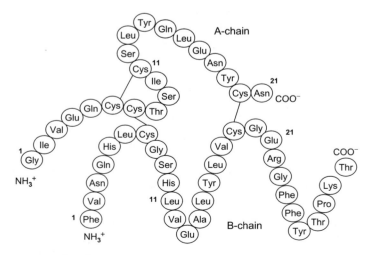

Figure 9.6 Chemical structure of insulin.

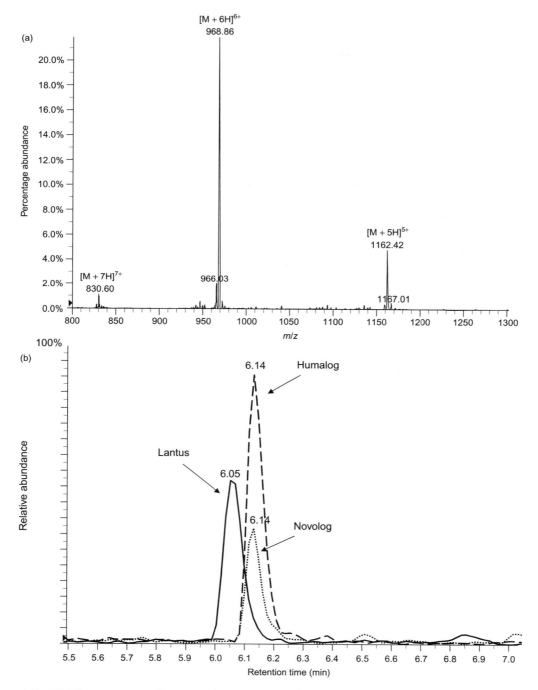

Figure 9.7 ESI full-scan spectrum of human insulin (a) generating the multiply charged molecules $[M + 5H]^{5+}$, $[M + 6H]^{6+}$ and $[M + 7H]^{7+}$ at m/z 1162.4, 968.9 and 830.6, respectively. Extracted ion chromatogram (b) of a plasma sample fortified with Humalog (m/z 1162), Novolog (m/z 1166) and Lantus (m/z 1213) at 10 pmol/mL. Insulins were analysed as intact proteins. (Reproduced from Thevis et al. [63] with permission.)

from the β chain (residue 42–60, 2059 Da). Doubly or triply charged ions of these peptides are suitable for LC-ESI-MS-MS screening. The assay is able to identify, confirm and quantify HBOCs in human and equine plasma samples. Respective assays were established using either QTOF or triple-quadrupole (QqQ) MS [64, 65]. Sample preparation requires a laborious combination of SPE, filtration of macromolecules and enzymatic digestion.

Summary

LC-MS has been introduced in different branches of doping analysis.

Qualitative progress in the analysis of highly polar substances (diuretics) could be achieved in combination with a significant reduction of sample preparation. Other low-mass pharmaceutical substances (stimulants, narcotics, β-blockers) may certainly be analysed by LC-MS, but well-established and validated GC assays are still dominant.

Progress in steroid analysis is mainly focused on substances with high proton affinity (polar substituents, conjugated double bonds, corticosteroids, steroid conjugates). The identification of relevant steroids by LC-MS has considerably advanced the classic GC-MS screening for anabolic steroids, but an extensive replacement of the conventional surveys is not likely.

Finally, there is increasing application of LC-MS to the analysis of doping-relevant proteins. Synthetic (e.g. insulin) or non-human (e.g. haemoglobin) peptides may be positively discriminated from endogenous analogues, whereas recent approaches to an unequivocal identification of abuse of recombinant EPO or growth hormone are promising, but so far insufficiently sensitive and reliable for routine application.

References

1. World Anti-Doping Agency. *The Prohibited List 2004. International Standard*. Montreal: WADA, 2004.

2. Houghton E, Dumasia MC, Wellby JK. The use of combined high performance liquid chromatography negative ion chemical ionization mass spectrometry to confirm the administration of synthetic corticosteroids to horses. *Biomed Mass Spectrom* 1981; 8: 558–564.

3. Garbis SD, Hanley L, Kalita S. Detection of thiazide-based diuretics in equine urine by liquid chromatography/mass spectrometry. *J AOAC Int* 1998; 81: 948–957.

4. Bowers LD. Analytical advances in detection of performance-enhancing compounds. *Clin Chem* 1997; 43: 1299–1304.

5. Bowers LD, Fregien K. HPLC/MS confirmation of peptide hormones in urine: an evaluation of detection limits. In: Donike M, Geyer H, Gotzmann A, et al., eds, *Recent Advances in Doping Analysis 6*. Cologne: Sport und Buch Strauß, 1994: 175–184.

6. Thieme D, Große J, Lang R, et al. Detection of mesocarb metabolite by LC-TS/MS. In: Donike M, Geyer H, Gotzmann A, et al., eds, *Recent Advances in Doping Analysis 2*. Cologne: Sport und Buch Strauß, 1995: 275–284.

7. Ventura R, Fraisse D, Becchi M, et al. Approach to the analysis of diuretics and masking agents by high-performance liquid chromatography-mass spectrometry in doping control. *J Chromatogr* 1991; 562: 723–736.

8. Krummen M, Hilkert AW, Juchelka D, et al. A new concept for isotope ratio monitoring liquid chromatography/mass spectrometry. *Rapid Commun Mass Spectrom* 2004; 18: 2260–2266.

9. Greig MJ, Bolanos B, Quenzer T, et al. Fourier transform ion cyclotron resonance mass spectrometry using atmospheric pressure photoionization for high-resolution analyses of corticosteroids. *Rapid Commun Mass Spectrom* 2003; 17: 2763–2768.

10. Wu SL, Choudhary G, Ramstrom M, et al. Evaluation of shotgun sequencing for proteomic analysis of human plasma using HPLC coupled with either ion trap or Fourier transform mass spectrometry. *J Proteome Res* 2003; 2: 383–393.

11. Magnusson MO, Sandstrom R. Quantitative analysis of eight testosterone metabolites using column switching and liquid chromatography/tandem mass spectrometry. *Rapid Commun Mass Spectrom* 2004; 18: 1089–1094.

12. Kawasaki N, Ohta M, Itoh S, et al. Usefulness of sugar mapping by liquid chromatography/mass

spectrometry in comparability assessments of glycoprotein products. *Biologicals* 2002; 30: 113–123.

13. Berges R, Segura J, Ventura R, *et al.* Discrimination of prohibited oral use of salbutamol from authorized inhaled asthma treatment. *Clin Chem* 2000; 46: 1365–1375.

14. Nelson RE, Grebe SK, DJ OK, *et al.* Liquid chromatography-tandem mass spectrometry assay for simultaneous measurement of estradiol and estrone in human plasma. *Clin Chem* 2004; 50: 373–384.

15. Kauppila TJ, Kostiainen R, Bruins AP. Anisole, a new dopant for atmospheric pressure photoionization mass spectrometry of low proton affinity, low ionization energy compounds. *Rapid Commun Mass Spectrom* 2004; 18: 808–815.

16. Ventura R, Segura J. Detection of diuretic agents in doping control. *J Chromatogr B* 1996; 687: 127–144.

17. World Anti-Doping Agency. *Technical Document TD2004 MRPL*. Montreal: WADA, 2004: 1–2.

18. Thieme D, Grosse J, Lang R, *et al.* Screening, confirmation and quantification of diuretics in urine for doping control analysis by high-performance liquid chromatography-atmospheric pressure ionisation tandem mass spectrometry. *J Chromatogr B* 2001; 757: 49–57.

19. Deventer K, Van Eenoo P, Delbeke FT. Simultaneous determination of beta-blocking agents and diuretics in doping analysis by liquid chromatography/mass spectrometry with scan-to-scan polarity switching. *Rapid Commun Mass Spectrom* 2004; 19: 90–98.

20. Deventer K, Delbeke FT, Roels K, *et al.* Screening for 18 diuretics and probenecid in doping analysis by liquid chromatography-tandem mass spectrometry. *Biomed Chromatogr* 2002; 16: 529–535.

21. Goebel C, Trout G, Kazlauskas R. Rapid screening method for diuretics in doping control using automated solid phase extraction and liquid chromatography-electrospray tandem mass spectrometry. *Anal Chim Acta* 2003; 502: 65–74.

22. Lehner AF, Harkins JD, Karpiesiuk W, *et al.* Clenbuterol in the horse: confirmation and quantitation of serum clenbuterol by LC-MS-MS after oral and intratracheal administration. *J Anal Toxicol* 2001; 25: 280–227.

23. Thevis M, Opfermann G, Schanzer W. Liquid chromatography/electrospray ionization tandem mass spectrometric screening and confirmation methods for beta2-agonists in human or equine urine. *J Mass Spectrom* 2003; 38: 1197–206.

24. Segura J, Ventura R, Jurado C. Derivatization procedures for gas chromatographic-mass spectrometric determination of xenobiotics in biological samples, with special attention to drugs of abuse and doping agents. *J Chromatogr B* 1998; 713: 61–90.

25. Thevis M, Opfermann G, Schanzer W. High speed determination of beta-receptor blocking agents in human urine by liquid chromatography/tandem mass spectrometry. *Biomed Chromatogr* 2001; 15: 393–402.

26. Leinonen A, Kuuranne T, Kotiaho T, *et al.* Screening of free 17-alkyl-substituted anabolic steroids in human urine by liquid chromatography-electrospray ionization tandem mass spectrometry. *Steroids* 2004; 69: 101–109.

27. Leinonen A, Kuuranne T, Kostiainen R. Liquid chromatography/mass spectrometry in anabolic steroid analysis – optimization and comparison of three ionization techniques: electrospray ionization, atmospheric pressure chemical ionization and atmospheric pressure photoionization. *J Mass Spectrom* 2002; 37: 693–698.

28. Guo T, Chan M, Soldin SJ. Steroid profiles using liquid chromatography-tandem mass spectrometry with atmospheric pressure photoionization source. *Arch Pathol Lab Med* 2004; 128: 469–475.

29. Catlin DH, Sekera MH, Ahrens BD, *et al.* Tetrahydrogestrinone: discovery, synthesis, and detection in urine. *Rapid Commun Mass Spectrom* 2004; 18: 1245–1249.

30. Draisci R, Palleschi L, Marchiafava C, *et al.* Confirmatory analysis of residues of stanozolol and its major metabolite in bovine urine by liquid chromatography-tandem mass spectrometry. *J Chromatogr A* 2001; 926: 69–77.

31. Draisci R, Palleschi L, Ferretti E, *et al.* Confirmatory analysis of 17beta-boldenone, 17alpha-boldenone and androsta-1,4-diene-3,17-dione in bovine urine by liquid chromatography-tandem mass spectrometry. *J Chromatogr B* 2003; 789: 219–226.

32. Buiarelli F, Cartoni GP, Coccioli F, *et al.* Determination of trenbolone and its metabolite in bovine fluids by liquid chromatography-tandem mass spectrometry. *J Chromatogr B* 2003; 784: 1–15.

33. Van Poucke C, Van Peteghem C. Development and validation of a multi-analyte method for the

detection of anabolic steroids in bovine urine with liquid chromatography-tandem mass spectrometry. *J Chromatogr B* 2002; 772: 211–217.

34. Wang C, Catlin DH, Demers LM, *et al.* Measurement of total serum testosterone in adult men: comparison of current laboratory methods versus liquid chromatography-tandem mass spectrometry. *J Clin Endocrinol Metab* 2004; 89: 534–543.

35. Reilly CA, Crouch DJ. Analysis of the nutritional supplement 1AD, its metabolites, and related endogenous hormones in biological matrices using liquid chromatography-tandem mass spectrometry. *J Anal Toxicol* 2004; 28: 1–10.

36. Shackleton CH, Chuang H, Kim J, *et al.* Electrospray mass spectrometry of testosterone esters: potential for use in doping control. *Steroids* 1997; 62: 523–529.

37. Kim JY, Choi MH, Kim SJ, *et al.* Measurement of 19-nortestosterone and its esters in equine plasma by high-performance liquid chromatography with tandem mass spectrometry. *Rapid Commun Mass Spectrom* 2000; 14: 1835–1840.

38. Borts DJ, Bowers LD. Direct measurement of urinary testosterone and epitestosterone conjugates using high-performance liquid chromatography/tandem mass spectrometry. *J Mass Spectrom* 2000; 35: 50–61.

39. Bowers LD, Sanaullah. Direct measurement of steroid sulfate and glucuronide conjugates with high-performance liquid chromatography-mass spectrometry. *J Chromatogr B* 1996; 687: 61–68.

40. Kuuranne T, Kotiaho T, Pedersen-Bjergaard S, *et al.* Feasibility of a liquid-phase microextraction sample clean-up and liquid chromatographic/mass spectrometric screening method for selected anabolic steroid glucuronides in biological samples. *J Mass Spectrom* 2003; 38: 16–26.

41. Kuuranne T, Aitio O, Vahermo M, *et al.* Enzyme-assisted synthesis and structure characterization of glucuronide conjugates of methyltestosterone (17 alpha-methylandrost-4-en-17 beta-ol-3-one) and nandrolone (estr-4-en-17 beta-ol-3-one) metabolites. *Bioconjug Chem* 2002; 13: 194–199.

42. Deventer K, Delbeke FT. Validation of a screening method for corticosteroids in doping analysis by liquid chromatography/tandem mass spectrometry. *Rapid Commun Mass Spectrom* 2003; 17: 2107–2114.

43. Fluri K, Rivier L, Dienes-Nagy A, *et al.* Method for confirmation of synthetic corticosteroids in doping urine samples by liquid chromatography-electrospray ionisation mass spectrometry. *J Chromatogr A* 2001; 926: 87–95.

44. Antignac JP, Le Bizec B, Monteau F, *et al.* Study of natural and artificial corticosteroid phase II metabolites in bovine urine using HPLC-MS/MS. *Steroids* 2002; 67: 873–882.

45. Appolonova SA, Shpak AV, Semenov VA. Liquid chromatography-electrospray ionization ion trap mass spectrometry for analysis of mesocarb and its metabolites in human urine. *J Chromatogr B* 2004; 800: 281–289.

46. Maralikova B, Weinmann W. Simultaneous determination of delta9-tetrahydrocannabinol, 11-hydroxy-delta9-tetrahydrocannabinol and 11-nor-9-carboxy-delta9-tetrahydrocannabinol in human plasma by high-performance liquid chromatography/tandem mass spectrometry. *J Mass Spectrom* 2004; 39: 526–531.

47. Thevis M, Opfermann G, Krug O, *et al.* Electrospray ionization mass spectrometric characterization and quantitation of xanthine derivatives using isotopically labelled analogues: an application for equine doping control analysis. *Rapid Commun Mass Spectrom* 2004; 18: 1553–1160.

48. Todi F, Mendonca M, Ryan M, *et al.* The confirmation and control of metabolic caffeine in standardbred horses after administration of theophylline. *J Vet Pharmacol Ther* 1999; 22: 333–342.

49. Tang FP, Leung GN, Wan TS. Analyses of quaternary ammonium drugs in horse urine by capillary electrophoresis-mass spectrometry. *Electrophoresis* 2001; 22: 2201–2209.

50. Liu C, Bowers LD. Mass spectrometric characterization of the beta-subunit of human chorionic gonadotropin. *J Mass Spectrom* 1997; 32: 33–42.

51. Jacoby ES, Kicman AT, Laidler P, *et al.* Determination of the glycoforms of human chorionic gonadotropin beta-core fragment by matrix-assisted laser desorption/ionization time-of-flight mass spectrometry. *Clin Chem* 2000; 46: 1796–1803.

52. Sonksen PH. Insulin, growth hormone and sport. *J Endocrinol* 2001; 170: 13–25.

53. de Kock SS, Rodgers JP, Swanepoel BC. Growth hormone abuse in the horse: preliminary assessment of a mass spectrometric procedure for IGF-1 identification and quantitation. *Rapid Commun Mass Spectrom* 2001; 15: 1191–1197.

54. Tsushima T, Katoh Y, Miyachi Y, et al. Serum concentrations of 20K human growth hormone in normal adults and patients with various endocrine disorders. Study Group of 20K hGH. *Endocr J* 2000; 47(Suppl): S17–S21.
55. Wu Z, Bidlingmaier M, Dall R, et al. Detection of doping with human growth hormone. *Lancet* 1999; 353: 895.
56. Wu SL, Jardine I, Hancock WS, et al. A new and sensitive on-line liquid chromatography/mass spectrometric approach for top-down protein analysis: the comprehensive analysis of human growth hormone in an *E. coli* lysate using a hybrid linear ion trap/Fourier transform ion cyclotron resonance mass spectrometer. *Rapid Commun Mass Spectrom* 2004; 18: 2201–2207.
57. Wu SL, Amato H, Biringer R, et al. Targeted proteomics of low-level proteins in human plasma by LC/MS^n: using human growth hormone as a model system. *J Proteome Res* 2002; 1: 459–465.
58. Lasne F, Martin L, Crepin N, et al. Detection of isoelectric profiles of erythropoietin in urine: differentiation of natural and administered recombinant hormones. *Anal Biochem* 2002; 311: 119–126.
59. Che FY, Shao XX, Wang KY, et al. Characterization of derivatization of sialic acid with 2-aminoacridone and determination of sialic acid content in glycoproteins by capillary electrophoresis and high performance liquid chromatography-ion trap mass spectrometry. *Electrophoresis* 1999; 20: 2930–2937.
60. Kawasaki N, Haishima Y, Ohta M, et al. Structural analysis of sulfated *N*-linked oligosaccharides in erythropoietin. *Glycobiology* 2001; 11: 1043–1049.
61. Caldini A, Moneti G, Fanelli A, et al. Epoetin alpha, epoetin beta and darbepoetin alfa: two-dimensional gel electrophoresis isoforms characterization and mass spectrometry analysis. *Proteomics* 2003; 3: 937–941.
62. Stanley SM, Poljak A. Matrix-assisted laser-desorption time-of flight ionisation and high-performance liquid chromatography-electrospray ionisation mass spectral analyses of two glycosylated recombinant epoetins. *J Chromatogr B* 2003; 785: 205–218.
63. Thevis M, Ogorzalek Loo RR, et al. Mass spectrometric identification of synthetic insulin. In: Schänzer W, Geyer H, Gotzmann A, et al., eds, *Recent Advances in Doping Analysis 11*. Cologne: Sport und Buch Strauß, 2003: 227–237.
64. Guan F, Uboh C, Soma L, et al. Unique tryptic peptides specific for bovine and human hemoglobin in the detection and confirmation of hemoglobin-based oxygen carriers. *Anal Chem* 2004; 76: 5118–5126.
65. Thevis M, Ogorzalek Loo RR, Loo JA, et al. Doping control analysis of bovine hemoglobin-based oxygen therapeutics in human plasma by LC-electrospray ionization-MS/MS. *Anal Chem* 2003; 75: 3287–3293.

10

Pesticide analysis using LC-MS

Félix Hernández, Juan V. Sancho and Oscar J. Pozo

General overview

Pesticides have been widely used since the early to mid-twentieth century to protect agricultural and horticultural crops against damage. They are also used in the home and workplace to provide a pest-free environment. Around 1000 active ingredients have been employed so far, formulated in thousands of different commercial products. They include a wide range of compounds used in pest control, e.g. insecticides, herbicides, fungicides, rodenticides and molluscicides, with very different physicochemical characteristics, and large differences in their polarity, volatility and persistence, among various other parameters.

Many public health benefits have been gained from the use of synthetic pesticides; however, in spite of the obvious advantages, the potential adverse impact on the environment and public health is substantial. Once in the environment, currently marketed pesticides are relatively labile and do not tend to persist for long periods of time. However, with the widespread use of pesticides, it is virtually impossible to avoid exposure at some level [1] and, in order to understand fully their impact on human health, it is necessary to investigate the concentration levels of pesticides and their metabolites in human samples. The biological monitoring of exposure involves the measurement of a biomarker of exposure (normally the pesticide or its metabolites) in human blood, serum, urine or tissues [1, 2].

Occupational exposure occurs mainly in the mixing and loading of equipment, and also in the spraying and application of pesticides. Absorption resulting from dermal exposure is the most common route of uptake for workers, although inhalation and ingestion can also contribute to pesticide uptake [3]. However, most poisonings by pesticides occur as a result of misuse or accidental over-exposure. Overall, pesticides account for a considerable number of human intoxication cases, which are not always well diagnosed or documented. The diagnosis of acute intoxication is expected to be supported by analytical investigations using quantitative, specific and selective methods [4].

The choice of method in analytical toxicology depends on the problem to be solved, and reliable analytical methods are always required for the correct identification and quantification of analytes. As a general rule, two approaches can be considered when pesticide analysis in biological samples is required – target analysis, when the compounds to be determined have been previously established, or general unknown screening (GUS), where there are no hints on the xenobiotic contaminant. Target analysis is by far the most widely employed option and requires the pre-selection of analytes (e.g. based on the pesticide use in the geographical area investigated or on the clinical symptoms in cases of intoxication). However, what clinical and forensic toxicologists expect from a GUS procedure is the detection and unambiguous identification of the xenobiotic(s) involved in intoxication cases, even if they have no clues to guide the search [5].

Coupling of chromatographic separations to mass spectrometry (MS) as a detection system is currently the technique of choice for identification and quantification of most of the xenobiotics encountered by analytical toxicologists. The complementary use of gas chromatography (GC)-MS and liquid chromatography (LC)-MS offers the possibility to determine a wide range of compounds with different physicochemical characteristics. Thus, GC-MS is the 'golden standard' for detection and quantification of drugs and poisons volatile under GC conditions [6], whereas LC-MS, particularly with robust, modern atmospheric pressure ionisation (API), enables the determination of most of the compounds that are not amenable to GC, i.e. thermolabile, non-volatile, highly polar, ionic compounds.

In the field of pesticide analysis, chemists face a variety of compounds, including not only parent pesticides, but also their metabolites, that have to be determined normally at very low concentration levels in complex matrices. Optimised chromatographic methods have to be employed, and coupling with MS techniques is compulsory to satisfy the sensitivity and selectivity requirements.

In the past, the bulk of published work dealing with the analysis of pesticides in biological samples has been related to the determination of persistent organochlorine pesticides (OCPs), including polychlorinated biphenyls (PCBs), in whole blood, serum, plasma and other lipid-rich matrices, such as adipose tissue or breast milk [1, 7]. Since these pesticides are inherently lipophilic, they tend to bioaccumulate in the fat stores of the body. Although some OCP metabolites are determined in urine, they are typically measured as the intact parent pesticide in lipid-rich matrices using GC-based methods.

However, currently used pesticides are non-persistent, do not tend to bioaccumulate, and are typically metabolised and excreted from the body within a matter of hours or days. For this reason, the target substrate to monitor their exposure is more frequently urine than blood. Although some compounds are excreted mostly as intact pesticides (e.g. the herbicides 2,4-dichlorophenoxyacetic acid [2,4-D], 2-methyl-4-chlorophenoxyacetic acid [MCPA], diquat and paraquat), exposure to most compounds has to be assessed by measuring their biotransformation products in urine. Because of their rapid metabolism, determination in blood is appropriate only in the case of recent exposure to high doses (e.g. acute intoxication) [7].

The majority of pesticide metabolites are non-volatile, polar and ionic compounds. These characteristics make LC-MS the technique of choice for their determination. Advantages of this technique, e.g. direct analysis of aqueous samples, no derivatisation required and high sample throughput, are widely recognised. In addition, the ability of LC-MS to analyse intact drug conjugates (i.e. glutathione, and glucurono and sulphonate conjugates) avoids the hydrolysis step applied in analytical toxicology to release free metabolites [8]. The limitation in this case comes more from the commercial availability of analytical reference standards and this becomes a critical factor in many analyses related to pesticide metabolites. In spite of the above-mentioned advantages of LC-MS, recent reviews on analytical methods used in biological monitoring of pesticide exposure [1, 7] show that LC-MS has not yet been widely applied in the field of pesticide residue analysis. As with other fields of application, the main reason for this delay must undoubtedly be attributed to the difficulties and delay in coupling LC with MS. However, there are also other reasons that apply specifically to pesticide analysis in biological samples. Among them, the fact that environmental and toxicological analysis is in most cases a non-commercial application, not attracting large investments of money, probably plays an important role. This is corroborated by reviews on applications of LC-MS in analytical toxicology dated from the late 1990s [6, 8]. In our recent review on the use of LC-MS for pesticide residue analysis in biological samples [9], we showed that most applications of LC-MS in this field have appeared from 2000 onwards. In fact, more than half the papers have been published in the last 3 years. Due to the inherent and widely accepted advantages of this technique, a remarkable increase in the number of applications is expected in the very near future.

Table 10.1 shows an overview of LC-MS applications to the analysis of pesticides and metabolites in human samples. As reported in Table 10.1, both single and tandem MS (MS-MS) have been applied, although the majority of recent applications deal with the use of MS-MS.

The main applications of LC-MS for pesticide analysis in biological samples will be discussed in this chapter. Sample processing required for the different types of analytes investigated will be considered, with emphasis on their quantification,

Table 10.1

Analytes	Compound type/pesticide	Matrix	Interface	MS	Observations	Ref/year
6 parathion and fenitrothion metabolites		Urine	ESI	QqQ	Semiquantitative method for conjugates and DAPs. Two MS/MS transitions used for confirmation	(10) 2004
– PNP	Metab/parathion ethyl, methyl					
– MPNP	Metab/fenitrothion					
– DAPs (DMP, DMTP)	Metab/several OPPs					
– Conjugated 4-nitrophenol (sulphate and glucuronid)	Metab/parathion ethyl, methyl					
Acetamiprid	Parent	Urine, air Clothes	ESI	QqQ		(43) 2004
7 OPPs metabolites		Urine	APCI ESI	QqQ	Signal enhancement for carboxylic acids in urine. Pseudo-MS/MS for TCPY	(26) 2004 (44) 2004 (25) 2003
– CIT	Metab/isazophos methyl, ethyl					
– CMHC	Metab/coumaphos					
– DEAMPY	Metab/pyrimiphos methyl					
– IMPY	Metab/diazinon					
– MDA	Metab/malathion					
– PNP	Metab/parathion methyl, ethyl					
– TCPY	Metab/chlorpyrifos methyl, ethyl					
6 Herbicides and/or metabolites						
– atrazine mercapturate	Metab/atrazine					
– acetochlor mercapturate	Metab/acetochlor					
– alachlor mercapturate	Metab/alachlor					
– metolachlor mercapturate	Metab/metolachlor					
– 2,4-D; 2,4,5-T	Parent herbicides					
1 insect repellant (DEET)	Parent					
5 Pyretroids metabolites						
– Cis-DBCA	Metab/deltametrin					
– Cis-DCCA	Metab/cyfluthrin, permethrin, cypermethrin					
– Trans-DCCA						
– 3-PBA	Several pyretroids					
– 4-F-3-PBA	Metab/cyfluthrin					

Table 10.1 Continued

Analytes	Compound type/pesticide	Matrix	Interface	MS	Observations	Ref/year
ETU	Metab/ethylenebisdithiocarbamates	Urine	ESI	QqQ		(12) 2003
Screening of drugs, metabolites and pesticides	Little information available on pesticides	Urine	ESI	TOF	Toxicological screening (substance data base for 637 compounds) Exact mass, retention time and metabolite pattern are required for unequivocal identification	(29) 2003
10 OPPs pesticides/metabolites - acephate, methamidophos - 7 OPPs metabolites (see Olsson, 2004) - BTA	Parent OPP Metab/OPPs Metab/azinphos methyl	Urine	ESI	QqQ	Pseudo-MS/MS for TCPY	(25) 2003
2 triazines (atrazine, simazine) - 2-chloro-4,6-diamino-1,3,5-triazine	Parent herbicides Metab/triazines	Urine	ESI	Q		(45) 2003
238 drugs including the pesticides malathion and methyl parathion	Parent OPPs	Serum	ESI	QqQ	Screening method	(21) 2003
PNP	Metab/parathion methyl, ethyl	Urine	APCI	QqQ	Two transitions used for confirmation Around 16,000 samples analyzed	(42) 2002
	Metab/several OPPs	Urine	ESI	QqQ	Ion-pairing chromatography Two MS/MS transitions used for confirmation Poor reproducibility in retention times	(14) 2002
4 DAPs (DEP, DETP, DEDTP, DMDTP)						
PNP MPNP	Metab/parathion methyl, ethyl Metab/fenitrothion	Urine	ESI	QqQ	Two MS/MS transitions used for confirmation	(33) 2002

Table 10.1 Continued

Analytes	Compound type/pesticide	Matrix	Interface	MS	Observations	Ref/year
Alachlor DEA, CDEPA	Parent herbicide Metab/alachlor	Plasma Urine (rat)	ESI	IT		(15) 2002
14 polar pesticides - 11 Carbamates - 3 Benzimidazoles	Parent pesticides	Serum	ESI	Q	Two ions for confirmation Method developed to investigate cases of acute intoxication	(4) 2001
3 azole fungicides (triadimenol, triadimefon, tetraconazol)	Parent fungicides	Urine	ESI	Q		(46) 2001
2 OPPs metabolitesUrine - IMPY - MDA 2 Herbicides/metabolites - atrazine mercapturate - 2,4-D 1 Pyrethroid metabolite (3PBA)	Metab/diazinon Metab/malathion Metab/atrazine Parent herbicide Metab/several pyrethroids		APCI	QqQ		(47) 2000
Chlorpyrifos TCPY	Parent OPP Metab/chlorpyrifos	Urine Serum	ESI	QqQ	Pseudo-MS/MS for TCPY Several transitions used for confirmation	(16) 2000
2,4-D	Parent herbicide	Urine	APCI	QqQ	LC-MS/MS used for confirmation of immunoassay	(35) 2000
2 carbamates (aldicarb, carbofuran) 3 carbamates metabolites - aldicarb sulphoxide, aldicarb sulphone - 3-hydroxycarbofuran	Parent carbamates Metab/aldicarb	Urine	ESI	Q	LC-MS used for confirmation	(36) 2000
2,4-D MDA Atrazine mercapturate	Parent herbicide Metab/malathion Metab/atrazine	Urine	APCI	QqQ		(48) 1999

Table 10.1 Continued

Analytes	Compound type/pesticide	Matrix	Interface	MS	Observations	Ref/year
21 OPPs 8 carbamates	Parent OPPs Parent carbamates	Urine Serum	APCI	Magnetic Sector		(22) 1996
Metolachlor mercapturate	Metab/metolachlor	Urine	APCI	QqQ		(49) 1997
Alachlor mercapturate	Metab/alachlor	Urine	APCI	QqQ		(30) 1996
Malathion 5 metabolites – MO, MDA, α- and β-MCA, DM-MA	Parent OPP Metab/malathion	Urine Blood	TSP	Q	Paper in Japanese n.a.	(17) 1994
8 methyl carbamates	Parent	Serum	APCI	Magnetic Sector		(50) 1993
21 OPPs	Parent	Blood	APCI	Magnetic Sector	Approach developed for determining substances in patients with acute agricultural chemical toxicity	(28) 1992
ETU	Metab/ethylenebisdithiocarbamates	Urine	TSP	Q		(11) 1992
Methomyl	Parent carbamate	Blood	TSP	n.a.	LC-MS/MS used for confirmation	(37) 1991

identification and confirmation. Matrix effects (MEs), which can severely affect the quantification of analytes using LC-MS, and the importance of chromatographic separation, even using the power of the MS technique, will be discussed. Finally, applications of the different types of mass analysers in both single and tandem mode will be reviewed.

Sample matrices and analytes investigated

Various analytical methods have been used to measure pesticides and metabolites in a variety of matrices including urine, serum, breast milk, saliva or postpartum meconium.

The choice of biological matrix depends upon a number of variables, including the pharmacokinetics of the toxicant, the availability of the matrix, the ease of matrix manipulation and the sensitivity of the analytical method. As stated previously, the monitoring of metabolites in urine is the most common approach; however, in cases of acute intoxication, the parent compound and metabolites have also been measured in blood, serum, gastric content or body tissues. Other matrices such as hair or adipose tissue might be useful to monitor long-term exposure to pesticides. More details related to biological monitoring of exposure to pesticides can be found in the excellent work performed by Barr *et al.* [1, 2].

In general, urine presents some disadvantages over serum or blood as it is not a regulated body fluid and dilution correction is required. Knowledge of the biotransformation pathways of the compound is also needed in order to establish the molecular species to be monitored. Yet, urine offers the advantages of a (usually) large available volume and easy, non-invasive sampling, which seems to lead the interest towards this substrate, as can be deduced from Table 10.1.

The physicochemical characteristics of the analyte and of possible interferents present in the matrix are important factors to be considered when developing LC-MS analytical methods. In the case of serum, most matrix interferents can be minimised by protein precipitation, which is normally followed by an additional clean-up step based on liquid–liquid partition or extraction (LLE) or solid-phase extraction (SPE). Urine presents the inconvenience of containing high amounts of polar compounds such as urea, salts, glucuronides and sulphates, which can interfere with the analyte's determination. This is one of the main problems when developing quantitative urinary methods, as will be discussed later.

With regard to the analyte's characteristics, in general, ionic, polar or medium-polar compounds have been determined by LC-MS. Some polar intact pesticides have been measured in urine, e.g. the organophosphorous insecticides metamidophos and acephate, the herbicides 2,4-D and 2,4,5-trichlorophenoxyacetic acid (2,4,5-T), or the insect repellent *N,N*-diethyl-*m*-toluamide (DEET). However, most work has been done on metabolites, such as: specific hydrolysis products of OPPs like chlorpyrifos, coumaphos, diazinon, isazophos, malathion, parathion, pirimiphos, fenitrothion, azinphos or their methyl derivatives; specific and common piretroid metabolites of cyfluthrin, deltamethrin, permethrin and cypermethrin; and mercapturic acid conjugates of herbicides like atrazine, acetochlor, alachlor and metolachlor. Some LC-MS methods have been proposed for the simultaneous measurement of several pesticide metabolites, demonstrating the suitability of this technique for multi-residue analysis. A few methods have also been developed for the determination of non-polar and medium-polar pesticides in urine, such as chlorpyrifos, acetamiprid, simazine and alachlor, demonstrating the versatility and broad scope of the LC-MS technique (for references, *see* Table 10.1).

The direct determination of conjugates can be performed by LC-MS, although it may present some inconveniences, including the lower sensitivity in comparison to the analysis carried out after hydrolysis, together with the already mentioned lack of commercially available standards. This occurs because only one conjugated metabolic product is measured, whereas after hydrolysis the free and conjugated fraction are summed up and measured together. Another reason for this loss of sensitivity is related to their high polarity and low retention in reversed-phase (RP) mode. As a consequence, they are early eluting compounds in a chromatographic fraction where there is a large amount of interferences and ion suppression is very likely to occur. Normally, high percentages of water would be required in the mobile phase for better retention, which is not optimal for effective nebulisation/ionisation in the ion source. The above statement does not apply for normal-phase or hydrophilic interaction chromatography (HILIC) where higher retention and organic content are usually involved. Moreover, the low specificity of fragmentation using LC-MS-MS and the difficulty of performing a selective clean-up may also represent important limitations in the determination of conjugated metabolites. The MS-MS behaviour of most of the conjugates involves the fragmentation of the conjugated group, being $[M - 176]^-$, and ions at mass-to-charge (m/z) ratios 113 and 175, characteristic for glucuronides, and $[M - 80]^-$ characteristic for sulphates, all of them measured in the negative-ion mode, or $[M - 176 + H]^+$, measured in the positive-ion mode. As conjugation is a common metabolic process for many substances, these fragments/transitions can be shared with other conjugates of the same nominal mass and their monitoring may not provide sufficient selectivity of detection. Additionally, the presence of high amounts of other glucuronide and sulphate conjugates in urine with physicochemical properties similar to those of the analyte makes an efficient clean-up more difficult [10].

For these reasons, the most common analytical approach is still the determination of the total polar metabolite released after an enzymatic (or acid) hydrolysis step.

Ethylenethiourea (ETU) is a common metabolite and degradation product of ethylene bisdithiocarbamate fungicides, such as maneb, mancozeb and ziram, and is also used as an accelerant in rubber production [1, 7]. Due to its high polarity, ETU is excreted free in urine and is suitable for direct LC-MS analysis in this substrate. An early method, based on the use of a thermospray (TSP) interface, offered increased selectivity for ETU in urine in comparison to LC with conventional detectors, although sensitivity was not satisfactory [11]. However, the use of the more efficient electrospray ionisation (ESI) source coupled to tandem MS enabled the improvement of the limit of detection (LOD) down to 0.5 µg/L [12].

Some metabolites may originate from more than one pesticide, which inhibits specific identification of the source of exposure. This is the case of dialkylphosphate metabolites (DAPs), which may be derived from a variety of OPPs and ETU, which is a common metabolite for ethylenebisdithiocarbamates. The monitoring of DAPs, excreted in urine as sodium or potassium salts, therefore provides information about exposure to a class of pesticides instead of a single compound. Given the high polarity of these compounds and the lack of chemical groups that can be measured by conventional LC detectors such as UV, LC-MS is an attractive option for their determination in urine at low concentration levels. However, this determination is problematic, as these small ionic compounds exhibit low ionisation efficiency with ESI [13] and, in addition, it is difficult to achieve a satisfactory chromatographic separation. Ion chromatography or ion-pairing chromatography using RP stationary phases is required, but this implies the use of mobile phase modifiers, which may affect sensitivity by LC-MS. The use of tetrabutylammonium (TBA) as an ion-pairing reagent in the mobile phase allows the LC-ESI-MS-MS determination of diethyl phosphate (DEP), diethyl dithiophosphate (DEDTP), diethyl thiophosphate (DETP) and dimethyl thiophosphate (DMTP) with a Discovery C_{18} column (Supelco, Bellefonte, PA, USA) [14]. However, TBA was not suitable for the determination of DMP due to the ion-suppression effect produced by the continuous introduction of the ion-pairing modifier in the ion source. The removal of TBA from the mobile phase and the introduction of the ion-pairing reagent into the sample vial resulted in good chromatographic retention with minimal ion suppression [10]. In this way, LC-MS-MS proved to be an advantageous alternative to the GC-MS methodology normally applied for the trace-level determination of DAPs, which requires tedious and time-consuming derivatisation steps.

The strong potential of LC-MS for the determination of metabolites in urine is nowadays widely accepted. However, LC-MS has been scarcely used for the determination of pesticides in other matrices (serum or blood samples) in order to control exposure or to diagnose acute poisoning. As can be seen in Table 10.1, a few methods involving serum/blood matrices have been used to determine parent pesticides and also for the determination of metabolites of pesticides such as alachlor [15], chlorpyrifos [16] or malathion [17].

The determination of pesticides in other human matrices such as hair or adipose tissue can be a helpful tool to evaluate long-term exposure to pesticides. The most widely studied pesticides for this purpose are OCPs due to their persistence and lipophilic character. However, the absence of any API ionisable centre limits their determination by LC-MS. These matrices have also been analysed for the evaluation to long-term exposure of OPPs or carbamates. Thus, different OPPs such as diazinon, chlorpyrifos and malathion have been determined in hair samples for this purpose [18]. Russo et al. [19] developed a GC-MS method for the multi-residue determination of OPPs in different human tissues, demonstrating the applicability of these matrices for long-term pesticide exposure evaluation. Despite the potential applicability of LC-MS in this field, published methods normally make use of GC-MS. The high sensitivity and specificity of LC-MS together with the possibility for direct injection of aqueous/methanolic extracts could allow us to simplify the sample pre-treatment and to improve the method characteristics. For example, the use of LC-MS would avoid the solvent partition usually performed in hair analysis [18] by directly injecting the methanolic extract, and would improve the sensitivity and selectivity for thermally labile analytes, such as methomyl, compared with the LC-UV determination [20].

As adipose tissue analysis is normally used to detect long-term exposure to lipophilic compounds, e.g. OCPs, which are efficiently determined by GC much better than by LC-MS, this matrix is not subjected to analysis by this technique.

Sample preparation

Suitable sample preparation is a prerequisite for the chromatography of biosamples. Typically, conventional methods require the isolation of analytes after cleavage of conjugates by enzymatic or chemical hydrolysis and, in the case of GC separation, also the derivatisation of polar functions. However, sample processing prior to LC-MS can be considerably reduced. Minimisation of sample treatment is further favoured in LC-MS-MS procedures because of the excellent sensitivity and selectivity offered by tandem MS. This reason, as well as the compatibility of aqueous samples with RP-LC systems and the ability of LC-MS to detect the majority of pesticide metabolites without derivatisation, makes LC-MS(-MS) one of the most powerful analytical tools currently available in pesticide residue analysis.

A two-step solvent extraction (butyl acetate from basic medium, followed by dichloromethane:isopropanol from acidic medium) gave satisfactory results for the simultaneous screening of 238 drugs (two OPPs included, malathion and methyl parathion) in blood by LC-ESI-MS-MS [21]. Extraction of serum with dichloromethane has also been used for the determination of propoxur by LC-MS with atmospheric pressure chemical ionisation (APCI) [22]. Deproteinisation of serum with acetonitrile (ACN) provides simple sample preparation for the trace-level determination of chlorpyrifos and its main metabolite 3,5,6-trichloro-2-pyridinol in serum. After protein precipitation with ACN, the supernatant can be injected directly in an LC-LC-MS-MS system, reaching LODs of 1.5 µg/L without any observed ion suppression (recoveries higher than 87%) due to the efficiency of coupled-column LC (LC-LC) for removing interferents [16].

SPE is an excellent way to isolate analytes from the biological substrate and there is a considerable interest in the development of on-line SPE-LC procedures in order to simplify the handling of biological samples. A variety of selective sorbents is available, based on different retention mechanisms, such as RP, ion-pair and ion-exchange, restricted-access materials, immunoaffinity sorbents and molecular imprinted polymers [23, 24].

On-line SPE-LC using selective sorbents, especially when coupled to MS(-MS), is an interesting approach for sample preparation in the determination of organic analytes in biological fluids, as it satisfies the demands of selectivity, sensitivity, reliability and rapidity [24]. However, there are still very few applications of SPE-LC-MS-MS in pesticide analysis of biological samples in the literature and a considerable increase in the number of applications is expected in the next few years.

Oasis (Waters, Milford, MA, USA) MCX mixed-mode SPE cartridges have been used in off-line mode for a general LC-ESI-MS unknown screening for basic drugs and toxic compounds in serum [5], as well as for extraction of several pesticides from serum samples prior to LC-MS detection [4]. In this latter paper the authors demonstrated the complementary role of GC-MS and LC-MS in detection and quantification of 61 pesticides of toxicological significance in humans. Two different cartridges were used depending on the acid/base properties of the analytes – Oasis HLB for non-polar compounds (OCPs and OPPs) prior to GC-MS and Oasis MCX prior to LC-MS determination of 14 basic polar compounds (carbamates and benzimidazoles). The cationic-exchange properties of this material allow an optimal extraction by pH modification.

In order to determine the sum of free and conjugated metabolite, an enzymatic (or acid) hydrolysis step previous to LC-MS is often carried out, particularly when the pure standards of conjugates are not available. The hydrolysis step has proved critical for determination of the OPP metabolites *p*-nitrophenol (PNP) and 3,5,6-trichloro-2-pyridinol (TCPY). In the case of TCPY, the free form could not be detected, whereas traces of unbound PNP were detected in a large proportion of the urine samples analysed without hydrolysis [25]. Between 45 and 99% of the total PNP was conjugated, in accordance with other studies where 81% of the total PNP was detected as sulphate conjugated. At higher concentrations, a larger fraction of the total PNP was found in the conjugated form. The differences in the relative amount of free metabolite between PNP and TCPY could be due to the different metabolism of the two compounds.

An illustrative example of the application of LC-MS in the biological monitoring of pesticide exposure through the determination of pesticide metabolites in urine is the work performed by the National Centre for Environmental Health (Atlanta, GA, USA) [26]. In this laboratory, LC-MS-MS is routinely applied to the determination in human urine of up to 19 markers of commonly used pesticides. All of them are metabolites, except for the parent herbicides 2,4-D and 2,4,5-T, and the insect repellent DEET. The analytical procedure is based on the addition of isotopically labelled internal standards (ISs) to a 2-mL volume of urine, and on enzymatic hydrolysis of glucuronide- or sulphate-conjugated metabolites. After a simple clean-up step on SPE cartridges (Oasis HLB), the sample extract is separated in two fractions and analysed in two different LC-MS-MS set-ups – one using ESI (five pyrethroid metabolites analysed in negative-ionisation mode) and the other using APCI (15 pesticides and metabolites analysed both in positive-ion mode and, in a subsequent injection, in negative-ion mode) (Figure 10.1). 3-Phenoxybenzoic acid (3-PBA), which, being a metabolite of over half of the pyrethroid insecticides registered in the USA, was considered as a good marker for general pyrethroid exposure, could be determined with both ion sources. The high throughput, excellent sensitivity (most LODs below 0.5 µg/L) and the multi-residual character of this method demonstrate its suitability for biological monitoring of pesticide exposure.

In some instances, and provided that ion suppression is evaluated, sample preparation can be simplified, thus minimising the analytical errors associated with this step. As an example, direct injection of urine into a LC-MS-MS gives satisfactory results for the determination of several DAPs (LODs between 1 and 2 µg/L) with the only requirement of adding a volatile ion-pairing reagent, such as TBA, to obtain adequate chromatographic separation. The addition of TBA also allowed the direct (semi-quantitative) determination of the anionic species PNP-glucuronide and PNP-sulphate in urine [10, 14]. In this case, a considerable ME was observed for both glucuronide and sulphate conjugates, obtaining a signal suppression of around 50% for glucuronide and a signal enhancement of about 160% for sulphate. For this reason, the methods applicability was restricted to a semi-quantitative approach.

The use of LC-LC is an effective way to reduce

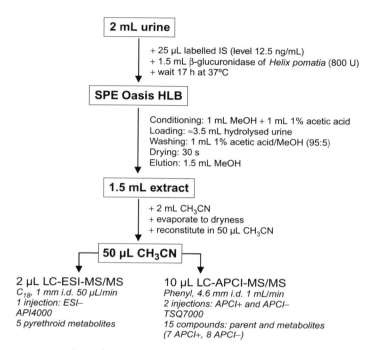

Figure 10.1 Sample preparation scheme for the LC-MS-MS determination of 19 markers of commonly used pesticides in human urine. Determination includes three parent compounds (two herbicides and one insect repellent) and 16 metabolites (seven from OPPs, five from pyrethroids and four from herbicides). (Data from Olsson et al. [26].)

sample handling. LC-LC-MS-MS is a powerful hyphenated technique that has been successfully applied in our laboratory for the sensitive and selective determination of several OPP metabolites in urine and serum samples [10, 16] (see below).

LC-MS

Table 10.2 shows selected LC-MS configurations applied for the determination of pesticides and metabolites in biological samples. The most commonly used stationary phases in LC-MS methods for pesticide residue analysis are RP C_{18} columns (around 80% of reported methods). For specific applications, phenyl columns [25] or polar-embedded columns [16] have been used. With regard to column dimensions, a shift to shorter lengths (50–100 mm) and smaller internal diameters (1–2 mm) is observed, mainly due to the recent increase in ESI-based methods.

As previously stated, multi-dimensional LC (coupled column) can be an excellent way of performing efficient and automated on-line clean-up in complex matrices, leading to rapid, selective and sensitive methods. LC-LC has been widely used in our laboratory to solve a variety of analytical problems in pesticide residue analysis, and its versatility and applications have been discussed recently [27]. LC-LC allows the direct injection of urine samples into the LC-LC-ESI-MS-MS system. Centrifuged urine samples are injected into a first separation column where early eluting interferents are sent to waste during an optimised time, known as the clean-up time. When the first analyte begins to elute, a high-pressure valve is activated and the fraction containing the analytes is transferred to a second analytical column during the so-called transfer time. When the high-pressure valve returns to initial conditions, the late-eluting interferents are

Table 10.2 Selected examples showing representative LC-MS configurations used for pesticide residue analysis in biological samples

LC characteristics				Interface/ionisation mode	MS characteristics			Reference (year)
Injection volume (µL)	Column dimensions (mm)	Stationary phase	Mobile phase		Analyser	Mode	No. of ions/transitions	
20	C-1[a]: 50 × 2.1 C-2: 100 × 2.1	Discovery C_{18}, 5 µm ABZ+, 5 µm	M-1: 25% ACN/75% formic acid 0.01% M-2: 65% ACN/35% H_2O 200 µL/min	ESI−	QqQ	SRM	2 transitions	10 (2004)
	100 × 2.1	Kromasil C_{18}, 5 µm	5 → 55% methanol, formic acid 0.01% 10 min; 300 µL/min					
2	100 × 1	Betasil C_{18}, 5 µm	49% ACN/51% acetic acid 0.1% 50 µL/min	ESI−	QqQ	SRM	1 transition	44 (2004)
5	100 × 1	Betasil Phenyl, 5 µm	30% ACN/70% acetic acid 0.15% 40 µL/min	ESI±	QqQ	SRM	1 transition (2 pseudo MS-MS for TCPY)	25 (2003)
2	150 × 1	Nucleosil C_{18}, 5 µm	30 → 80% ACN, 2 mM NH_4^+ pH 3 30 min; 50 µL/min	ESI+	Q	SIM	2 ions	4 (2001)
10	C-1: 50 × 2.1 C-2: 100 × 2.1	Discovery C_{18}, 5 µm ABZ+, 5 µm	M-1[b]: 65% ACN/35% formic acid 0.01% M-2: 80% ACN/20% formic acid 0.0025%	ESI±	QqQ	SRM	2 transitions (3 pseudo MS-MS for TCPY)	16 (2000)
20	250 × 4.6	Whatman ODS-3, 5 µm	60% ACN/40% acetic acid 0.2% 1 mL/min	APCI+	QqQ	SRM	2 transitions	48 (1999)
50	250 × 4.6	Whatman ODS-3, 5 µm	95% methanol/5% acetic acid 0.1% 1 mL/min	APCI+	QqQ	precursor-ion scan product-ion scan SRM	2 transitions	30 (1996)
100	300 × 3.9	µBondapak C_{18}, 30 µm	70 → 90% methanol 20 min; 1 mL/min	APCI±	magnetic sector	scan SIM	1 ion	28 (1992)

[a] Two different LC configurations used in this paper: LC-LC for nitrophenols, and LC for DAP and conjugates.
[b] M-1: 35% ACN/65% formic acid 0.01% in the case of TCPY.
C-1, C-2: first and second column in an LC-LC configuration; M-1, M-2: mobile phase in C-1 and C-2; Q: single-quadrupole instrument; QqQ: triple-quadrupole instrument. For other abbreviations, see Abbreviations, page xv.

washed out from the first column, while the analytes are separated by the second column, which preferably has a different retention mechanism to take full advantage of multi-dimensional LC. This combination makes the co-elution of matrix interferents and analytes more difficult, minimising the signal suppression often observed in LC-MS methods.

LC-LC in combination with MS-MS allowed the determination at sub-parts per billion levels of free metabolites (4-nitrophenol and 3-methyl-4-nitrophenol, metabolites of parathion and fenitrothion, respectively) or total metabolites after a simple enzymatic hydrolysis [10]. LC-LC-MS-MS has been also applied to the direct determination of chlorpyrifos and its main metabolite TCPY in urine, and to their trace-level determination in serum after protein precipitation with ACN [16] (Table 10.2).

Apart from a few early methods based on the use of a TSP interface [11, 17], most of the methods devoted to pesticide analysis in biosamples make use of the more recent APCI (about one-third of the reported methods since the early work of Kawasaki et al. [28] screening different OPPs in blood) and ESI (around 50% of methods) interfaces.

With the exception of early applications using double-focusing sector instruments [22, 28] or sporadic applications using ion trap [15] or time-of-flight [29] analysers, the majority of the published methods dealing with pesticide analysis in biosamples are based on the use of quadrupole mass filters. The inherent increase in sensitivity and selectivity achieved with triple quadrupole instrumental configurations in the selected-reaction monitoring (SRM) mode, even in spite of the high purchase cost, were noticed early, judging from the first application reported in 1996 by Dryskell [30]. Figure 10.2 shows, as an example, the difference between LC-MS and LC-MS-MS chromatograms for the determination of diazinon in human urine. The selection of an appropriate transition (m/z 305 → 169) in the MS-MS mode allows the determination of diazinon with increased selectivity and sensitivity in

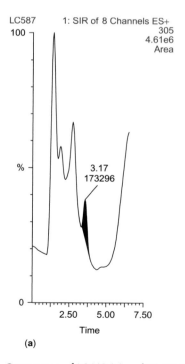
(a)

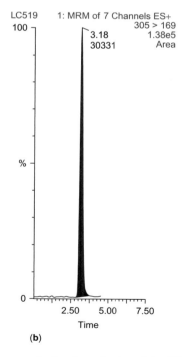
(b)

Figure 10.2 Comparison of LC-MS (a) and LC-MS-MS (b) chromatograms for the determination of the OPP diazinon in human urine (blank urine fortified at 25 μg/L).

comparison to the use of single MS mode (m/z 305), due to a much more favourable signal-to-noise (S/N) ratio.

Quite often, the transitions monitored in LC-MS-MS are based on the selection of $[M + H]^+$ or $[M - H]^-$ as precursor ions due to the relatively high acidity or basicity of most pesticides and metabolites. For some compounds the formation of other charged species has been described, such as clusters with solvents (ACN $[M + 42]^+$ and methanol $[M + 33]^+$), or adducts with sodium ($[M + 23]^+$) or ammonium ($[M + 18]^+$). However, the use of clusters or adducts as precursor ions may lead to problems such as poor reproducibility. Moreover, the fragments observed might present a low selectivity when working with solvent adducts or lower sensitivity when dealing with alkaline adducts that present a difficult fragmentation. Therefore, it is advisable to use the protonated or deprotonated ion as the precursor ion whenever it is feasible.

$[M + H]^+$ or $[M - H]^-$ ions can easily be formed in the ion source, depending on the presence of basic and acidic centres in the analyte molecules. As an example, Figure 10.3a (top) shows the full-scan spectrum in positive-ion mode of chlorpyrifos (molecular weight 349), an insecticide exhibiting two basic centres, a pyridine nitrogen and a thiophosphate, with protonation in an ESI ion source that is favoured ($[M + H]^+$ ion at m/z 350) [16].

In order to maximise sensitivity in the SRM mode, the cone voltage has to be optimised to favour the entrance of the ion chosen as the precursor into the mass analyser. In the case of chlorpyrifos, the abundance of the $[M + H]^+$ ion (precursor) was found to maximise at a voltage of 30 V, despite the occurrence of in-source fragmentation to give the ion at m/z 198. The full-scan spectrum also shows the typical isotopic pattern due to the presence of three chlorine atoms in the molecule (see peaks at m/z 352 and 354).

Once the precursor ion has been selected and the cone voltage optimised, the parameters for optimal SRM analysis (e.g. selection of product ions, collision energies, collision gas pressure) have to be tuned. In the chlorpyrifos example, the product ion spectrum (Figure 10.3a, middle) presented two abundant ions at m/z 198 and 125, with intensity maximised for both fragments at a collision energy of 20 eV. At the optimised collision energy the precursor ion almost disappeared, demonstrating efficient fragmentation. This fragmentation can be explained by the loss of an ethene molecule after a McLafferty rearrangement followed by the loss of the generated phosphothiolate (Figure 10.3a, bottom). Both fragments give excellent specificity as they are highly related to the original molecule. Therefore, two sensitive and selective transitions can be monitored. Additionally, the presence of abundant isotopic ions in the mass spectrum means that two different precursor ions (m/z 350 and 352) could also be selected to increase the number of transitions available.

Another example is shown in Figure 10.3b (top), corresponding to a molecule that can be easily deprotonated and is therefore suitable for ESI-MS in negative-ion mode. The full-scan spectrum shown corresponds to DEP, a DAP metabolite of several OPPs including chlorpyrifos, which is present in urine in the deprotonated form, thus reaching the ion source as $[M - H]^-$.

At a cone voltage of 20 V, the abundance of the $[M - H]^-$ signal (precursor) was found to maximise, although a small in-source fragmentation appeared at m/z 125. MS-MS experiments led to optimum collision energies of 12 and 20 eV for the two fragments at m/z 125 and 79, respectively (Figure 10.3b, middle). The abundance of the precursor ion in the daughter-ion scan spectrum is still important, revealing less efficient fragmentation in this case. Product ions can be explained by the loss of ethene after a McLafferty rearrangement (m/z 125) and the subsequent loss of ethanol (m/z 79) as shown in Figure 10.3b (bottom). These fragmentations correspond to the neutral loss of common molecules and are less specific than in the case of chlorpyrifos. The absence of isotopic pattern in the molecule (no bromine or chlorine atoms present) impedes the selection of several precursor ions in order to increase the number of available transitions. In spite of this, two transitions could be selected allowing the confirmation of this metabolite in the urine samples analysed.

A challenging situation occurs when the precursor ion exhibits poor MS-MS fragmentation

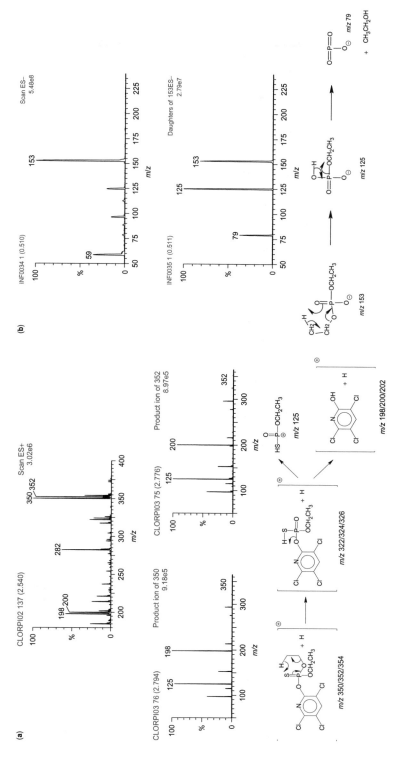

Figure 10.3 Full-scan spectrum (top), product-ion spectra (middle) and fragmentation pathway (bottom) for chlorpyrifos (a) and one of its DAP metabolites, DEP (b). (Modified from Hernández et al. [14] and Sancho et al. [16].)

due to its chemical characteristics. In this case, better sensitivity, at the expense of selectivity, can be obtained using a 'pseudo MS-MS' transition, where collision energy is set low enough to avoid fragmentation of the precursor ion in the collision cell and high enough to break up other ions with the same nominal m/z as the precursor, but less stability. Under these conditions, the same ion is monitored as precursor and product (so-called 'surviving-ion' technique). As an example, the strategy adopted in our laboratory for the determination of another chlorpyrifos metabolite, TCPY, is illustrated. Due to the presence of a phenolic group, which can be deprotonated to generate an abundant $[M - H]^-$ at m/z 196 (Figure 10.4a), TCPY is well suited for ESI-MS in the negative-ion mode. The isotopic ions at m/z 198 and 200 are due to the presence of three chlorine atoms in the molecule, as for the parent compound.

By increasing the collision energy to 20 eV, the product-ion spectrum of the TCPY precursor ion m/z 198 (Figure 10.4b, left) shows only an important fragment that corresponds to the chloride ion. At higher collision energies chloride was still the only fragment observed, but with lower sensitivity. Moreover, this transition is not very selective, as obviously it can be shared with other compounds containing chlorine. Additionally, the high amount of precursor ion still present shows low fragmentation efficiency, resulting in poor sensitivity for the transition m/z 198 → 35. Both the low sensitivity and low specificity of the product ion prompted us to acquire a 'pseudo-MS-MS transition' [16]. Figure 10.4 (middle, right) shows the product ion spectrum of TCPY obtained with a collision energy of 5 eV where no fragmentation is evident. As already mentioned, the use of 'pseudo-MS-MS' has limitations in selectivity compared with conventional MS-MS, but it allows a somewhat increased selectivity than ordinary single-stage MS analysis [25, 26]. In the case of TCPY, the use of additional 'pseudo-MS-MS' transitions due to the isotopic pattern given by the presence of chlorine (Table 10.3) allowed further improvements in the confirmation capability of the method [16]. The optimised MS-MS conditions for the determination of chlorpyrifos and its metabolites DEP and TCPY are summarised in Table 10.3.

Quantification and confirmation of analytes

LC separation coupled to tandem MS is considered among the most selective analytical techniques available nowadays and apparent interferences in the form of extra chromatographic peaks or peak shoulders are rarely encountered. Ion suppression resulting from inorganic salts and other components of complex matrices, such as urine, is the most common problem in quantitative LC-MS applications using ESI ion sources, whereas in APCI applications it is not regarded as a major problem [31]. The interference remains undetected because of the superior selectivity of the MS-MS detection, but it may lead to serious errors in the quantification of analytes if it is not appropriately corrected or minimised. The extent to which the analyte signal is suppressed can be simply determined from the signal difference between direct injection of standard solutions and sample extracts at the same concentration of analyte [32]. However, this approach provides no information about the chemical nature of interferents or the steps necessary to eliminate the interference. One possibility to visualise the interferents is to perform a full-scan acquisition, although the information obtained is biased to compounds ionisable in the current ion source and relevant interferents can remained unrevealed.

Apart from the unacceptable errors in quantification, ion suppression can be the cause for high LODs and false-negative results; however, less frequently, MEs can also produce signal enhancement for some analytes.

The use of isotopically labelled analytes as ISs in isotope-dilution-based methods is the best solution to compensate for MEs. The native and labelled analytes exhibit a virtually identical chemical behaviour. Therefore, it is expected that native and labelled analytes undergo equivalent MEs. Thus, the ratio of native to labelled species can be used to compensate for the variable recoveries [26]. In spite of this correction, strong ion-suppression effects can lead to poor sensitivity because of the decrease in mass spectrometric signal, forcing removal of interferences to obtain

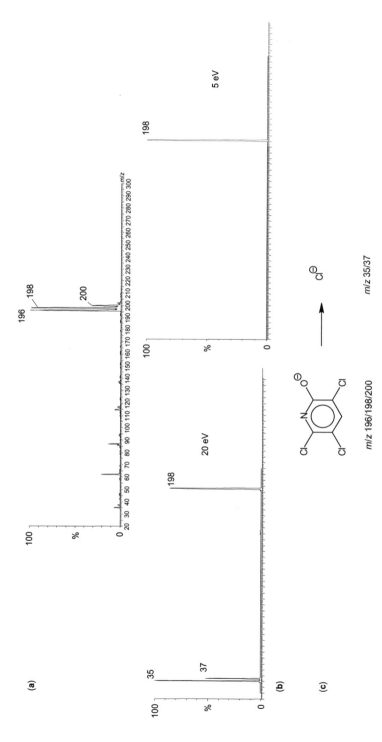

Figure 10.4 Full-scan spectrum (a), product-ion spectra (b) and fragmentation pathway (c) for the main chlorpyrifos metabolite, TCPY. (Modified from [16].)

Table 10.3 MS-MS optimised conditions for the determination of the OPP chlorpyrifos and some of its main metabolites in human urine

Compound	Structure	Ionisation mode	Precursor ion (m/z)	Cone voltage (V)	Product ion (m/z)	Collision energy (eV)
Chlorpyrifos		ESI+	350	30	198	20
					125	20
			352	30	200	20
					125	20
DEP		ESI–	153	20	125	12
					79	20
TCPY		ESI–	196	30	196	5
			198	30	198	5
			200	30	200	5

adequate LODs and limits of quantification (LOQs). Unfortunately, isotope-labelled analytes are not always available and this situation is even more frequent in the field of pesticide residue analysis. In such cases, the choice of an adequate IS is problematic [33].

Calibration in the matrix is always recommended when dealing with biological samples, but it requires the use of adequate blank extracts. Normally, pools of blank samples collected from a non-exposed population are used to prepare matrix-matched calibrators and quality control materials. The main problem of this approach is to assure the reproducibility for all samples, as the ME can be substantially different depending on the sample.

In some cases, another limitation is to find a true blank, which can be difficult when determining some compounds at the residue level. This is the case of PNP, where background levels of around 1 µg/L are often measured in the urine of non-exposed subjects [10, 34]. It is important to have knowledge of the background concentrations in humans in order adequately to estimate the degree of exposure in the examined population. Background data are available for some OPP metabolites, such as PNP, 2-isopropyl-6-methyl-pyrimidin-4-ol (IMPY), TCPY or malathion dicarboxylic acid (MDCA), which allow the prevalence and concentrations in the general population to be established [25].

In the absence of adequate labelled IS, the application of an efficient clean-up for minimising MEs is mandatory in order to obtain satisfactory results. Normally, SPE and LLE are the methods of choice for this purpose. As illustrated previously, an interesting alternative option is the use of LC-LC, efficient, automated, on-line clean-up. This technique, in combination with MS-MS, provides satisfactory quantification for several OPP metabolites in urine even without the need to use ISs [10, 16, 33].

The excellent characteristics of LC-MS detectors have led some authors to use this technique for the confirmation of pesticide-positive samples obtained by conventional analytical techniques

[35–37]. However, the selectivity of LC-MS detection may be overestimated in biological matrices due to their inherent complexity. Different confirmation criteria have been proposed, and some of them with particular reference to environmental samples [38]. An interesting criterion has been established in a recent decision of the European Union [39], based on the collection of identification points (IPs). The number of IPs required for confirmation is established as four and three depending on whether the analyte is a banned or non-banned compound, respectively. This means that at least three single ions have to be measured with LC-MS single instruments (three IPs) or two transitions monitored when using LC-MS-MS methods (four IPs). The measurement of one ion ratio within previously established tolerances is also necessary. This EU decision was originally set up for the determination of organic contaminants in food samples, although it has also been used recently for the confirmation of positive findings in other matrices such as environmental samples [38, 40, 41].

The monitoring of only a single ion in LC-MS-based methods or one transition in LC-MS-MS methods would lead to just one or two and a half IPs, respectively, thus being insufficient for positive confirmation. However, the approach of using a single ion or transition has been quite frequently used in the recent past, possibly due to the widespread idea that LC-MS(-MS) is by itself selective enough to confirm positive findings. When establishing the number of ions/transitions monitored, it should be kept in mind that this affects the sensitivity of the method. Additionally, in multi-residue methods for a large number of analytes, the use of two or more transitions per analyte could lead to poor definition of the chromatographic peak and, consequently, to poor quantitative performance of the method. One of the problems of recording two transitions for each analyte is that usually the second transition (used for confirmation) is less sensitive than the quantitative one. To be consistent, the LOD reported must be the lowest concentration to which the presence of the analyte can be detected and confirmed.

Together with the number of ions/transitions monitored, their specificity should also be taken into account. For example, the presence of an abundant isotopic ion in the molecule (e.g. in chlorine- or bromine-containing substances) would increase the attainable number of IPs, although other potential interferents exhibiting the same isotopic pattern could share the ions or transitions pre-selected for the analyte. Moreover, the use of non-selective losses in MS-MS-based methods, such as water or carbon dioxide, should be avoided in order to minimise the risk of reporting false positives or even false negatives.

In only few of the methods reviewed in this chapter has more than one ion or transition been monitored to assure positive findings. Lacassie *et al.* [4] used the pseudo-molecular ion or the most abundant characteristic ion for quantification in selected-ion monitoring (SIM) mode, and, when possible, two fragment ions for confirmation. Peak confirmation was made by matching of relative retention times and relative abundance of qualifier and quantifier ions. This approach was applied to confirm the presence of 11 carbamates and 3 benzimidazols in the serum of patients investigated in several intoxication cases.

In the reviewed LC-MS-MS methods applied to pesticide analysis in biosamples, measurement of two transitions has rarely been applied. As an example, monitoring of two transitions was used for the quantification and confirmation of the metabolite PNP [10, 42], and for the determination of alkylphosphates [14]. In the last case, the use of two transitions was also required to minimise possible errors due to the low reproducibility in retention times. Figure 10.5 shows the MS-MS chromatograms obtained for the determination of chlorpyrifos and several of its metabolites in serum and urine of a grower who applied this insecticide in citrus orchards. Both the parent compound and in its main metabolite TCPY was detected in serum collected the day after application. In addition, several metabolites was detected in urine, mainly TCPY and the DAPs, DEP and DETP. The use of three pseudo-MS-MS transitions for TCPY and two MS-MS transitions for chlorpyrifos and DAPs allowed us to confirm the presence of these compounds in the samples analysed [14].

For a thorough discussion on identification and confirmation criteria for LC-MS, *see* Chapter 5.

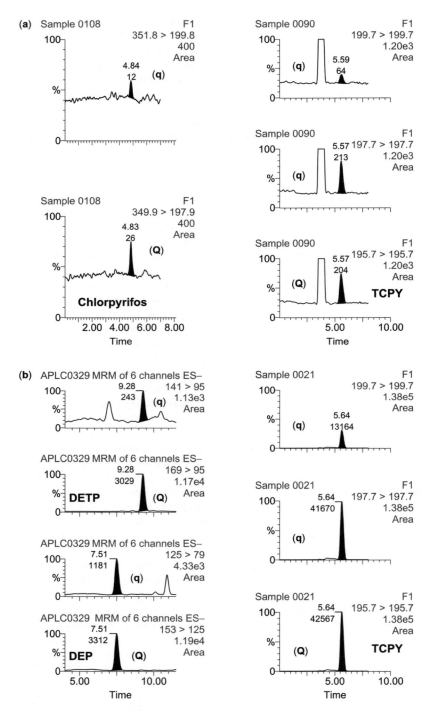

Figure 10.5 MS-MS chromatograms for chlorpyrifos and some of its metabolites determined in serum (a) and urine (b) of a grower who applied this insecticide for citrus pest control. Q: quantification; q: confirmation. (Urine chromatograms modified from Hernández et al. [14].)

Selected applications

Monitoring of exposure

Most published LC-MS methods for the determination of pesticides in biological samples have been developed for monitoring pesticide exposure – a field where high sensitivity is needed. In these cases, the analysis of a high number of samples is usually required and therefore the sample pre-treatment reduction achievable by LC-MS can help the analyst to reach this objective. Moreover, the wide polarity range of compounds ionisable by API sources allows determination by this technique of most of the pesticides (except highly apolar ones such as OCPs) and metabolites facilitating the development of multi-residue methods. These methods can be useful for general population exposure monitoring where the contamination source is not always controlled and a high number of potential pesticides have to be considered.

Olsson *et al.* [26] proposed a method for the efficient and rapid simultaneous determination of 19 markers of commonly used pesticides in human urine. More than 100 urine samples were analysed and the results were compared with those obtained by a previously reported GC-MS-MS method [34], giving comparable results even though the samples were analysed separated in time and with different standard solutions for calibration curves. The use of LC-MS-MS decreased the total sample preparation time by 75%, demonstrating the suitability of this technique in the general population exposure monitoring of pesticides.

A different approach is to develop specific methods in order to monitor the exposure to a particular pesticide or family of pesticides. This is, for example, the case of monitoring OPP exposure in growers applying these insecticides. An indirect measurement as the determination of cholinesterase activity in serum is the most commonly used indicator of OPP exposure. In addition to some already known problems associated with this determination, the invasive nature of blood collection makes it difficult to perform this control as efficiently and frequently as desirable. The use of LC-MS-MS can circumvent this problem by determining DAPs in urine in order to control OPP exposure [10, 14]. DAPs were determined in urine of growers who applied chlorpyrifos and the results compared with cholinesterase activity in serum. Additionally, the specific metabolite of chlorpyrifos, TCPY, was also determined in urine (Table 10.4). Except for a grower (identified as 'c' in Table 10.4) who seemed to have little exposure to this insecticide, the concentrations obtained for TCPY (concentration range 130–6300 µg/L) were much higher than for DAPs (concentration range 8–99 µg/L).

However, at lower concentrations, DAPs were found in all samples, due to the excellent sensitivity of the method, confirming the exposure to OPPs even in cases where the cholinesterase activity was at the normal level. In general, a good correlation was obtained between urinary metabolites and cholinesterase activity. This example shows the utility and advantages of LC-MS-MS urinary methods to measure the general exposure to OPP. The chromatograms in Figure 10.5b show the high sensitivity and specificity of the technique.

Acute poisoning cases

Although a number of fatalities due to pesticides have been reported, little quantitative toxicological data have been found in the literature based on the use of LC-MS. Only Lacassie *et al.* [4] reported a few cases of intoxication to demonstrate the suitability of LC-MS for diagnosis, fulfilling the quantification, specificity and

Table 10.4 Levels of cholinesterase activity in serum and chlorpyrifos metabolites in urine from several growers exposed to this insecticide

Grower	Cholinesterase activity (U/L)	DEP (µg/L)	DETP (µg/L)	TCPY (µg/L)
a	2592	99	40	6300
b	3581	65	47	1300
c	6368	11	8	16
d	4739	62	24	800
e	6707	21	8	130

selectivity requirements of diagnostic methods. Moreover, LC-MS methods avoid laborious, time-consuming extraction and derivatisation steps that are troublesome in emergency cases of severe poisoning. These authors developed a LC-MS method for 14 pre-selected polar pesticides (carbamates and benzimidazoles) in serum samples, previously extracted using mixed-mode Oasis MCX cartridges. This method was applied in some intoxication cases and positive results were found for two carbamates, carbofuran and aldicarb.

In the case of carbofuran intoxication, the blood samples of the patient showed a low serum cholinesterase activity (1720 U/L). A treatment with pralidoxime was first administrated and then stopped as soon as the intoxication by a carbamate pesticide instead of OPP was diagnosed by LC-MS. Carbofuran was detected at high concentrations (1.3–1.5 mg/L).

In the case of aldicarb intoxication, there were no initial hints about the possible toxicant, but the patient's symptoms were characteristic of an OPP intoxication, which was first confirmed by a significantly decreased blood cholinesterase activity. However, aldicarb has recently been identified as most likely to be responsible for this intoxication when gastric contents were analysed by LC-MS. Aldicarb was not quantified, as the method was validated only in serum and potential MEs from gastric components were not addressed.

In these examples, an LC-MS method with a single quadrupole was developed, but the benefits of developing LC-MS-MS methods for acute intoxication diagnosis have not yet been described. Among potential advantages, MS-MS detection allows us to put less emphasis on chromatographic separation, thus increasing the sample throughput and achieving the quick response desirable in acute poisoning cases. The possibility of a rapid detection of DAPs in urine makes it easier to assign the intoxication to either OPPs or carbamates — a useful diagnostic test for an adequate treatment as shown in the example of carbofuran. Another potential benefit is the inherent sensitivity of the technique which, although not necessary in the diagnosis of intoxication cases where high concentration levels are expected, may help in accelerating and simplifying the analysis. Thus, almost no sample treatment would be required and reliable quantification could be easily performed by simple high dilution of the sample with LC-grade water (e.g. ×10, ×50 or even higher) in order to overcome undesirable MEs.

Concluding remarks

LC-MS, after long-term development of adequate interfaces, is now an appropriate technique for application in the field of pesticide residue analysis. Determination of low concentration levels of pesticides and their metabolites in biosamples is necessary to investigate the relationship between pesticide exposure and health effects. In contrast to persistent organochlorine compounds, currently used pesticides are non-persistent and do not tend to accumulate in the environment or in lipid-rich tissues. They are typically metabolised and excreted from the body within a few days. Thus, the determination of polar metabolites in urine is the most common way to assess the exposure to currently used pesticides — an analytical field where LC-MS is the technique of choice. As a result of the further improved sensitivity and selectivity, LC-MS-MS provides a straightforward advantageous approach for the determination of polar, non-volatile and thermolabile compounds, like the majority of pesticide metabolites. In addition, aqueous samples or extracts can be directly injected and polar/ionic analytes determined without any derivatisation step (as typically required in GC methods). All these characteristics allow the minimisation of sample treatment in comparison to GC-MS-based methodology.

Until now, the main applications of LC-MS have been focused on the determination of pesticides and/or metabolites in urine in order to evaluate the exposure to pesticides. However, LC-MS presents a great potential for other applications, such as acute poisoning diagnosis or monitoring of long-term exposure, particularly in the case of medium-polar or polar target analytes. Measurement of intact pesticides in whole blood, serum, urine or body tissues may be performed to confirm acute poisoning cases, where the reduction of sample treatment achieved by LC-MS might help to minimise the time of analytical response. Due to the inherent advantages of LC-MS in this field, it is expected that the number of applications will drastically increase in the near future.

References

1. Barr DB, Needham LL. Analytical methods for biological monitoring of exposure to pesticides: a review. *J Chromatogr B* 2002; 778: 5–29.

2. Barr JR, Driskell WJ, Hill RH, *et al*. Strategies for biological monitoring of exposure for contemporary-use pesticides. *Toxicol Ind Health* 1999; 15: 169–180.

3. Anwar WA. Biomarkers of human exposure to pesticides. *Environ Health Perspect* 1997; 105(Suppl 4): 801–806.

4. Lacassie E, Marquet P, Gaulier JM, *et al*. Sensitive and specific multiresidue methods for the determination of pesticides of various classes in clinical and forensic toxicology. *Forensic Sci Int* 2001; 121: 116–125.

5. Saint-Marcoux F, Lachatre G, Marquet P. Evaluation of an improved general unknown screening procedure using liquid chromatography-electrospray-mass spectrometry by comparison with gas chromatography and high performance liquid chromatography-diode array detection. *J Am Soc Mass Spectrom* 2003; 14: 14–22.

6. Maurer HH. Liquid chromatography-mass spectrometry in forensic and clinical toxicology. *J Chromatogr B* 1998; 713: 3–25.

7. Aprea C, Colosio C, Mammone T, *et al*. Biological monitoring of pesticide exposure: a review of analytical methods. *J Chromatogr B* 2002; 769: 191–219.

8. Hoja H, Marquet P, Verneuil B, *et al*. Applications of liquid chromatography-mass spectrometry in analytical toxicology: a review. *J Anal Toxicol* 1997; 21: 116–126.

9. Hernández F, Sancho JV, Pozo OJ. Critical review of the application of liquid chromatography/mass spectrometry to the determination of pesticide residues in biological samples. *Anal Bioanal Chem* 2005; 382: 934–946.

10. Hernández F, Sancho JV, Pozo OJ. An estimation of the exposure to organophosphorus pesticides through the simultaneous determination of their main metabolites in urine by liquid chromatography-tandem mass spectrometry. *J Chromatogr B* 2004; 808: 229–239.

11. Kurttio P, Vartiainen K, Auriola S. Measurement of ethylenethiourea using thermospray liquid chromatography-mass spectrometry. *J Anal Toxicol* 1992; 16: 85–87.

12. Sottani C, Bettinelli M, Fiorentino ML, *et al*. Analytical method for the quantitative determination of urinary ethylenethiourea by liquid chromatography/electrospray ionization tandem mass spectrometry. *Rapid Commun Mass Spectrom* 2003; 17: 2253–2259.

13. Cech NB, Enke CG. Practical implications of some recent studies in electrospray ionization fundamentals. *Mass Spectrom Rev* 2001; 20: 362–387.

14. Hernández F, Sancho JV, Pozo OJ. Direct determination of alkyl phosphates in human urine by liquid chromatography/electrospray tandem mass spectrometry. *Rapid Commun Mass Spectrom* 2002; 16: 1766–1773.

15. Zang LY, DeHaven J, Yocum A, *et al*. Determination of alachlor and its metabolites in rat plasma and urine by liquid chromatography-electrospray ionization mass spectrometry. *J Chromatogr B* 2002; 767: 93–101.

16. Sancho JV, Pozo OJ, Hernández F. Direct determination of chlorpyrifos and its main metabolite 3,5,6-trichloro-2-pyridinol in human serum and urine by coupled-column liquid chromatography/electrospray-tandem mass spectrometry. *Rapid Commun Mass Spectrom* 2000; 14: 1485–1490.

17. Katagi M, Tatsuno M, Tsuchihashi H. Determination of malathion metabolites in biological samples by LC-MS. *Japan J Toxicol Environ Health* 1994; 40: 357–364.

18. Tsatsakis A, Tutudaki M. Progress in pesticide and POPs hair analysis for the assessment of exposure. *Forensic Sci Int* 2004; 145: 195–199.

19. Russo M. V, Campanella L, Avino P. Determination of organophosphorus pesticides residues in human tissues by capillary gas chromatography-negative chemical ionization mass spectrometry analysis. *J Chromatogr B* 2002; 780: 431–441.

20. Tsatsakis AM, Tutudaki MI, Tzatzarakis MN, *et al*. Pesticide deposition in hair: Preliminary results of a model study of methomyl incorporation into rabbit hair. *Vet Hum Toxicol* 1998; 40: 200–203.

21. Gergov M, Ojanpera I, Vuori E. Simultaneous screening for 238 drugs in blood by liquid chromatography-ionspray tandem mass spectrometry with multiple-reaction monitoring. *J Chromatogr B* 2003; 795: 41–53.

22. Itoh H, Kawasaki S, Tadano J. Application of liquid chromatography-atmospheric pressure chemical ionization mass spectrometry to pesticide analysis. *J Chromatogr A* 1996; 754: 61–76.

23. Hennion MC. Solid-phase extraction: method development, sorbents, and coupling with liquid chromatography. *J Chromatogr A* 1999; 856: 3–54.

24. Hernández F, Sancho JV. Liquid chromatography: multidimensional. In: Worsfold PJ, Townshend A, Poole CF, eds, *Encyclopedia of Analytical Science 5*, 2nd edn. Oxford: Elsevier, 2005: 197–294.

25. Olsson AO, Nguyen JV, Sadowski MA, *et al.* A liquid chromatography/electrospray ionisation-tandem mass spectrometry method for quantification of specific organophosphorus pesticide biomarkers in human urine. *Anal Bioanal Chem* 2003; 376: 808–815.

26. Olsson OA, Baker SE, Nguyen JV, *et al.* A liquid chromatography-tandem mass spectrometry multiresidue method for quantification of specific metabolites of organophosphorus pesticides, synthetic pyrethroids, selected herbicides, and DEET in human urine. *Anal Chem* 2004; 76: 2453–2461.

27. Hogendoorn EA, van Zoonen P, Hernández F. The versatility of coupled-column LC. *LC-GC Europe* 2003; 16: 44–51.

28. Kawasaki S, Ueda H, Itoh H, *et al.* Screening for organophosphorous pesticides using liquid chromatography-atmospheric pressure chemical ionisation mass spectrometry. *J Chromatogr* 1992; 595: 193–202.

29. Pelander A, Ojanpera I, Laks S, *et al.* Toxicological screening with formula-based metabolite identification by liquid chromatography/time-of-flight mass spectrometry. *Anal Chem* 2003; 75: 5710–5718.

30. Driskell WJ, Hill RH, Shealy DB, *et al.* Identification of a major human urinary metabolite of alachlor by LC-MS/MS. *Bull Environ Contam Toxicol* 1996; 56: 853–859.

31. Matuszewski BK, Constanzer ML, Chavez-Eng CM. Matrix effect in quantitative LC/MS/MS analyses of biological fluids: a method for the determination of finasteride in human plasma at picogram per milliliter concentrations. *Anal Chem* 1998; 70: 882–889.

32. Buhrman DL, Price PI, Rudewicz PJ. Quantitation of SR 27417 in human plasma with electrospray liquid chromatography tandem mass spectrometry: a study of ion suppression. *J Am Soc Mass Spectrom* 1996; 7: 1099–1105.

33. Sancho JV, Pozo OJ, López FJ, *et al.* Different quantitation approaches for xenobiotics in human urine samples by liquid chromatography/electrospray tandem mass spectrometry. *Rapid Commun Mass Spectrom* 2002; 16: 639–645.

34. Hill RH, Lead SL, Baker S, *et al.* Pesticide residues in urine of adults living in the United States: reference range concentrations. *Environ Res* 1995; 71: 99–108.

35. Lyubimov AV, Garry VF, Carlson RE, *et al.* Simplified urinary immunoassay for 2,4-D: validation and exposure assessment. *J Lab Clin Med* 2000; 136: 116–124.

36. Martínez Fernández J, Parrilla Vázquez P, Martínez Vidal JL. Analysis of N-methylcarbamates and some of their main metabolites in urine with liquid chromatography using diode array detection and electrospray mass spectrometry. *Anal Chim Acta* 2000; 412: 131–139.

37. Driskell WJ, Groce DF, Hill RH. Methomyl in the blood of a pilot who crashed during aerial spraying. *J Anal Toxicol* 1991; 15: 339–340.

38. Reetsma T. Liquid chromatography-mass spectrometry and strategies for trace-level analysis of polar organic pollutants. *J Chromatogr A* 2003; 1000: 477–501.

39. Commission of the European Communities. *Commission Decision 2002/657/EC* of 12 August 2002, implementing Council Directive 96/23/EC concerning the performance of analytical methods and the interpretation of results, 2002.

40. Hernández F, Ibáñez M, Sancho JV, *et al.* Comparison of different mass spectrometric techniques combined with liquid chromatography for confirmation of pesticides in environmental water based on the use of identification points. *Anal Chem* 2004; 76: 4349–4357.

41. Stolker AAM, Dijkman E, Niesing W, *et al.* Identification of residues by LC/MS/MS according to the new European Union guidelines: application to the trace analysis of veterinary drugs and contaminants in biological and environmental matrices.

In: Ferrer I, Thurman EM, eds, *Liquid Chromatography/Mass Spectrometry, MS/MS and Time-of-Flight MS: Analysis of Emerging Contaminants (ACS Symposium Series 850)*. American Chemical Society: New York, 2003.

42. Barr DB, Turner WE, DiPietro E, *et al*. Measurement of *p*-nitrophenol in the urine of residents whose homes were contaminated with methyl parathion. *Environ Health Perspect* 2002; 110: 1085–1091.

43. Marin A, Martínez Vidal JL, Egea González FJ, *et al*. Assessment of potential (inhalation and dermal) and actual exposure to acetamiprid by greenhouse applicators using liquid chromatography-tandem mass spectrometry. *J Chromatogr B* 2004; 804: 269–275.

44. Baker SE, Olsson AO, Barr DB. Isotope dilution high-performance liquid chromatography-tandem mass spectrometry method for quantifying urinary metabolites of synthetic pyrethroid insecticides. *Arch Environ Contam Toxicol* 2004; 46: 281–288.

45. Pozzebon JM, Vilegas W, Jardim ISCF. Determination of herbicides and a metabolite in human urine by liquid chromatography-electrospray ionisation mass spectrometry. *J Chromatogr A* 2003; 987: 375–380.

46. Martínez JM, Martínez Vidal JL, Vazquez PP, *et al*. Application of restricted-access media column in coupled-column RPLC with UV detection and electrospray mass spectrometry for determination of azole pesticides in urine. *Chromatographia* 2001; 53: 503–509.

47. Baker SE, Barr DB, Driskell WJ, *et al*. Quantification of selected pesticide metabolites in human urine using isotope dilution high-performance liquid chromatography/tandem mass spectrometry. *J Expo Anal Environ Epidemiol* 2000; 10: 789–798.

48. Beeson MD, Driskell WJ, Barr DB. Isotope dilution high-performance liquid chromatography/tandem mass spectrometry method for quantifying urinary metabolites of atrazine, malathion, and 2,4-D-dichloropenoxyacetic acid. *Anal Chem* 1999; 71: 3526–3530.

49. Driskell WJ, Hill RH. Identification of a major human urinary metabolite of metholachlor by LC-MS/MS. *Bull Environ Contam Toxicol* 1997; 58: 929–933.

50. Kawasaki S, Nagumo F, Ueda H, *et al*. Simple, rapid and simultaneous measurement of eight different types of carbamate pesticides in serum using liquid chromatography-atmospheric pressure chemical ionization mass spectrometry. *J Chromatogr* 1993; 620: 61–71.

11

Peptide analysis using LC-MS

Isabel Riba Garcia and Simon J. Gaskell

Analysis of peptides

Introduction

Over recent years, mass spectrometry (MS) has become one of the major growth techniques for biomolecular analysis. Its applications cover a variety of different areas in biology, drug discovery, forensic science, chemistry, etc., with toxicology being one of the latest additions. Although numerous toxicological studies have been performed using gas chromatography T(GC-MS), the development of novel ionisation techniques, such as atmospheric pressure ionisation (API) and matrix-assisted laser desorption ionisation (MALDI), which allow the analysis of different classes of compounds, has extended the areas of application in toxicological research from drug abuse and postmortem toxicology to oxidative stress, studies of teratogens, cancer research, apoptosis and many others.

Mass spectrometry

The development of so-called 'soft' ionisation techniques, i.e. electrospray ionisation (ESI) and MALDI, permitted the analysis of thermally labile compounds. These ionisation methods have now become conventional, and research efforts have more recently become focused upon the use of analysers and detectors to improve the sensitivity and accuracy of analysis. With all this, MS has become one of the most important techniques for the analysis of peptides and proteins both in vivo and in vitro; therefore, the analysis of modified peptides and proteins that are putatively toxic for organisms can be studied. At the same time, abnormal levels of endogenous compounds that could produce toxicological effects can also be studied in vivo.

Matrix-assisted laser desorption ionisation

MALDI was firstly introduced by Karas and Hillenkamp [1], working in parallel with Tanaka *et al.* [2], and since then has been developed to a point where it is among the major techniques currently in use for the analysis of biopolymers.

Gas-phase ions are generated by application of laser pulses, generally using nitrogen lasers, with an emission wavelength of 337 nm. The sample is mixed with an excess of a 'matrix' – a compound with an absorption maximum similar to the laser wavelength and which bears an acidic proton (for positive-ion formation) for transfer to the analyte. The matrix–analyte mixture is allowed to dry on the MALDI target to form a heterogeneous distribution of crystals. When the laser is fired, excitation of the matrix molecules decreases the binding forces to the target, and a 'plume' of matrix and analyte neutrals and ions is released. Although the ionisation process is not fully understood, the significant net result is proton transfer to the analyte [3–5]. Using this ionisation process, singly charged ions are mainly observed. As in the case of ESI, the amount of energy transferred to the analyte is small and little or no fragmentation is observed without further excitation of the ions; accordingly, the technique is also described as 'soft' ionisation. MALDI has found wide application in peptide and protein analysis, particularly in combination with time-of-flight (TOF) MS.

Post-source decay (PSD) is an extension of

MALDI-TOF in which the ions formed in the ion source gain internal energy during the desorption process and can release this energy by fragmenting in the field-free (flight tube) region. The product ions formed have lower kinetic energies, but maintain the same velocity as their precursor ions. The reflectron achieves separation according to kinetic energy and therefore according to the mass-to-charge (m/z) ratio in this instance. In practice, satisfactory focusing of a wide range of product m/z values requires sequential data acquisition at several reflectron voltages, with post-acquisition accumulation of data into a single spectrum. This requirement can be obviated if a curved-field reflectron is used [6, 7]. The PSD technique becomes a genuine tandem MS method if an ion gate is incorporated for selection (albeit generally with modest resolution) of precursor ions. For the analysis of peptides (see below), the fragmentation of singly charged ions, as produced by MALDI, is commonly less informative than the fragmentation of multiply charged ions; this, together with the generally modest yield of fragment ions and the time-consuming data acquisition, has contributed to the infrequent uptake of PSD as a routine tool.

TOF-TOF

A refinement of TOF-based tandem MS may be achieved by the coupling of two separate analyser regions with an intermediate decomposition region [8]. The incorporation of an ion gate within the first analyser region allows selection of the precursor species (albeit with modest resolution), which may then be activated for decomposition by collision with neutral gas prior to the introduction of surviving precursor and products into the second TOF analyser (generally a reflectron). If the decomposition region is held at elevated potential, the range of relative kinetic energies of the product ions is reduced, allowing the acquisition of complete product-ion spectra using a single set of reflectron potentials.

Peptide analysis using mass spectrometry

MS is one of the most useful techniques for the analysis of peptides. Using MS, not only can the mass of the peptide be obtained, but also sequence information and possible modifications. Although the ion fragmentation pathways are complex, understanding is sufficient to allow unequivocal identification of at least a partial sequence in many cases. Peptide fragmentation nomenclature is widely accepted. Cleavage of peptide bonds can yield ions containing the N-terminus of the peptide, known as b-ions, and fragment ions containing the C-terminus, known as y-ions [9].

Both ESI and MALDI ionisation are widely used for peptide analysis. Whilst MALDI provides the simpler approach for molecular weight determination, ESI is more widely used for quantification purposes and, in conjunction with tandem MS, for sequencing.

Liquid chromatography

Isolation of peptides from biological matrices is a difficult task because they are present in a complex mixture with other components such as acids, amino acids, sugars, salts and others. Prior to MS analysis, some degree of peptide purification is therefore generally required. Although techniques like capillary electrophoresis and micellar electrokinetic chromatography (MEKC) are gaining in application, the most widely used techniques for peptide and protein isolation and analysis are reversed-phase liquid chromatography (RP-LC), ion-exchange chromatography (IEC) and size-exclusion chromatography (SEC), achieving separation according to hydrophobicity, net charge and molecular size, respectively.

Reverse-phase liquid chromatography

RP-LC is the most commonly used separation technique for peptides. The separation is based on differences in the hydrophobicity of analytes, allowing partition between a stationary phase (contained within a column) and a mobile phase which flows through the stationary phase. A basic principle of 'like interacts with like' applies, where non-polar analytes will be preferentially retained by the non-polar stationary phase, whereas polar components will pass through the column relatively unhindered, preferring to remain in the polar mobile phase.

Improved RP-LC separation of peptides has

been demonstrated by the development of surface-alkylated polystyrene monolithic columns, suitable also for capillary LC-MS. The advantage of these columns is that they consist of a single polymeric or silica-based monolith. No frits to retain the packed bed are required; moreover, they provide a stable chromatography bed, adjustable pore diameter and low column back-pressure under high eluent flows [10].

Another type of packing material that has been reported to be very effective for the separation of peptides in complex mixtures is porous graphitic carbon [11]. This packing material is very effective for the separation of small peptides from phenolic compounds.

Ion-exchange chromatography

Retention in IEC is based on the presence of charged groups covalently attached to the stationary phase – anion-exchange columns carry a positive charge and cation-exchange columns carry a negative charge. An increase in salt or buffer concentration results in decreased retention and the effect is greater for more highly charged compounds. The ionic strength of the mobile phase is varied to control sample retention.

Capillary liquid chromatography

Requirements for the analysis of very low amounts of compounds (down to attomoles) have led to developments in LC. The most commonly used columns, with internal diameters (i.d.) in the range 3.2–4.6 mm and operated at mobile-phase flow rates in excess of 200 µL/min, have been substituted by capillary columns with 150–500 µm i.d. (and flow rates of 1–10 µL/min) or nano-columns with 10–150 µm i.d. (and flow rates from 10 to 1000 nL/min) [12] for analysis of peptides and proteins from biological samples. The main advantages of using smaller-diameter columns are higher sensitivity, improved resolution leading to higher peak capacity and superior separation efficiencies (compared with conventional-scale LC). The sensitivity increase obtained using smaller-diameter columns arises as a result of a reduction in the volume in which the components elute. The main advantage of using capillary and nano-LC for the analysis of biological samples is that dilute or low concentrated samples can be concentrated on the LC column prior to elution.

Two-dimensional liquid chromatography

Multi-dimensional separation may be necessary in some cases. This approach exploits two or more independent physical properties of the analytes to achieve a higher resolution and loading capacity than can be achieved by a single dimension alone. Different strategies are therefore possible to provide complete separation of analytes from a sequential combination of two one-dimensional (1-D) approaches that, used independently, would not suffice [13]. Some of the most commonly used properties are anion- or cation exchange, RP, hydrophobic interaction or SEC and all their combinations. The combination of strong cation exchange in the first dimension followed by RP in the second dimension seems to have several advantages. First, it provides orthogonality as the first dimension separates by charge and the second by hydrophobicity. Second, the elution solvents from the first dimension may be solvents that are compatible with RP-LC, resulting in focusing of the chromatographic peaks in the second dimension. Finally, the second-dimension elution solvent is compatible with the mass spectrometric techniques normally used for the ionisation of the analytes [14].

A combination of solid-phase extraction followed by two-dimensional (2-D) LC-MS has been used for the characterisation of neuropeptides in rat brain tissue [14]. A strong cation-exchange (SCX) column in the first dimension followed by an RP column in the second dimension was used. On injection, the positively charged components of the samples are trapped on the SCX column; the different components then elute in different molarities of ammonium formate from a step gradient. The fractions eluting from this first dimension were diluted with water containing 0.1% trifluoroacetic acid and trapped onto a C_{18} pre-column which was back-flushed onto the RP analytical column (second dimension) where the peptides elute in a water/acetonitrile gradient. Two RP trap pre-columns (operating in parallel) were used to speed up the process.

An on-line 2-D-LC system with integrated sample treatment has also been described for the analysis of proteins and peptides with a molecular mass below 20 kDa [15].

On-line and off-line 2-D-LC systems are now very popular for peptide separations during proteomics analysis, with or without gel separation of proteins (*see below*).

LC-MS

Coupling LC and MS is now straightforward; the solvents and flow rates used in LC are compatible with ESI analysis. The advantages of using nano-flow LC systems have been augmented by developments in low-flow ESI [16, 17]. Wilm and Mann [18] demonstrated that the diameter of the droplets emitted by the Taylor cone was closely related to the diameter of the tip of the cone where the droplets are produced. The low-flow ESI provides increased sensitivity and signal stability compared with higher-flow sources. The improvement in sensitivity is mainly a result of improvements in ionisation efficiency due to the formation of smaller droplets and improved ion-sampling efficiency into the MS as a result of positioning of the spray tip closer to the MS inlet [18, 19].

LC-MALDI

When ESI-MS-MS is used with on-line LC, analysis of analytes must occur within their chromatographic residence times; this limits the analysis to a window of typically 10–30 s. As a result, many co-eluting peptides, particularly those of low abundance, may go undetected. In contrast, if chromatographic fractions are collected off-line and spotted onto a MALDI target, re-analysis of the sample can occur as many times as required.

One of the disadvantages of off-line analysis is the sample loss; as the fractions are collected from a low-flow system (with subsequent addition of matrix), recovery of the sample for target spotting may not be complete. To overcome this problem, different systems for on-line MALDI spotting have been developed. The use of a heated capillary nebuliser that provides for rapid evaporation of the eluent has been described by Gostick *et al.* [20]. Alternative approaches include a non-contact liquid deposition system which uses a controlled pulsed electric field [21], a heated droplet interface [22], and an oscillating capillary nebuliser [23]; assisted fractionation by a liquid sheath flow and solvent evaporation by a gas sheath flow [24] have also been described.

Although both techniques, LC-ESI and LC-MALDI, have their advantages and disadvantages, comparisons between LC-ESI-MS-MS and LC-MALDI-MS-MS, showed that neither technique is invariably superior, but rather they are complementary [25, 26].

Capillary electrophoresis

Electrophoresis is the migration of ions through solution in the presence of an applied electric field. In the early 1980s, Jorgenson and Lukacs [27, 28] demonstrated capillary electrophoresis (CE) using 75-µm i.d. fused silica capillaries. A phenomenon of CE is electro-osmotic flow (EOF), which is the net movement of solution through a buffer-filled capillary under the influence of an electric field. The magnitude and direction of the flow depends on the applied potential and the surface properties of the capillary. EOF requires careful control if high-efficiency separations with good resolution are to be achieved.

Although EOF produces a bulk flow in one direction leading to reduced analysis times and increased separation efficiency (sharper peaks due to less diffusion), resolution will often be less than that obtained in a system where EOF is suppressed as the components have less time to separate. Although there are several 'modes' of CE (including isoelectric focusing and isotachophoresis), the most commonly used is capillary zone electrophoresis (CZE) which separates the components into discrete bands or zones as they migrate through the capillary.

CE-MS

Coupling CE with ESI is not straightforward, as consideration needs to be taken of the voltages applied to provide the separation and the voltage required for the ESI process. Further the low or non-existent mobile phase flow (EOF) means that standard approaches for coupling LC to ES interfaces are not suitable.

Several interfaces for CE-MS have been developed.

- *The coaxial sheath flow interface [29].* This uses three concentric capillaries. The inner capillary (fused silica) is used for the electrophoretic separation and is inserted into a narrow metal tube which corresponds to the ESI capillary. The ESI capillary (stainless steel) is placed concentrically around the fused silica capillary and delivers the sheath liquid, which is necessary to complete the electrical circuit. The third concentric tube (stainless steel) delivers the gas flow (nebulising gas) which helps the ES process and provides cooling for the CE capillary.
- *The sheathless interface [30].* The CE capillary is sharpened by etching the wall with hydrofluoric acid or mixtures of sulphuric and phosphoric acids. The outside surface of the capillary is coated with a layer of conductive metal to complete the electric circuit.
- *The liquid junction interface [31].* The CE capillary is partially decoupled from the ESI circuit. Independent optimisation of ESI and CE is one of the major advantages. The junction is made of a low dead-volume T-part. A make-up flow is admitted through the side arm of the T and introduced to the sprayed liquid via the gap between the CE capillary and the ESI emitter to complete the circuit.

Applications of peptide analysis in toxicology

MS has been widely used in toxicology studies, mainly by using GC-MS(-MS) in the analysis of small molecules. More recently, ESI interfaced with LC (LC-MS and LC-MS-MS) has provided the method of choice for screening and quantification of peptides and proteins with therapeutic or toxic properties. Compounds such as toxins from marine, plant or animal origin (venoms), hormones, cytokines and peptides or proteins produced by genetic engineering are suitable for this type of analysis. One of the problems in the analysis of peptides and proteins by MS is the difficulty in their extraction and separation. However, as mentioned early in this chapter, different methods have been developed to successfully separate peptides and proteins from complex mixtures for mass spectrometric analyses. Nevertheless, the analysis of peptides from biological matrices remains complex and the analyses discussed in this chapter correspond to the few examples present in the toxicology literature.

Forensic toxicology

An important objective of clinical and forensic toxicological analysis is the identification of initially undetermined substances in biological material summarised as 'general unknown screening (GUS)'. Moreover, the sample volume of blood or urine available is generally low and the need for accurate, rapid identification of components may be urgent. Looking for an *a priori* unknown poison in cases of suspected poison intoxication by MS is a challenge since only limited numbers of mass spectra of drugs or pesticides are available [32]. In order to develop a comprehensive method for systematic screening in postmortem analysis, appropriate steps in the complex process of sample work-up, isolation of crude extracts, differentiation and separation between heterogeneous phases, concentration of intermediate and final fractions, and the detection and identification of key components have to be selected and carried out in a systematic way (*see* Chapter 6). This approach must therefore be well planned. In addition to the diversity of specimens, the forensic toxicologist has to consider the limited stability of postmortem specimens which is caused by human and bacterial degradation enzymes, and which can result in sample artefacts and contamination. When working with tissue, samples require homogenisation and it may be necessary to concentrate the sample by protein precipitation. The precipitants required, such as strong acids or bases, may result in the loss of relevant substances [33] or co-precipitation of contaminating artefacts.

With all these considerations in mind and knowing the low stability of peptides and proteins, it is not surprising that few reports of peptide or protein analyses by MS in forensic toxicology are available [34].

Analysis of endogenous and exogenous peptides

Analysis of peptides and proteins in urine from Fanconi syndrome [35, 36]

It is known that urine from healthy patients does not contain many peptides and proteins; however, urine from Fanconi syndrome patients is rich in both types of components. MS analysis by LC-ESI-MS-MS of the fractions with and without SEC separation was carried out as part of a study of toxicological effects of different endogenous and exogenous compounds. Potentially bioactive peptides and novel proteins were identified in urine of Fanconi syndrome patients that were not present in normal urine samples. This method can also be applied to different toxicological studies where renal malfunction is observed or expected. However, correct interpretation of the available data is essential both to understand normal renal physiology and to assess hypotheses that implicate proteinuria or peptiduria as early indicators of renal diseases.

Orexin-A: quantification of peptides administered as drugs

One of the earliest papers on peptide quantification came from a group at SmithKline Beecham [37]. Orexin-A, an endogenous hypothalamic peptide that is localised in areas of the brain and spinal cord, and is associated with nociceptive processing, was quantified using LC-MS-MS.

Pharmacokinetics studies, oral bioavailability and steady-state brain penetration were investigated in the conscious rat. For brain-penetration studies, rats were sacrificed and the brains homogenised. Orexin-A was given orally or intravenously to rats to study the benefits of the compound as a drug. Blood and brain homogenate samples were treated identically for LC-MS-MS analysis. However, one of the major problems encountered by the authors was the recovery of the peptide from the biological matrices, as co-extracted components affected the analysis of the peptide. Different separation techniques were used for peptide purification, with protein precipitation being the most effective. At the same time, LC separation together with tandem MS using multiple-reaction monitoring (MRM gave enough sensitivity and selectivity for the analysis.

Analysis of neuropeptides

Analyses of the levels of neuropeptides have been successfully performed using solid-phase extraction (SPE) followed by LC-MS-MS analysis. These techniques have been successfully used in the study of compounds such as substance P derivatives (SPDs), and the bradykinin antagonists CU201 and B201 (Figure 11.1).

SPDs were the first compounds reported to inhibit signal transduction via multiple peptide effects and to inhibit the growth of many small cell lung cancer (SCLC) cell lines [38]. Subsequent studies showed that SPDs induced apoptosis in SCLC cell lines. Similar to SPDs, the peptide dimer CU201 acts as a biased agonist by inhibiting $G_{\alpha q}$-protein activation, and subsequent downstream events, and by stimulating $G_{\alpha 12,13}$-proteins (G_{α}-proteins are guanidine nucleotide-binding proteins) and its associated components, resulting in the induction of apoptosis and inhibition of growth. CU201 also inhibits the growth of a number of non-SCLC, breast cancer and prostate cancer cell lines. These neuropeptides and their receptors are logical targets for novel therapeutic strategies. These observations led to the development of several specific peptide inhibitors, including monoclonal antibodies to the peptide, receptor and peptide antagonists [39]. Stability and preliminary pharmacokinetics of another bradykinin antagonist polypeptide, B201, have also been performed [40].

Also, in the area of peptidomics, Desiderio et al. [41] have used ultra-filtration followed by SPE and nano-LC-MS-MS for the analysis of human cerebrospinal fluid (CSF). CSF offers a unique window to the diagnosis of central nervous system (CNS) disorders and to clarify the basic molecular mechanisms of CNS disorders. Their studies, so far, provide the proof of principle for the determination and identification of peptides and modified peptides in CSF (Figure 11.2). Although many peptides were identified using the above analysis alone, confirmation of the sequence of some of the larger peptides ($M_r >$ 3 kDa) required tryptic hydrolysis after SPE. Also, peptides were observed and identified even when

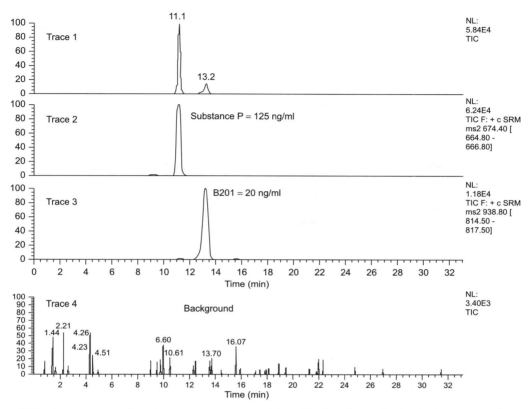

Figure 11.1 Reconstructed selected-reaction monitoring (SRM) profiles from known levels of substance P and B201 in mouse plasma. The transitions used for the analysis were: 674.4–665.0 Th for substance P and 938.9–816.0 Th for B201. Trace 1 shows the total-ion chromatogram of mouse plasma spiked with authentic standards, substance P and B201; traces 2 and 3 are the SRM profiles for substance P and B201 respectively; and trace 4 shows the total-ion chromatogram corresponding to the mouse plasma without the addition of authentic standards. (Reproduced from Feng et al. [40] with permission.)

their concentrations differed by several orders of magnitude. They conclude that a method for the analysis of CSF is feasible and effective, but in order to detect a greater number of peptides, especially low abundance peptides and those diseases that are related, multi-dimensional LC would have to be used.

Changes in the amounts and modifications of the peptides reflect an aberrant processing of the protein precursors and with this the pathology of the related biological system could be revealed.

Different research groups are at the moment concentrating their investigations in the analysis of endogenous peptides from CSF and brain tissue by nano-LC-MS-MS, especially using quadrupole TOF-type instruments as the resolution, mass accuracy, sensitivity and fragmentation data are sufficient for the identification of most of the peptides, including possible modifications [42–46].

A peptide bank of circulating human peptides obtained using human haemofiltrate (HF) as a source of purification was generated by Forssmann et al. [47] in 1997. Samples were obtained from patients suffering from chronic renal failure. The peptides from the HF were separated by cation exchange followed by RP. The SCX/RP fractions represent a peptide bank containing bioactive, desalted and lyophilised peptides from blood.

250 Chapter 11 • Peptide analysis using LC-MS

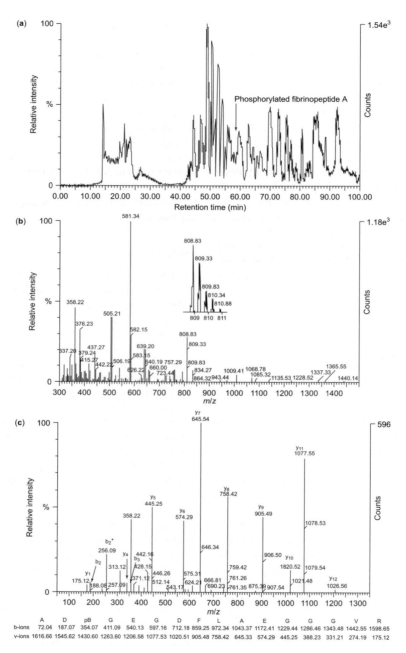

Figure 11.2 LC-MS-MS characterisation of phosphorylated fibrinopeptide A. (a) Base-peak total-ion chromatogram of the CSF compounds that have $M_r < 5$ kDa. The retention time of phosphorylated fibrinopeptide A is indicated by the arrow. (b) Mass spectrum that includes the phosphorylated fibrinopeptide A $[M + 2H]^{2+}$ precursor ion; the inset of the $[M + 2H]^{2+}$ precursor ion at m/z 808.83 of phosphorylated fibrinopeptide A demonstrates the double charge. (c) Product-ion mass spectrum of the $[M + 2H]^{2+}$ precursor ion at m/z 808.83 of phosphorylated fibrinopeptide A. The amino acid sequence of that peptide is listed beneath the spectrum. The mass of each b- and y-ion is also indicated. The underlined numbers correspond to the ions detected by LC-MS-MS. (Reproduced from Yuan and Desiderio [41] with permission.)

Analysis of endogenous intact insulin in blood samples [48]

Even if insulin is normally considered as a small protein due to its molecular weight, it is included in the peptide analysis section because, using this method, the intact analyte is detected by MS.

Measurements of insulin levels in human blood have been recently reported. Although this study was not a 'toxicology study' as such, the method is applicable for analysis of abnormal levels of insulin. The method involves purification and concentration of plasma insulin by SPE. The advantage of the application of SPE and LC-MS for the quantification of insulin is the high specificity compared with other techniques such as radioimmunoassay (RIA) [48]. Moreover, the developed LC-MS method is not subject to interferences associated with RIA, e.g. arising from haemolysis, and is more robust and reliable.

Analysis of proteins

Toxicoproteomics

The first step in many proteomics analyses is the identification of proteins from sodium dodecyl-sulphate–polyacrylamide gel electrophoresis (SDS–PAGE) gels (one- or two-dimensional electrophoresis); this normally requires the in-gel digestion of the proteins of interest with an enzyme, commonly trypsin, followed by MS analysis. In many cases, these experiments are performed using MALDI to generate a peptide mass fingerprint (PMF). However, the requirement of a peptide sequence for complete and confident protein identification leads to the use of more complex techniques such as LC-MS-MS where the peptides are separated by LC, ionised by ESI and fragmented in the collision cell using, for example, QTOF-type instruments or ion traps. Protein recognition is achieved by comparison of the masses and/or sequences of the proteolytic peptides with those expected for known or predicted protein sequences [49–52].

The most common way to study toxicological changes or the effect of endogenous or exogenous stimuli is by relative quantification of the proteins present in the different conditions. Using these methods the absolute amount of protein is not calculated, but rather the ratio of the quantities of a given protein under the conditions to be compared. The approaches include densitometric analysis of protein spots separated by 2-D gel electrophoresis (2-DE) or differential labelling using fluorescent dyes prior to 2-DE (2-DIGE). Alternatively, differential isotope labelling of samples to be compared can be followed by MS or MS-MS analysis. The labelling can be achieved during cell growth in culture [53], or by derivatisation at the protein [54] or proteolytic peptide levels [55].

Densitometry

In order to achieve the necessary separation and resolution required for quantification by densitometry or differential labelling, samples must be separated by 2-DE. Many modern advances such as narrow range immobilised pH gradient (IPG) strips and solubilisation buffers have improved the resolution and spot detection capabilities of this technique. However, often pairs of 2-D gel images cannot be directly superimposed and a number of computer-assisted image analysis software packages have been developed to allow such comparisons to be performed. The first step of the computerised analysis involves the conversion of the gel image into digital data. The matching process then involves three steps: processing of the 2-D gel images, spot detection and 2-D protein pattern matching. Image warping and the removal of both background noise and streak artefacts are also required to limit gel-to-gel variation which can affect the accuracy of the analysis and quantification [56].

Differential labelling using fluorescent dyes

This approach enables the relative quantification of proteins from two different samples using a single 2-DE gel [57], thereby greatly improving reproducibility. Each sample is covalently labelled prior to 2-DE separation with succinimidyl esters of cyanine dyes (Cy2, Cy3 or Cy5). The N-hydroxy-succinimidyl ester undergoes a nucleophilic substitution reaction with the lysine ε-amino group to give an amide. Samples

A and B are labelled with Cy3 and Cy5, respectively, and an internal standard (IS), typically an equal amount of sample A and B, is labelled with Cy2. The samples are mixed together and separated on the same gel, allowing accurate relative quantification of both samples in a single gel via fluorescence detection of the three dyes. The Cy2 dye, or IS, is needed because expression levels of different samples are compared in the same gel. The IS is a pooled sample comprising equal amounts of each sample to be compared; this ensures that all proteins occurring in the samples are represented, allowing inter- and intra-gel matching. The spot volumes from the labelled samples are compared with the IS which gives standardised abundances, so variations in gel running are taken into consideration. This method also has its limitations [57]; it requires that only 1–2% of the lysine residues are derivatised to maintain the solubility of the proteins and to avoid affecting the protein migration on the 2-DE. Moreover, it is usually advisable to perform a second experiment in which the labelling of the two preparations is reversed, to allow for differential derivatisation; clearly, this increases the cost and time required per analysis. This staining method is significantly less sensitive than silver staining, so many of the low-abundance proteins are not visualised or quantified; however, new developments using saturated labelling dyes have improved the detection limits of the technique [58].

Stable isotope labelling coupled with mass spectrometry

A more versatile technique for precise relative quantification involves the labelling of each set of proteins (or peptides derived from them) from two different cell states or types with light and heavy isotopes. The differentially labelled samples are then combined and analysed by MS. The heavy and light analogues will appear as signal doublets, and the ratio between them will indicate the molar ratio between the proteins in the different preparations.

One approach to differential isotopic labelling is the incorporation of label during cell culture. Oda et al. [53], for example, cultured cells in ^{14}N- and ^{15}N-labelled media before combination, proteolytic digestion and analysis by LC-MS. The kinetics of isotope labelling during yeast cell culture has been used by Beynon et al. [59] in order to measure turnover rates for individual proteins.

A second general approach to relative quantification involves differential isotope labelling of recovered protein samples, an example being the popular isotope-coded affinity tag (ICAT) approach [54]. Cysteine-containing proteins are derivatised using a reagent incorporating a heavy or light label and a biotin moiety. The labelled samples are combined and cleaved using trypsin before the hydrolysate is separated on an avidin-affinity column. Cysteine-containing peptides are recovered and analysed by LC-MS or LC-MS-MS. Comparison of the abundances of isotopic variants of the tryptic peptides yields molar ratios for the proteins from which they are derived. A disadvantage of ICAT is that at least one cysteine residue must be present in proteins to be studied, but the method benefits from a lack of reliance on 2-D gels, with consequent applicability to proteins not amenable to such analysis (such as those of extreme pI values).

A final general approach to differential isotope labelling for relative quantification of proteins involves derivatisation with light or heavy tags of proteolytic peptides. One example of many such approaches is the use of guanidination of lysine residues (using [$^{14}N_2$]- or [$^{15}N_2$]O-methylisourea) to yield homoarginine. An attendant benefit is the enhancement in response during MALDI MS that follows this derivatisation [60]. Estimates of the abundance ratios of heavy:light variants are obtained for multiple peptides derived from the same protein. If tandem MS is performed for identification, y-series ions are observed as isotopic doublets, again providing multiple estimates of abundance ratios for the light and heavy variants. The benefits of tandem MS analysis are evident also in the use of iTRAQ™ (isotope tags for relative and absolute quantification) reagents [61]. Labelling of primary amines in tryptic peptides is achieved by incorporation of one of four isobaric tags, each of which incorporates an isotopically distinct 'reporter' group and an isotopically distinct 'balance' group, the masses of which add up to a common value. A four-way comparison of protein digests may then be performed by separate derivatisation, followed by mixing and LC-MS-MS analysis. Fragmentation of the individual protonated peptides yields isotopically distinct reporter ions, the relative abundances of which reflect the molar ratios in the four samples of the protein from which they are derived.

Proteomics technologies have been applied to different toxicological studies, such as the study of the effect of bromobenzene on the liver [62]. Here, hepatotoxicity in rats was induced by bromobenzene, and its effects on the expression of genes and proteins studied by transcriptomics and proteomics, respectively. The general objective of such studies is to compare pathophysiological effects with changes at the RNA or protein level, with a view to early detection and prediction of toxicity.

Another application of toxicoproteomics is in cancer research. Analysis of different preparations by proteomics technologies of patients suffering from cancer is routinely used at this time [63, 64]. Proteins are separated by 2-D PAGE, digested with trypsin and the peptides analysed by MS. Interesting proteins will be those that appear in one preparation, but not the other, or those with levels that change from one preparation to the other.

Modified proteins

The interaction with proteins of the product of lipid oxidation, 4-hydroxy-2-nonenal (HNE), provides an example of the use of MS for the detection and determination of modified protein structures. HNE may react with proteins at histidine, cysteine or lysine residues. Precursor ion scanning during tandem MS of the tryptic hydrolysate of oxidatively modified (in vitro) low-density lipoprotein has been used to screen for peptides that fragment to give a specific fragment ion at m/z 266 corresponding to the modified histidine immonium ion [65]. Peptides so identified were then fully characterised using the product-ion scanning mode of tandem MS.

Summary

Modern MS methods allow the determination of the sequences of peptides and proteins, including the elucidation of structural modifications. Mixture analysis is facilitated by the application of combined LC-MS and of tandem MS. While the MS analysis of intact proteins is feasible, the determination of sequence is normally performed via enzyme-catalysed digestion and analysis of the resulting peptides. The use of proteolytic peptides as analytical surrogates for the protein from which they are derived applies also to quantitative analyses using MS, where differential isotope labelling allows the comparison of protein expression in separate biological samples. Applications of these techniques in toxicological studies remain few, but are likely to increase as the methods become increasingly routine.

References

1. Karas M, Hillenkamp F. Laser desorption ionization of proteins with molecular masses exceeding 10,000 daltons. *Anal Chem* 1988; 60: 2299–2301.

2. Tanaka K, Waki H, Ido Y, Akita S, *et al*. Protein and polymer analyses up to m/z 100,000 by laser ionization time-of-flight mass spectrometry. *Rapid Commun Mass Spectrom* 1988; 2: 151–153.

3. Zenobi R, Knochenmuss R. Ion formation in MALDI mass spectrometry. *Mass Spectrom Rev* 1998; 17: 337–366.

4. Gluckmann M, Karas M. The Initial ion velocity and its dependence on matrix, analyte and preparation method in ultraviolet matrix-assisted laser desorption ionization. *J Mass Spectrom* 1999; 34: 467–565.

5. Karas M, Gluckmann M, Schafer J. Ionization in matrix-assisted laser desorption/ionization: singly charged molecular ions are the lucky survivors. *J Mass Spectrom* 2000; 35: 1–12.

6. Cornish TJ, Cotter RJ. A curved-field reflectron for improved energy focusing of product ions in time-of-flight mass spectrometry. *Rapid Commun Mass Spectrom* 1993; 7: 1037–1040.

7. Cornish TJ, Cotter RJ. A curved field reflectron time-of-flight mass-spectrometer for the simultaneous

focusing of metastable product ions. *Rapid Commun Mass Spectrom* 1994; 8: 781–785.

8. Medzihradszky KF, Campbell JM, Baldwin MA, *et al*. The characteristics of peptide collision-induced dissociation using a high-performance MALDI-TOF/TOF tandem mass spectrometer. *Anal Chem* 2000; 72: 552–558.

9. Biemann K. Contribution of mass spectrometry to peptide and protein structure. *Biomed Environ Mass Spectrom* 1988; 16: 99–111.

10. Huang X, Zhang S, Schultz GA, *et al*. Surface-alkylated polystyrene monolithic columns for peptide analysis in capillary liquid chromatography-electrospray ionization mass spectrometry. *Anal Chem* 2002; 74: 2336–2344.

11. Desportes C, Charpentier M, Duteurtre B, *et al*. Liquid chromatographic fractionation of small peptides from wine. *J Chromatogr A* 2000; 893: 281–291.

12. Chervet JP, Ursem M, Salzmann JP. Instrumental requirements for nanoscale liquid chromatography. *Anal Chem* 1996; 68: 1507–1512.

13. Delahunty C, Yates JR, III. Protein identification using 2-D-LC-MS/MS. *Methods* 2005; 35: 248–255.

14. Holm A, Storbraten E, Mihailova A, *et al*. Combined solid-phase extraction and 2D LC-MS for characterization of the neuropeptides in rat-brain tissue. *Anal Bioanal Chem* 2005; 382: 751–759.

15. Wagner K, Miliotis T, Marko-Varga G, *et al*. An automated on-line multidimensional HPLC system for protein and peptide mapping with integrated sample preparation. *Anal Chem* 2002; 74: 809–820.

16. Emmett MR, Caprioli RM. Micro-electrospray mass spectrometry: ultra-high-sensitivity analysis of peptides and proteins. *J Am Soc Mass Spectrom* 1994; 5: 605–613.

17. Andren PE, Emmett MR, Caprioli RM. Micro-electrospray: zeptomole/attomole per microliter sensitivity for peptides. *J Am Soc Mass Spectrom* 1994; 5: 867–869.

18. Wilm MS, Mann M. Electrospray and Taylor-cone theory, Dole's beam of macromolecules at last? *Int J Mass Spectrom Ion Proc* 1994; 136: 167–180.

19. Wilm M, Mann M. Analytical properties of the nanoelectrospray ion source. *Anal Chem* 1996; 68: 1–8.

20. Wall DB, Berger SJ, Finch JW, *et al*. Continuous sample deposition from reversed-phase liquid chromatography to tracks on a matrix-assisted laser desorption/ionization precoated target for the analysis of protein digests. *Electrophoresis* 2002; 23: 3193–3204.

21. Ericson C, Phung QT, Horn DM, *et al*. An automated noncontact deposition interface for liquid chromatography matrix-assisted laser desorption/ionization mass spectrometry. *Anal Chem* 2003; 75: 2309–2315.

22. Zhang B, McDonald C, Li L. Combining liquid chromatography with MALDI mass spectrometry using a heated droplet interface. *Anal Chem* 2004; 76: 992–1001.

23. Fung KY, Askovic S, Basile F, *et al*. A simple and inexpensive approach to interfacing high-performance liquid chromatography and matrix-assisted laser desorption/ionization-time of flight-mass spectrometry. *Proteomics* 2004; 4: 3121–3127.

24. Mirgorodskaya E, Braeuer C, Fucini P, *et al*. Nanoflow liquid chromatography coupled to matrix-assisted laser desorption/ionization mass spectrometry: sample preparation, data analysis, and application to the analysis of complex peptide mixtures. *Proteomics* 2005; 5: 399–408.

25. Bodnar WM, Blackburn RK, Krise JM, *et al*. Exploiting the complementary nature of LC/MALDI/MS/MS and LC/ESI/MS/MS for increased proteome coverage. *J Am Soc Mass Spectrom* 2003; 14: 971–979.

26. Zhen Y, Xu N, Richardson B, *et al*. Development of an LC-MALDI method for the analysis of protein complexes. *J Am Soc Mass Spectrom* 2004; 15: 803–822.

27. Jorgenson JW, Lukacs KD. Capillary zone electrophoresis. *Science* 1983; 222: 266–272.

28. Jorgenson JW, Lukacs KD. Free-zone electrophoresis in glass capillaries. *Clin Chem* 1981; 27: 1551–1553.

29. Tetler LW, Cooper PA, Powell B. Influence of capillary dimensions on the performance of a coaxial capillary electrophoresis-electrospray mass spectrometry interface. *J Chromatogr A* 1995; 700: 21–26.

30. Wahl JH, Gale DC, Smith RD. Sheathless capillary electrophoresis-electrospray ionization mass spectrometry using 10 µm I.D. capillaries: analysis of tryptic digests of cytochrome c. *J Chromatogr A* 1994; 659: 217–222.

31. Lee ED, Muck W, Henion JD, et al. On-line capillary electrophoresis-ion spray tandem mass spectrometry for the determination of dynorphins. *J Chromatogr* 1988; 458: 313–321.

32. Stimpfl T, Vycudilik W. Automatic screening in postmortem toxicology. *Forensic Sci Int* 2004; 142: 115–125.

33. Spiehler V, ed. *Proceedings of the 1994 Joint TIAFT/SOFT International Meeting*, Tampa, FL, 1995.

34. Swatton JE, Prabakaran S, Karp NA, et al. Protein profiling of human postmortem brain using 2-dimensional fluorescence difference gel electrophoresis (2-D DIGE). *Mol Psychiatry* 2004; 9: 128–143.

35. Cutillas PR, Norden AG, Cramer R, et al. Detection and analysis of urinary peptides by on-line liquid chromatography and mass spectrometry: application to patients with renal Fanconi syndrome. *Clin Sci (Lond)* 2003; 104: 483–490.

36. Norden AG, Sharratt P, Cutillas PR, et al. Quantitative amino acid and proteomic analysis: very low excretion of polypeptides >750 Da in normal urine. *Kidney Int* 2004; 66: 1994–2003.

37. Bingham S, Davey PT, Babbs AJ, et al. Orexin-A, an hypothalamic peptide with analgesic properties. *Pain* 2001; 92: 81–90.

38. Langdon SP, Sethi T, Richie A, et al. Broad spectrum neuropeptide agonists inhibit the growth of small cell lung cancer *in vivo*. *Cancer Res* 1992; 52: 4554–4557.

39. Chan DC, Gera L, Stewart JM, et al. Bradykinin antagonist dimer, CU201, inhibits the growth of human lung cancer cell lines *in vitro* and *in vivo* and produces synergistic growth inhibition in combination with other antitumor agents. *Clin Cancer Res* 2002; 8: 1280–1287.

40. Feng WY, Chan KK, Covey JM. Electrospray LC-MS/MS quantitation, stability, and preliminary pharmacokinetics of bradykinin antagonist polypeptide B201 (NSC 710295) in the mouse. *J Pharm Biomed Anal* 2002; 28: 601–612.

41. Yuan X, Desiderio DM. Human cerebrospinal fluid peptidomics. *J Mass Spectrom* 2005; 40: 176–181.

42. Eriksson U, Andren PE, Caprioli RM, et al. Reversed-phase high-performance liquid chromatography combined with tandem mass spectrometry in studies of a substance P-converting enzyme from human cerebrospinal fluid. *J Chromatogr A* 1996; 743: 213–220.

43. Andren PE, Caprioli RM. Determination of extracellular release of neurotensin in discrete rat brain regions utilizing *in vivo* microdialysis/electrospray mass spectrometry. *Brain Res* 1999; 845: 123–129.

44. Zhang H, Stoeckli M, Andren PE, et al. Combining solid-phase preconcentration, capillary electrophoresis and off-line matrix-assisted laser desorption/ionization mass spectrometry: intracerebral metabolic processing of peptide E *in vivo*. *J Mass Spectrom* 1999; 34: 377–383.

45. Williams K, Wu T, Colangelo C, et al. Recent advances in neuroproteomics and potential application to studies of drug addiction. *Neuropharmacology* 2004; 47(Suppl 1): 148–166.

46. Che FY, Fricker LD. Quantitative peptidomics of mouse pituitary: comparison of different stable isotopic tags. *J Mass Spectrom* 2005; 40: 238–249.

47. Schulz-Knappe P, Schrader M, Standker L et al. Peptide bank generated by large-scale preparation of circulating human peptides. *J Chromatogr A* 1997; 776: 125–132.

48. Darby SM, Miller ML, Allen RO, et al. A mass spectrometric method for quantitation of intact insulin in blood samples. *J Anal Toxicol* 2001; 25: 8–14.

49. Zhang W, Chait BT. ProFound: an expert system for protein identification using mass spectrometric peptide mapping information. *Anal Chem* 2000; 72: 2482–2489.

50. Yates JR, III, Eng JK, Clauser KR, et al. Search of sequence databases with uninterpreted high-energy collision-induced dissociation spectra of peptides. *J Am Soc Mass Spectrom* 1996; 7: 1089–1098.

51. Mann M, Hojrup P, Roepstorff P. Use of mass spectrometric molecular weight information to identify proteins in sequence databases. *Biol Mass Spectrom* 1993; 22: 338–345.

52. Perkins DN, Pappin DJ, Creasy DM, et al. Probability-based protein identification by searching sequence databases using mass spectrometry data. *Electrophoresis* 1999; 20: 3551–3567.

53. Oda Y, Nagasu T, Chait BT. Enrichment analysis of phosphorylated proteins as a tool for probing the phosphoproteome. *Nat Biotechnol* 2001; 19: 379–382.

54. Gygi SP, Rist B, Gerber SA, et al. Quantitative analysis of complex protein mixtures using isotope-coded affinity tags. *Nat Biotechnol* 1999; 17: 994–999.

55. Brancia FL, Butt A, Beynon RJ, *et al.* A combination of chemical derivatisation and improved bioinformatic tools optimises protein identification for proteomics. *Electrophoresis* 2001; 22: 552–559.

56. Veeser S, Dunn MJ, Yang GZ. Multiresolution image registration for two-dimensional gel electrophoresis. *Proteomics* 2001; 1: 856–870.

57. Tonge R, Shaw J, Middleton B, *et al.* Validation and development of fluorescence two-dimensional differential gel electrophoresis proteomics technology. *Proteomics* 2001; 1: 377–396.

58. Shaw J, Rowlinson R, Nickson J, *et al.* Evaluation of saturation labelling two-dimensional difference gel electrophoresis fluorescent dyes. *Proteomics* 2003; 3: 1181–1195.

59. Pratt JM, Robertson DH, Gaskell SJ, *et al.* Stable isotope labelling *in vivo* as an aid to protein identification in peptide mass fingerprinting. *Proteomics* 2002; 2: 157–163.

60. Brancia FL, Oliver SG, Gaskell SJ. Improved matrix-assisted laser desorption/ionization mass spectrometric analysis of tryptic hydrolysates of proteins following guanidination of lysine-containing peptides. *Rapid Commun Mass Spectrom* 2000; 14: 2070–2073.

61. Pappin DJC, Barlet-Jones M. Methods, mixtures, kits and compositions pertaining to analyte determination. *Patent WO 2004/070352*, 2004.

62. Heijne WH, Stierum RH, Slijper M, *et al.* Toxicogenomics of bromobenzene hepatotoxicity: a combined transcriptomics and proteomics approach. *Biochem Pharmacol* 2003; 65: 857–875.

63. Shen J, Person MD, Zhu J, *et al.* Protein expression profiles in pancreatic adenocarcinoma compared with normal pancreatic tissue and tissue affected by pancreatitis as detected by two-dimensional gel electrophoresis and mass spectrometry. *Cancer Res* 2004; 64: 9018–9026.

64. Cheung PK, Woolcock B, Adomat H, *et al.* Protein profiling of microdissected prostate tissue links growth differentiation factor 15 to prostate carcinogenesis. *Cancer Res* 2004; 64: 5929–5933.

65. Bolgar MS, Gaskell SJ. Determination of the sites of 4-hydroxy-2-nonenal adduction to protein by electrospray tandem mass spectrometry. *Anal Chem* 1996; 68: 2325–2330.

12

LC-MS in forensic chemistry

Jehuda Yinon

Introduction

Forensic science is best defined as the application of science to the processes of the law in order to achieve justice. The term 'science' in the definition is a broad one and encompasses several disciplines, from chemistry, toxicology, odontology, jurisprudence through to engineering sciences and even psychiatry/behavioural science.

The forensic scientist generally has expertise in one of these disciplines together with a thorough understanding of his or her role in the judiciary and in relation to other fellow forensic scientists when trying to piece together a solution to a particular problem or puzzle. Hence, forensic laboratories can be involved in a multitude of tasks, using multiple disciplines and techniques in order to obtain answers from evidence gathered. In more general terms, the various tasks of the forensic laboratory can be summarised in three major activities: identification, comparison and individualisation/authentication of evidence.

- *Identification*: includes the identification of drugs, explosives, hair, paint, glass, fibres and ignitable materials
- *Comparison*: in order to link evidence from the scene of a crime to a suspect, e.g. the comparison of paint, hair, fibre, glass, soil, bullets, fingerprints or bloodstains (DNA)
- *Individualisation/authentication*: in cases of forgery, whether of ancient objects, pieces of art, precious stones, etc.; questioned documents also fall into this category (i.e. by comparison of inks) and, of course, currency.

Materials of forensic interest are therefore materials that can be found at crime scenes and/or on suspects that can help investigators to solve the crime, or materials that can serve as evidence in court. In most cases these materials are found in trace quantities and in complex matrices, which necessitates highly sensitive and selective analytical methods for their identification and characterisation.

We have seen in previous chapters that liquid chromatography-mass spectrometry (LC-MS) is a powerful tool in the analytical toxicologist's arsenal, providing complementary information to gas chromatography (GC)-MS. With the advent of MS^n detection systems, both these hyphenated techniques have rapidly become techniques of choice in forensic trace analysis. LC-MS is a widely used method in forensic analysis, particularly in applications where non-volatile, labile or high-molecular-weight compounds are being analysed. The following sections aim to give an overview of the applications of LC-MS to other forensic chemistry puzzles, highlighting recent developments and current practices.

Lacrimators

Lacrimators, or tear gas sprays, have been widely used by law enforcement agents to control crowds and riots, but are also marketed for personal protection [1]. Increasingly, these products have been used for self-defence, but also by criminals. The occurrence of finding gas spray residues as forensic evidence at crime scenes has become more frequent.

Representative active ingredients in tear gas sprays include:

- 2-chloroacetophenone (CN)
- o-chlorobenzylidene malononitrile (CS)
- oleoresin capsicum (OC) – a hot pepper extract.

The most active ingredients in OC are capsaicinoids. There are five naturally occurring capsaicinoids: capsaicin, dihydrocapsaicin, nordihydrocapsaicin, homocapsaicin and homodihydrocapsaicin [2]. Capsaicin is the most active and potent of these compounds. The synthetic analogue nonivamide exhibits the same pungency as capsaicin. Pepper spray weapons usually contain a 10% solution of OC diluted in a suitable solvent and a gaseous propellant.

The analysis of such compounds was previously carried out by GC-MS after derivatisation with trimethylsilyl due to their low volatility [1]. A recent method to identify and quantify directly the predominant capsaicinoid analogues in extracts of fresh peppers, in OC and in pepper sprays was successfully developed for LC-MS [3]. The analysis was performed using the following equipment:

- *Instrument*: Agilent series 1100 LC-MSD
- *Column*: MetaSil Basic reversed-phase LC column (100 × 3.0 mm, 3 μm particle size)
- *Mobile phase*: stepwise gradient of methanol: water, containing 0.1% (v/v) formic acid
- *Ion monitoring*: the following [M + H]$^+$ ions were detected:
  octanoyl-vanillamide (mass-to-charge ratio or m/z 280; used as internal standard)
  nonivamide (m/z 294)
  nordihydrocapsaicin (m/z 294)
  capsaicin (m/z 306)
  dihydrocapsaicin (m/z 308)
  homocapsaicin (m/z 320)
  homodihydrocapsaicin (m/z 322).

In the pepper spray products, capsaicin and dihydrocapsaicin constituted 80–90% of the total capsaicinoid concentration. Variability in product composition was observed not only in products from different manufacturers, but also in products from the same manufacturer.

A potential difficulty in detecting forensic evidence like capsaicinoids on clothing is the possibility that the clothing may be cleaned or washed in an attempt to remove the chemical evidence. Time and conditions of storage are also important factors and might affect the detection of the chemical residue. A study was designed to investigate the effects of storage and washing on the detection of pepper spray residues containing capsaicinoids on various fabrics, using positive-ion LC-MS with an electrospray ionisation (ESI) source [4]. The analyses were performed as per Reilly *et al.* [3].

A typical reconstructed ion mass chromatogram obtained from the analysis of a cotton fabric control sample containing a series of capsaicinoids is shown in Figure 12.1. Capsaicinoids persisted on a variety of fabrics after extended storage at different temperatures and even following attempts to remove them from the fabric. Over 85% of the original concentrations of nordihydrocapsaicin, nonivamide and capsaicin were detectable by LC-MS for up to 6 months after storage at room temperature. Retention of the capsaicinoids on the fabrics following washing was affected by fabric type and by the chemical properties of the individual capsaicinoid analogues.

A method based on chemical derivatisation of capsaicinoids using a diazonium salt was developed for the visualisation of colourless, UV-activated fluorescent dye-free pepper sprays on textiles [5]. Identification of both the capsaicinoids and their derivatives was confirmed by LC-MS.

- *Instrument*: Agilent 1100 binary pump and autosampler with a Thermo-Finnigan LCQ ion trap (IT) with an atmospheric pressure chemical ionisation (APCI) source
- *Column*: YMC Basic S-3 (50 × 2.0 mm)
- *Mobile phase*: gradient of 10 mmol/L ammonium formate:methanol
- *Ion monitoring*: ion chromatograms of a 3-μL injection of an MK4 First Defense colourless pepper spray extract included ions of:
  capsaicin (m/z 137 + 170 + 306)
  dihydrocapsaicin (m/z 137 + 172 + 308)
  capsaicin derivative (m/z 286 + 455)
  dihydrocapsaicin derivative (m/z 286 + 457).

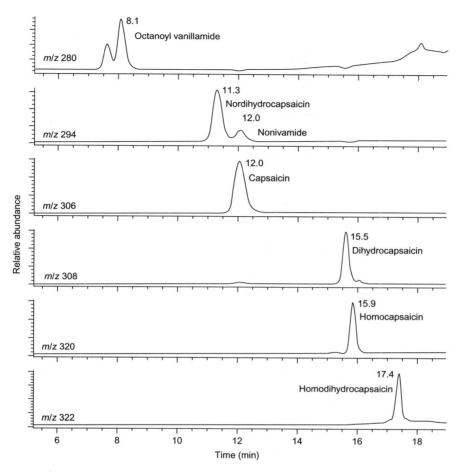

Figure 12.1 Mass chromatograms of a cotton fabric control sample containing a series of capsaicinoids. (Reproduced from Reilly et al. [4] with permission from ASTM.)

Inks

Chemical analysis of writing inks, in combination with other physical methods, is often used to establish whether a questioned document is authentic or forged. Such analyses deal mostly with the comparison of two inks on various documents to determine if they were written with similar inks. The most common types of inks in these investigations are ballpoint pen inks, which are composed mainly of solvents, ionic bases and acid dyes, resins, additives and other components. As with capsaicins, these components are non-volatile and hence LC-MS lends itself to the analysis of ink components, such as dyes.

A method based on the profiling of dye components by ESI-MS was developed [6]. The method involved benzyl alcohol extraction of ballpoint pen ink from paper, followed by ESI-MS. The ink extracts were analysed without chromatographic separation by flow injection.

- *Instrument*: Thermo-Finnigan LCQ-Duo system equipped with an ESI ion source, a Spectra-System P4000 gradient pump, a Spectra-System A3000 autosampler and a Thermo-Finnigan Xcalibur data system; the autosampler injector valve was directly connected to the inlet valve

of the LC-MS interface via a polyether-etherketone (PEEK™) tube
- *Mobile phase*: methanol:water:acetic acid (75:24.5:0.5) at a flow rate of 200 µL/min.

Typically, 2 and 5 µL of ink extracts was injected for analysis in positive and negative modes, respectively. Basic and acid dyes in the inks were detected in the positive- and negative-ion modes, respectively, with each dye producing one or two characteristic ions. In this study, a total of 44 blue inks, 23 black inks and 10 red inks were analysed, and their mass spectra were recorded to establish a searchable library to be used in forensic document investigations. The characteristic ions in the dyes of the above series of ink formulations are reported in Table 12.1.

Dyes in textile fibres

Textile fibres found at a crime scene can be used as physical evidence in cases that involve personal contact in which cross-transfers may occur between the clothing of the suspect and the victim. The value of fibres as evidence will depend on the forensic scientist's ability to narrow their origin to a limited number of sources (or even to a single source). The mass production of textiles makes this a difficult task. An important part of forensic fibre examination involves the characterisation of textile dyes. As these dyes are non-volatile, MS techniques used for their analysis included electron ionisation (EI) MS, field desorption (FD) MS and fast atom bombardment (FAB) MS [7], but not GC-MS.

After the development of the thermospray interface, LC-MS was also applied to characterise dyes extracted from textile fibres [8–10]. While in an earlier study, thermospray ionisation was used [9], ESI was the mode of ionisation in a more recent study [10].

- *Instrument*: Agilent 100 MSD quadrupole mass spectrometer equipped with an ESI source and interfaced to an Agilent 1100 LC apparatus (this has the versatility of being switched between positive- and negative-ion modes for the detection of dyes forming either positive or negative ions)
- *Column*: Zorbax Eclipse XDB-C18 (150 × 2.1 mm)
- *Mobile phase*: methanol:water gradient at a flow rate of 200 µL/min.

Dyes were extracted from fibres with a methanol:water (1:1) mixture in a sealed glass capillary heated to 100°C. A 20-µL aliquot of the methanol:water mixture was used to extract a 5-mm-long piece of thread. Figure 12.2 shows the ESI mass spectrum, in the negative-ion mode, of Direct Red 1 (molecular weight 627) extracted from a cotton fibre [10]. The highest mass ion observed was at m/z 582 due to $[M - 2Na + H]^-$.

Additional dyes extracted from textile fibres and analysed by LC-MS include Acid Red 151 (extracted from nylon), Basic Green 4 (extracted from acrylic), Disperse Blue 56 (extracted from polyester), Disperse Red 1 (extracted from acetate), Disperse Red 4 (extracted from polyester), and Direct Blue 1, Direct Blue 71, Direct Blue 75 and Direct Blue 90 (the last four extracted from cotton). A database of ESI mass spectra of dyes extracted from various textile fibres could be a useful tool for forensic analysts.

Chemical warfare agents

The analysis and identification of chemical warfare agents used in terrorist activities are a major forensic problem. Although GC-MS has been extensively used to identify these compounds, LC-MS has been found to be a more suitable analytical method in cases where the chemical agents are in aqueous solutions.

Packed capillary column LC-ESI-MS has been successfully used to detect and identify four organophosphorous chemical warfare agents in aqueous samples [11].

- *Instrument*: Applied Biosystems Model 140B dual syringe pump with a Micromass Autospec-Q tandem mass spectrometer equipped with an ESI source
- *Column*: Zorbax C_{18} SB packed fused-silica capillary (150 × 0.32 mm, 5 µm particle size)
- *Injector*: Rheodyne 8125 injector with a 5-µL sample loop
- *Mobile phase*: A (0.1% trifluoroacetic acid in

Table 12.1 Characteristic ions in dyes of a series of ballpoint pen inks (reproduced from Ng et al. [6] with permission from ASTM

Sample	Found components (m/z of major ions)		Components not found
	Positive mode	Negative mode	
Blue 1	Basic Violet 1 (m/z 372) Basic Violet 3 (m/z 358) Ditoyl guanidine (m/z 240)	Solvent Blue 38 (m/z 734)	
Blue 2	Basic Violet 1 (m/z 372) Basic Violet 3 (m/z 358) Basic Violet 10 (m/z 443) aryl guanidines (m/z 226, m/z 240, m/z 254, m/z 268)	Solvent Blue 38 (m/z 734) Acid Blue 9 (m/z 373.5, m/z 769) anti-oxidant (m/z 339) corrosion inhibitor (m/z 473)	Solvent Yellow 19
Blue 3	Basic Violet 1 (m/z 372) Basic Violet 3 (m/z 358) Solvent Blue 23 (m/z 516) aryl guanidines (m/z 226, m/z 240, m/z 254, m/z 268)	Solvent Blue 38 (m/z 734)	
Blue 4	Basic Violet 1 (m/z 372) Basic Violet 3 (m/z 358) Solvent Blue 23 (m/z 516)	Solvent Blue 38 (m/z 734) anti-oxidant (m/z 339) corrosion inhibitor (m/z 473)	
Blue 5	Basic Violet 1 (m/z 372) Basic Violet 3 (m/z 358) Solvent Blue 23 (m/z 516)	Solvent Blue 38 (m/z 734)	
Black 1	Basic Violet 1 (m/z 372) Basic Violet 3 (m/z 358) Nigrosine (m/z 530)		Solvent Orange 3
Black 2	Basic Violet 1 (m/z 372) Basic Violet 3 (m/z 358) Nigrosine (m/z 530)	Acid Blue 9 (m/z 373.5, m/z 769)	Solvent Orange 3
Black 3	Basic Violet 1 (m/z 372) Basic Violet 3 (m/z 358) Basic Red 1 (m/z 443) Nigrosine (m/z 530)	Acid Yellow 36 (m/z 352)	
Black 4	Basic Violet 1 (m/z 372) Basic Violet 3 (m/z 358) ditoyl guanidine (m/z 240)	Solvent Blue 38 (m/z 734)	Solvent Orange 3
Red 1	Basic Violet 10 (m/z 443)	Acid Yellow 36 (m/z 352)	Solvent Yellow 19
Red 2 Red 3	Basic Red 1 (m/z 443)	Solvent Orange 25 (m/z 304, m/z 312, m/z 457, m/z 468	
Purple 1	Basic Violet 1 (m/z 372) Basic Violet 3 (m/z 358) Basic Violet 10 (m/z 443)	Acid Blue 9 (m/z 373.5, m/z 769)	
Green 1	Solvent Green 1 (m/z 329) aryl guanidines (m/z 226, m/z 240, m/z 254, m/z 268)	Solvent Blue 38 (m/z 234) Sepisol Fast Yellow TN (467)	

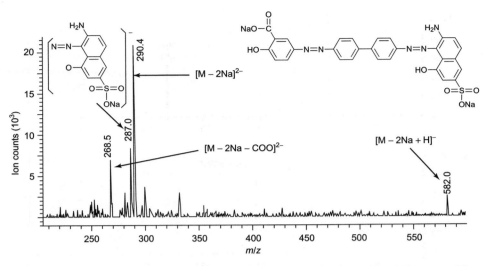

Figure 12.2 ESI mass spectrum in the negative-ion mode of Direct Red 1 (molecular weight 627), extracted from a cotton fibre. (Reproduced from Huang *et al.* [10] with permission from ASTM.)

water) and B (0.1% trifluoroacetic acid in acetonitrile [ACN]:water) (95:5); elution programme: 1–75% B over 30 min, flow rate 200 µL/min and split before the injector, so that the flow through the column was 5 µL/min.

Aqueous samples containing the four organophosphorous chemical warfare agents – isopropyl methylphosphonofluoridate (sarin or GB), O-ethyl-N,N-dimethylphosphoramidocyanidate (tabun or GA), cyclohexyl methylphosphonofluoridate (GF) and pinacolyl methylphosphonofluoridate (soman or GD) – at concentrations of 0.01–0.1 g/L were analysed. Figure 12.3 shows the ESI mass spectra of the four compounds from an aqueous mixture at a concentration of 0.1 g/L.

Snow and soil samples containing chemical warfare agents were analysed by LC-ESI-MS [12, 13].

- *Instrument*: Micromass LCT time-of-flight mass spectrometer equipped with a Z-spray ESI source; mass spectral data were acquired with a resolution of 5000 (50% valley definition)
- *Column*: Zorbax C_{18} SB packed fused-silica capillary (150 × 0.32 mm, 5 µm particle size)
- *Injector*: Rheodyne 8125 injector with a 5-µL sample loop
- *Mobile phase*: A (0.1% trifluoroacetic acid in water) and B (0.1% trifluoroacetic acid in ACN:water [95:5]; elution programme: 1–75% B over 30 min, flow rate 200 µL/min and split before the injector, so that the flow through the column was 16 µL/min.

A number of organophosphorous and organosulphur compounds were identified, based on acquired high-resolution ESI mass spectra, which contained significant $[M + H]^+$ ions.

A packed capillary LC-ESI-MS-MS method was developed for the identification of mustard hydrolysis products in aqueous extracts of soil [14].

- *Instrument*: Waters/Micromass qTOF Ultima tandem mass spectrometer equipped with a Z-spray ESI ion source; data were acquired in the continuum mode at a resolution of 9000 (50% valley definition)
- *Column*: Zorbax C_{18} SB packed fused-silica capillary (150 × 0.32 mm, 5 µm particle size)
- *Injector*: Rheodyne 8125 injector with a 5-µL sample loop
- *Mobile phase*: A (0.1% trifluoroacetic acid in water) and B (0.1% trifluoroacetic acid in ACN:water) (95:5); elution programme: 5–75% B over 30 min, flow rate 10 µL/min.

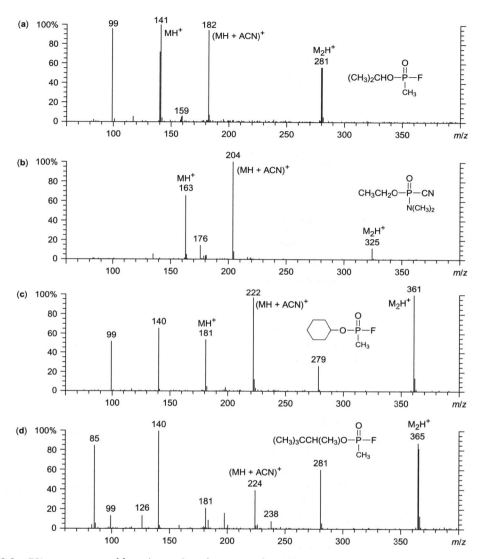

Figure 12.3 ESI mass spectra of four chemical warfare agents from an aqueous mixture at a concentration of 0.1 g/L. (a) Isopropyl methylphosphonofluoridate (sarin or GB); (b) O-ethyl-N,N-dimethylphosphoramidocyanidate (tabun or GA); (c) cyclohexyl methylphosphonofluoridate (GF); and (d) pinacolyl methylphosphonofluoridate (soman or GD). (Reproduced from D'Agostino et al. [11] with permission from Elsevier Science.)

High-resolution mass spectra were obtained for thiodiglycol, the hydrolysis product of mustard and a series of other sulphur-containing diols. MS-MS was used to characterise five longer-chain diols that could not be previously identified.

Explosives

The analysis of explosives has important applications in both forensic and environmental fields [15]. In forensics, the applications include analysis of post-explosion residues, and identification of traces of explosives on suspects' hands, clothing and other related items. The results of these analyses are not only necessary for the investigation of a bombing, but can also serve as evidence in court.

The methodologies for the analysis of explosives are mainly based on GC-MS and LC-MS. The thermal lability of many explosives, along with the requirements of high sensitivity, especially in the analysis of post-explosion residues, make LC-MS a method of choice for the forensic identification of explosives. Both ESI and APCI have been used, depending on the type of explosives encountered [16].

LC-ESI-MS-MS fragmentation processes of a series of nitroaromatic, nitramine and nitrate ester explosives were studied in the negative-ion mode [17]. The studied explosives included 2,4,6-trinitrotoluene (TNT), 1,3,5-trinitro-1,3,5-triazacyclohexane (RDX), 1,3,5,7-tetranitro-1,3,5,7-tetrazacyclooctane (HMX), 2,4,6-N-tetranitro-N-methylaniline (tetryl), pentaerythritol tetranitrate (PETN), nitroglycerin (NG) and ethylene glycol dinitrate (EGDN).

The main ions observed in the ESI mass spectrum of TNT were $[M]^-$ (m/z 227), $[M-H]^-$ (m/z 226), $[M-OH]^-$ (m/z 210) and $[M-NO]^-$ (m/z 197). MS-MS fragmentation of the molecular ion of TNT at m/z 227 yielded fragment ions at m/z 210 $[M-OH]^-$ and at m/z 197 $[M-NO]^-$. MS-MS fragmentation of the ion at m/z 197 $[M-NO]^-$ yielded an ion at m/z 167 $[(M-NO)-NO]^-$.

Several additives have been used in the LC-ESI-MS analysis of explosives in order to enhance the sensitivity of the method. Nitramine and nitrate ester explosives showed an enhanced response with ammonium nitrate additive by forming $[M+NO_3]^-$ adduct ions in the negative-ion mode [18, 19]. Figure 12.4 shows the LC-ESI-MS mass chromatograms of a 25 µg/L Semtex sample (a plastic explosive containing RDX and PETN) with post-column introduction of ammonium nitrate [19]. LC separation was

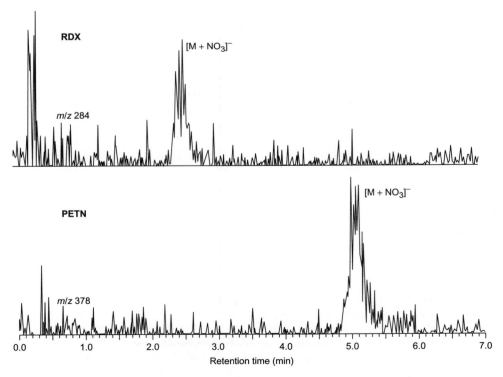

Figure 12.4 LC-ESI-MS mass chromatograms of a 25 µg/L Semtex sample – a plastic explosive containing RDX and PETN. (Reproduced from Zhao and Yinon [19] with permission from Elsevier Science.)

achieved with a C_{18} column (100 × 2.1 mm, 5 μm particle size), using an isocratic mobile phase of methanol:water (70:30, by vol.) at a flow rate of 150 μL/min.

The formation of RDX cluster ions in LC-MS and the origin of the clustering agents have been studied in order to determine whether the clustering anions originate from self-decomposition of RDX in the source or from impurities in the mobile phase [20]. Isotopically labelled RDX ($[^{13}C_3]$RDX and $[^{15}N_6]$RDX) were used in order to establish the composition and formation route of RDX adduct ions produced in ESI and APCI sources. Results showed that in ESI, RDX clusters with formate, acetate, hydroxyacetate and chloride anions, present in the mobile phase as impurities at parts per million levels. In APCI, parts of the RDX molecules decompose, yielding NO_2^- species, which in turn cluster with a second RDX molecule, producing abundant $[M + NO_2]^-$ cluster ions. Figure 12.5 shows the ESI mass spectrum of RDX. The ions $[M + 45]^-$, $[M + 59]^-$, $[M + 75]^-$, $[2M + 59]^-$ and $[2M + 75]^-$ are adduct ions formed as a result of the presence of these impurities, probably in the high-performance LC (HPLC)-grade methanol of the mobile phase.

LC-APCI-MS-MS in the positive-ion mode was used for trace analysis of triacetone triperoxide (TATP) [21].

- *Instrument*: Agilent 1100 LC instrument was coupled to a Thermo-Finnigan 'Navigator' bench-top quadrupole mass spectrometer equipped with an APCI source
- *Column*: $ProC_{18}$ (150 × 2.0 mm, 3 μm particle size)
- *Mobile phase*: methanol:water (70:30, by vol.) with 5 mmol/L ammonium acetate buffer, at flow rates between 0.1 and 0.2 mL/min.

Samples were injected as ACN solutions. The peaks observed in the LC-MS mass spectrum were at m/z 75, 89, 90, 91, 102, 107, 194, 240 and 252. The ion at m/z 240 is believed to be the $[M + NH_4]^+$ adduct ion, formed because of the use of

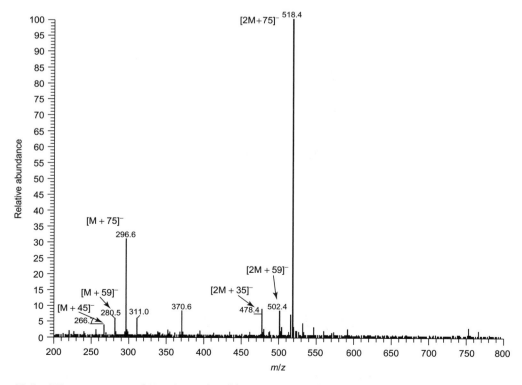

Figure 12.5 ESI mass spectrum of RDX. (Reproduced from Gapeev *et al.* [20] with permission from John Wiley & Sons.)

nitrogen drying gas. This ion was enhanced when using ammonium acetate buffer. A detection limit of 100 µg/L was obtained using MS-MS in the selected-reaction monitoring (SRM) mode (m/z 240 → 89).

In another study, the analysis of peroxide explosives by LC-APCI-MS-MS [22] was carried out using the following apparatus.

- *Instrument*: Waters 600-MS liquid chromatograph coupled to a Thermo-Finnigan 700 triple-stage quadrupole; the LC equipment was connected to the mass spectrometer by an upgraded Thermo-Finnigan API interface
- *Column*: Nova-Pack 4 µm C_{18} (150 × 3.9 mm)
- *Mobile phase*: methanol:water (75:25, by vol.) with 2.5 mmol/L ammonium acetate, at a flow rate of 0.4 mL/min.

The explosives studied included hexamethylenetriperoxidediamine (HMTD) and TATP. The base peaks in the mass spectra of TATP and HMTD were $[TATP + NH_4]^+$ at m/z 240 and $[HMTD - H]^+$ at m/z 207, respectively. The main daughter ions in the MS-MS spectrum of the TATP precursor ion at m/z 240 were at m/z 223 $[TATP + H]^+$ and at m/z 74 $[TATP/3]^+$. The main daughter ion in the MS-MS spectrum of the HMTD precursor ion at m/z 207 was at m/z 118 $[O_2(CH_2)_3N]^+$. Detection limits for TATP and HMTD were 3.3 and 0.26 ng on-column by LC-APCI-MS in full-scan mode and, respectively, 0.8 and 0.08 ng by LC-APCI-MS-MS in the multiple-reaction monitoring (MRM) mode.

LC-APCI-MS-MS in the negative-ion mode was used for identification of TNT, RDX and 2,4,6,8,10,12-hexanitro-2,4,6,8,10,12-hexaazaisowurtzitane (CL-20) [23].

- *Instrument*: Varian 1200 LC-MS with an APCI source
- *Column*: C_{18} (150 × 4.0 mm, 5 µm particle size)
- *Mobile phase*: water:isopropanol:methanol (60:30:10, by vol.) with 0.1% chloroform at an isocratic flow rate of 0.8 mL/min.

Base peaks at m/z 227 $[M]^-$, 257 $[M + Cl]^-$ and 473 $[M + Cl]^-$ were observed in the mass spectra of TNT, RDX and CL-20, respectively. These ions were used as precursor ions in MS-MS analysis. Main fragment ions observed in the daughter-ion spectra were at m/z 210 $[M - OH]^-$ and 197 $[M - NO]^-$ for TNT, 46 $[NO_2]^-$ for RDX, and 233 and 154 $[M - C_5H_5O_8N_9]^-$ for CL-20. Detection limits were in the low parts per billion range.

Identification and characterisation of explosives is of major importance in forensic analysis of post-explosion residues. In addition to the type of explosives, it is important to know their country of origin and, if possible, their manufacturer. Each manufacturer will produce explosives with characteristic differences in the type and amount of byproducts, impurities and additives, depending on the purity of the raw materials and solvents used and the type of manufacturing processes, thus resulting in a typical profile of byproducts, organic impurities and additives.

The byproducts of industrial TNT, including isomers of trinitrotoluene, dinitrotoluene, trinitrobenzene and dinitrobenzene, were investigated using LC-APCI-MS in negative-ion mode in order to build a profile for the characterisation of TNT samples from various origins [24]. MS-MS was used for further identification of some of the nitroaromatic isomers.

- *Instrument*: Thermo-Finnigan LCQ$_{DUO}$ IT mass spectrometer
- *Column*: C_{18} (150 × 3.2 mm, 5 µm particle size)
- *Mobile phase*: three mobile phase systems consisting of methanol:isopropanol:water or methanol:water were employed for different groups of standard mixtures at a flow rate of 0.4 mL/min.

Figure 12.6 shows the mass chromatograms of a standard mixture of six TNT, three DNB, four DNT and one TNB isomer(s) in methanol:water (50:50, by vol.) at a concentration of 1 mg/L each. The full-scan negative-ion APCI mass spectra of the TNT isomers provided limited information on their identification, due to lack of fragment ions, despite the different abundance ratios of m/z 226 $([M - H]^-)$ to 227 $([M]^-)$ observed for different TNT isomers. Therefore, MS-MS analyses were performed.

The product ions of m/z 227 of the individual TNT isomers are summarized in Table 12.2. LC-APCI-MS mass chromatograms of samples from different sources showed differences in their profiles, thus demonstrating the capability of this method to characterise TNT samples. A collection of standard TNT samples from manufacturers

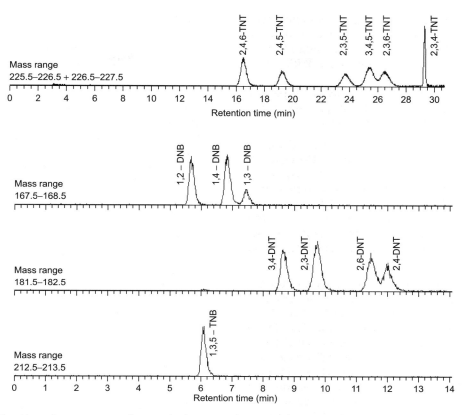

Figure 12.6 Mass chromatograms of a standard mixture of six TNT, three DNB, four DNT and one TNB isomer(s) in methanol:water (50:50, by vol.) at a concentration of 1 mg/L each. (Reproduced from Zhao and Yinon [24] with permission from Elsevier Science.)

could be the basis of a database of isomer by-product profiles, which would help forensic analysts in determining the origin of TNT found in post-explosion residues.

ESI-MS and ESI-MS-MS mass spectra of ammonium nitrate were recorded in both positive- and negative-ion modes, using a Thermo-Finnigan LCQ$_{DUO}$ IT mass spectrometer [25]. Sample solutions of 1 mmol/L, in methanol:water (50:50, by vol.), were introduced by syringe pump infusion into the mass spectrometer at a flow rate of 5 μL/min. It was found that at heated capillary temperatures in the range 55–150°C, in the positive-ion mode, cluster ions of the type $[(NH_4NO_3)_nNH_4]^+$ ($n = 1$–3) were dominant in the mass spectrum, which enables the characterisation and identification of the integral ammonium nitrate molecule. This was confirmed by using isotopically labelled ammonium nitrate ([^{15}N]ammonium, [^{15}N]nitrate, [^{18}O]nitrate) and deuterated water.

A study of a group of inorganic oxidisers was carried out with the same system [26]. Inorganic oxidisers are widely used as blasting agents in commercial explosives employed in mining and construction, and also in improvised home-made explosive devices. Studied compounds included sodium nitrate, potassium nitrate, ammonium sulphate, potassium sulphate, sodium chlorate, potassium chlorate, ammonium perchlorate and sodium perchlorate. Typical positive- and negative-ion clusters obtained were: $[(NaNO_3)_nNa]^+$ and $[(NaNO_3)_nNO_3]^-$ for sodium nitrate, $[(KNO_3)_nK]^+$ and $[(KNO_3)_nNO_3]^-$ for potassium nitrate, $[(K_2SO_4)_nK]^+$ and $[(K_2SO_4)_nKSO_4]^-$ for potassium sulphate, $[(NaClO_3)_nNa]^+$ and $[(NaClO_3)_nClO_3]^-$

Table 12.2 Precursor and product ions (with relative abundances) of TNT isomers (reproduced from Zhao and Yinon [24] with permission from Elsevier Science)

Precursor ion m/z (ion)	Product ions m/z (%)	Structure
2,4,6-TNT		
227 (M-)	210 (100)	[M – OH]-
	197 (48)	[M – NO]-
	181 (5)	[M – NO$_2$]-
	167 (4)	[M – 2NO]-
	151 (5)	[M – NO$_2$ – NO]-
	137 (13)	[M – 3NO]-
2,4,5-TNT		
227 (M-)	197 (100)	[M – NO]-
	181 (2)	[M – NO$_2$]-
2,3,5-TNT		
227 (M-)	197 (100)	[M – NO]-
3,4,5-TNT		
227 (M-)	197 (49)	[M – NO]-
	181 (100)	[M – NO$_2$]-
2,3,6-TNT		
227 (M-)	197 (100)	[M – NO]-
	181 (10)	[M – NO$_2$]-
2,3,4-TNT		
227 (M-)	197 (60)	[M – NO]-
	181 (100)	[M – NO$_2$]-
	151 (3)	[M – NO$_2$ – NO]-

for sodium chlorate, [(KClO$_3$)$_n$K]$^+$ and [(KClO$_3$)$_n$ClO$_3$]$^-$ for potassium chlorate, [(NH$_4$ClO$_4$)$_n$NH$_4$]$^+$ and [(NH$_4$ClO$_4$)$_n$ClO$_4$]$^-$ for ammonium perchlorate, and [(NaClO$_4$)$_n$Na]$^+$ and [(NaClO$_4$)$_n$ClO$_4$]$^-$ for sodium perchlorate.

An ESI mass spectrum of a Powermite sample (a commercially available slurry explosive) obtained at heated capillary temperature of 100°C, showed the presence of ammonium nitrate and sodium nitrate.

It was concluded that ESI-MS is well suited for the qualitative characterisation of the major inorganic oxidiser components in explosives.

References

1. Yinon J. Miscellaneous forensic applications of mass spectrometry. In: Yinon J, ed., *Forensic Mass Spectrometry*. Boca Raton, FL: CRC Press, 1987: 205–209.

2. Kataoka M, Seto Y, Tsuge K, *et al.* Stability and detectability of lachrymators and their degradation products in evidence samples. *J Forensic Sci* 2002; 47: 44–51.

3. Reilly CA, Crouch DJ, Yost GS. Quantitative analysis of capsaicinoids in fresh peppers, oleoresin capsicum and pepper spray products. *J Forensic Sci* 2001; 46: 502–509.

4. Reilly CA, Crouch DJ, Yost GS, *et al.* Detection of pepper spray residues on fabrics using liquid chromatography-mass spectrometry. *J Forensic Sci* 2002; 47: 37–43.

5. Cavett V, Waninger EM, Krutak JJ, *et al.* Visualization and LC/MS analysis of colorless pepper sprays. *J Forensic Sci* 2004; 49: 469–476.

6. Ng L-K, Lafontaine P, Brazeau L. Ballpoint pen inks: characterization by positive and negative ion-electrospray ionization mass spectrometry for the forensic examination of writing inks. *J Forensic Sci* 2002; 47: 1238–1247.

7. Betowski LD, Yinon, J, Voyksner RD. Mass spectrometry in the analysis of dyes in wastewater. In: Reife A, Freeman HS, eds, *Environmental Chemistry of Dyes and Pigments*. New York: Wiley, 1996: 255–292.

8. Yinon J, Betowski LD, Voyksner RD. LC-MS techniques for the analysis of dyes. In: Barcelo D, ed., *Applications of LC-MS in Environmental Chemistry*. Amsterdam: Elsevier, 1996: 187–218.

9. Yinon J, Saar J. Analysis of dyes extracted from textile fibers by thermospray high-performance liquid chromatography-mass spectrometry. *J Chromatogr* 1991; 586: 73–84.

10. Huang M, Yinon J, Sigman ME. Forensic identification of dyes extracted from textile fibers by liquid chromatography-mass spectrometry (LC-MS). *J Forensic Sci* 2004; 49: 238–249.

11. D'Agostino PA, Hancock JR, Provost LR. Packed capillary liquid chromatography-electrospray mass spectrometry analysis of organophosphorus chemical warfare agents. *J Chromatogr A* 1999; 840: 289–294.

12. D'Agostino PA, Chenier CL, Hancock JR. Packed

capillary liquid chromatography-electrospray mass spectrometry analysis of snow contaminated with sarin. *J Chromatogr A* 2002; 950: 149–156.

13. D'Agostino PA, Hancock JR, Chenier CL. Mass spectrometric analysis of chemical warfare agents and their degradation products in soil and synthetic samples. *Eur J Mass Spectrom* 2003; 9: 609–618.

14. D'Agostino PA, Hancock JR, Chenier CL. Packed capillary liquid chromatography-electrospray ionization (tandem) mass spectrometry of mustard hydrolysis products in soil. *J Chromatogr A* 2004; 1058: 97–105.

15. Yinon J, Zitrin S. *Modern Methods and Applications in Analysis of Explosives*. Chichester: Wiley, 1993.

16. Yinon J. Forensic analysis of explosives by LC/MS. *Forensic Sci Rev* 2001; 13: 19–28.

17. Yinon J, McClellan JE, Yost RA. Electrospray ionization tandem mass spectrometry collision-induced dissociation study of explosives in an ion trap mass spectrometer. *Rapid Comm Mass Spectrom* 1997; 11: 1961–1970.

18. Miller ML, Mothershead R, Leibowitz J, *et al.* The analysis of nitrated organic explosives by LC/MS: additive enhancement. In: *Proceedings of the 45th ASMS Conference on Mass Spectrometry and Allied Topics*, Palm Springs, CA, 1997: 52.

19. Zhao X, Yinon J. Identification of nitrate ester explosives by liquid chromatography-electrospray ionization and atmospheric pressure chemical ionization mass spectrometry. *J Chromatogr A* 2002; 977: 59–68.

20. Gapeev A, Sigman M, Yinon J. Liquid chromatography/mass spectrometric analysis of explosives: RDX adduct ions. *Rapid Commun Mass Spectrom* 2003; 17: 943–948.

21. Widmer L, Watson S, Schlatter K, *et al.* Development of an LC/MS method for trace analysis of triacetone triperoxide (TATP). *Analyst* 2002; 127: 1627–1632.

22. Xu X, van de Craats AM, Kok EM, *et al.* Trace analysis of peroxide explosives by high performance liquid chromatography-atmospheric pressure chemical ionization-tandem mass spectrometry (HPLC-APCI-MS/MS) for forensic applications. *J Forensic Sci* 2004; 49: 1230–1236.

23. Colorado A. Analysis of TNT, RDX and CL-20 by APCI LC/MS/MS. *Varian LC/MS Application Note 18*. Palo Alto, CA: Varian, 2004: www.varianinc.com/media/sci/apps/lcms18.pdf.

24. Zhao X, Yinon J. Characterization and origin identification of 2,4,6-trinitrotoluene through its by-product isomers by liquid chromatography-atmospheric pressure chemical ionization mass spectrometry. *J Chromatogr A* 2002; 946: 125–132.

25. Zhao X, Yinon J. Characterization of ammonium nitrate by electrospray ionization tandem mass spectrometry. *Rapid Commun Mass Spectrom* 2001; 15: 1514–1519.

26. Zhao X, Yinon J. Forensic identification of explosive oxidizers by electrospray ionization mass spectrometry. *Rapid Commun Mass Spectrom* 2002; 16: 1137–1146.

Index

abbreviations xv–xvii
ABT-518 structure 26
accuracy (bias)
  experiments and calculations 80–3
  lower limit of quantification 75
  validation 63, 79–80
acetylcodeine (AC) 155–6
6-acetylmorphine (6-AM) 55–7, 152
acquired immunodeficiency syndrome (AIDS) 74
alachlor 220
aldicarb 238
aldosterone antagonists 195–6
American Association of Clinical Chemists (AACC) 103
American Academy of Forensic Sciences (AAFS) 98
ammonia method, heroin production 151
amphetamine (AMP) 159–61
anabolic steroid 200–1
analogues, isotope-labelled 74
analysis of variance (ANOVA) 81–2, 90
  lack-of-fit test 77
Association of Official Racing Chemists (AORC) 206
atazanavir 74
atmospheric pressure chemical ionisation (APCI) 1, 5–6, 30, 52–3
atmospheric pressure ionisation (API) 1, 2, 23, 112
atmospheric pressure laser-ionisation (APLI) 1, 6
atmospheric pressure photo-ionisation (APPI) 1, 6, 30–1, 201–2
automation
  library searching 115–16
  sample preparation 46

background noise 118
barbituric acid 175–6
bench-top stability/in process stability 85
benzodiazepines 170–3
benzthiazide 196
(β) beta$_2$-agonists 197, 199
(β) beta$_2$-blockers 197–9
betamethasone 205
bias (accuracy)
  experiments and calculations 80–3
  lower limit of quantification 75
  validation 63, 79–80
biological samples analysis 152–9, 164
blood analysis 161–3
boldenone 100, 201
bosentan 38
buffers, ESI 3
buprenorphine (BP) (*Buprenex*) 56, 155–8

caffeine 195
calibration curves 58, 84
calibration model 75
calibrators 57
cannabidiol (CBD) 166
cannabinoids 168, 206
cannabinol (CBN) 166
cannabis 44, 164–7, 206
capillary electrophoresis 246
capillary electrophoresis-mass spectrometry (CE-MS) 246–7
capillary liquid chromatography 245
capsaicinoids 258–9
carbamates 221, 238
carbaryl 116
central nervous system (CNS) 248
cerebrospinal fluid (CSF) 248–9

channel electron multiplier (CEM) 11
chemical ionisation (CI) 5
chemical warfare agents 260–3
ChemINDEX 18
'China white' 152
2-chloroacetophenone (CN) 258
o-chlorobenzylidene malonitrile (CS) 258
D$_3$-chlorpromazine 58–9, 60–1
chlorpyrifos 223, 231, 234, 236–7
cholinesterase 237–8
chromatography
  conditions 47–9
  method development 33–6
  quantification 51–2
  separations 25, 124–5
Claviceps purpurea 168
clenbuterol 197
coaxial sheath flow interfaces 247
cocaine 86, 161, 163–5
codeine 55–7, 152–3
collision energy 12, 31–4
collision-induced dissociation (CID) 2, 14
  fragmentation 54
  in-source 113–14
  validation 74
comparison studies 132–4
concentration measurement 57–64
confirmation
  criteria 97–110
  identity 100
  pesticide analysis 232–6
corticosteroids 205–6
criteria, identification and confirmation 97–110
cross-contamination 50
cross talk 15–6, 119
curves, calibration 58, 84
cycloamidine derivatives 195–6

271

data-dependent acquisition (DDA) 16–18, 121–3
databases 99–100, 104–5
'date rape drugs' 170
daughter-ion-scan mode 12–6, 120–1
decision limits 105–7
deconvolution 17
densitometry 251
deprotonated molecular ions 2
desmethylsildenafil 177
detection, ion 11–2
deuterated analogues 58
diacetylmorphine (DAM) 154–5, 154–6
dialkylphosphates (DAPs) 220, 224, 226, 237–8
diazinon 229–30
diethyl phosphate (DEP) 230–1, 234–6
N,N-diethyl-*m*-toluamide (DEET) 219, 223, 226
diethylthiophosphate (DETP) 235–6
differential labelling 251–2
UV-diode array detection (DAD) 111
diquat 218
Direct Red 1 (dye) 262
diuretics 195–7
dopants 6
doping control 193–216
  historical and technical aspects 193–5
  large molecules 206–12
  small molecules 195–206
drugs
  immunoassays 111
  therapeutic 131–48
drugs of abuse 149–92
  amphetamine and ecstasy 159–63
  Cannabis 164–7
  cocaine 86, 161, 163–5
  illicit 151–70
  incapacitating 170
  LSD 44, 167–70
  opiate agonists 151–9
  screening 176–9
  street 152
dyes 260, 262
dynamic range 9–11, 60–2

Ecstasy 159–63, 175
electro-osmotic flow (EOF) 246
electron ionisation (EI) 113–14, 260

electron multipliers 11
electrophoresis, capillary 246
electrospray ionisation (ESI) 1–3, 29–30, 52–54, 201–2
endogenous steroids 201
energies, ionisation 30
'enhance procedure' 119
epimentendiol 201, 203
epitestosterone 202
erythropoetin (EPO) 208–9
Erythroxylon coca 163
esters, steroid 201–3
ethyl glucuronide (EtG) 4
ethyl sulphate (EtS) 4
ethylenethiourea (ETU) 220, 222, 224
EURACHEM 71, 80–2
European Union 235
experiments
  method validation 88–90
  stability 85–6
explosives 263–8
exposure monitoring 237

Fanconi syndrome 248
fast atom bombardment (FAB) 1, 260
fenitrothion metabolites 219
phosphorylated fibrinopeptide A 250
field desorption (FD) 260
The Fitness for Purpose of Analytical Methods (EURACHEM) 71, 80–2
flow-injection analysis (FIA) 14, 150
fluorescent dyes 251–2
forensic chemistry 257–69
  chemical warfare agents 260–3
  dyes in textile fibres 260, 262
  explosives 263–8
  inks 259–60
  lachrymators 257–8
Forensic Laboratory Guidelines 98
'Forget Pill' 170
Fourier-transform ion cyclotron resonance (FTICR) 10, 16, 195
fragmentation 54–7, 113
freeze/thaw stability 85
full scan 2–7, 12, 103, 120–1

gas chromatography (GC)-MS 98, 111
general unknown screening (GUS) 111–30, 217, 247

Gesellschaft für Toxikologische und Forensiche Chemie (GFTCh) 103
glucocorticosteroids 205–6
glucuronides, steroid conjugates 204
'Golden Triangle' 152
Grubbs test, statistics 76
guidelines, validation 71–2

haemoglobin-based oxygen carriers (HBOCs) 207, 210–12
hair analysis 172
hair, opiates and cocaine metabolites 86
hallucinogens 167–70
D_4-haloperidol 58–9, 60–1
Handbook of Chemometrics and Qualimetrics 71
Harmonized Guideline for Single-Laboratory Validation of Methods of Analysis 71–2
hashish 164–7
Helix pomatia 204
herbicides 219, 221
heroin 151–3
hexamethylenetriperoxidediamine (HMTD) 266
high resolution mass spectrometry (HR)-MS 99
homogeneity of variance 77
homoscedasticity 77
Humalog (insulin) 210–11
human chorionic gonadotrophin (hCG) 207
human growth hormone (hGH) 207–8
human immunodeficiency virus (HIV) 74
human insulin 210–11
hydrocodone 158
hydromorphone 158
hydrophilic interaction chromatography (HILIC) 224
γ-hydroxybutyrate (GHB) 173, 176
4-hydroxyl-2-nonenal (HNE) 253

ibogaine 170
identification
  confirmation 100
  criteria 97–110
  molecular 112–24
  pesticides 235
  unknown compounds 18
identification points (IPs) 99, 235
idoxifene 31

illicit drugs   151–70
immunoassays (IAs)   111, 131
in process stability   85
in-house spectral library   104
in-source collision-induced dissociation (CID)   2, 17, 113–4
incapacitating drugs   170
information-dependent acquisition (IDA)   18, 121–3
inks   259–61
insulin   210–11, 251
interfering signals   73–5
intermediate precision   80
internal standards (IS)
    calibration curves   61
    concentration   57
    method development   25–6
    samples   50
    therapeutic drugs   144
International Olympic committee (IOC)   102
ion detection   11–2
ion suppression *see* matrix effects
ion traps (ITs)
    ion separation   8–9
    mass spectrometers   54–6
    tandem mass spectrometry   16, 123
ion-exchange chromatography (IEC)   245
ionisation   1–6, 27
    energies   30
    analysis for therapeutic drugs   139
    quantification   52–4
    sources   1–6, 112
    steroids   201–2
ions
    cyclotron resonance   10
    detection   11–12
    separation   6–11
    transmission   31–3
2-isopropyl-6-methyl-pyrimidin-4-ol (IMPY)   234
isotope affinity tag (ICAT) approach   252
isotope tags for relative and absolute quantification (iTRAQ)   252
isotope-labelled analogues   26, 74

Jeffamine D-230   117
Journal of Chromatography B   80, 86

ketamine   172, 174
ketobemidone   158–9

labelled isotopes   26
lachrymators   257–8
Lantus (insulin)   210–11
large molecules, doping control   206–12
LC-MS-MS analysis
    data/information-dependent acquisition   16, 18, 121–3
    molecular identification   113–14
    pesticides   226–7, 230
    sample introduction   51
    separation systems   138–9
    systematic toxicology   126–7
    therapeutic drugs   132–44
libraries
    automated searching   115–16
    mass spectral   100–2
    searching   17
    systematic toxicological analysis   115
lime method, heroin production   151
limit of detection (LOD)
    decision limits   106
    drugs of abuse   149
    matrix effects   29
    validation   72
limit of quantification (LOQ)   29, 58
    upper limit of quantification (ULOQ)   84–5
    lower limit of quantification (ULOQ)   83–4
limits
    decision   105–7
    validation   83–5
linear ion traps (LITs)   9, 123
linearity   60–62, 75–79
liquid chromatography-matrix assisted laser desorption ionisation (LC-MALDI)   246
liquid junction interface   246–7
liquid-liquid extraction (LLE)   28, 37, 45–46
    pesticides   223
    therapeutic drugs   134
literature, validation   71–2
long-term stability of samples   85
loop diuretics   195–6
lower limit of quantification (LLOQ)
    definitions   83–4
    experimental design   90
    precision   75
    validation   72

lysergic acid diethylamide (LSD)   44, 167–70

McLafferty rearrangement   230
'magic mushrooms'   170
malathion   225
malathion dicarboxylic acid (MDCA)   234
mannitol   195
marijuana   164–7
mass accuracy   7
mass range   7, 113
mass resolution   7, 113
matrix effects (ME)   28–9
    chromatography   35
    ion suppression/enhancement   87–8
    ionisation   2, 23
    preliminary method validation   38–9
    sample preparation   46
    validation   72, 87–8
matrix-assisted laser desorption ionisation (MALDI)   243–4
    ionisation   1
    liquid chromatography   246
    peptide analysis   243–4
Merck Index   18
mesocarb   206
metabolite identification   18
methadone   158
methamphetamine (MAMP)   159–61
mehod development   23–39
methomyl   222
methyl carbamates   222
methyl parathion   225
3,4-methylenedioxyamphetamine (MDA)   159
3,4-methylenedioxyethylamphetamine (MDEA)   159
3,4-methylenedioxymethamphetamine (MDMA)   159
methylphenidate   173–4
D_3-mianserine   58–9, 60–1
midazolam   172
'Mind Eraser'   170
mobile phase   3, 5, 33–6 114
modafinil   206
modified proteins   253
molecular identification   112–24
molecular structure   98
monitoring exposure   237
morphine   55–7, 152–3
MS-MS *see* tandem mass spectrometry

multiple reaction monitoring (MRM)   14–15, 56, 198
mycophenolic acid   74

naltrexone   159–60
narcotics   206
natural opiates   152–5
neuropetides   248–50
neutral loss scan   15
nitroglycerin (NG)   264
p-nitrophenol   220, 226, 234
norbuprenorphine (NBP)   155
noscapine   151–2
Novolog (insulin)   210–11

octadecasilane   45
oleoresin capsicum (OC)   258
opiates
 agonists   151–2
 natural and semi-synthetic   152–5
 stability   86
opioids   154, 158–9
opium   151–2
optimisation   23–42
Orexin-A   248
organochlorine pesticides (OCPs)   218, 225, 237
organophosphorus pesticides (OPPs)   219–22, 225, 237–8
oxandrolone 201

Papaver somniferum   151
papaverine   151–2
paraquat   218
parathion metabolites   219
partial scans   103
particle beam ionisation (PBI)   1
peak fronting   52
peptide analysis   243, 244
peptides   2, 243–56
per se laws   149
pesticide analysis   217–42
phenobarbital   178
Pihkal: A Chemical Love Story   159
poisoning, acute   237–8
polarity   5
polychlorinated biphenyls (PCBs)   218
polyetherketone (PEEK)   50
poppies   151–2
positive identification criteria   115
post mortem   43, 247
post source decay (PSD)   243–4
post-column addition   3, 4
post-column infusion   28–9

post-extraction addition   28–9, 87–88
potassium-sparing diuretics   195–6
Powermite   268
pre-natal exposure
 cocaine   164
 opiates   159
precision
 experiments and calculations   80–3
 intermediate   80
 lower limit of quantification (LLOQ)   75, 83
 validation   63, 80
prednisolone   205
preliminary method validation   38–9
product (daughter)-ion-scan mode   12–6, 120–1
proteins
 analysis   251–3
 identification tools   18–19
 modified   253
 precipitation   36–7
proton transfer   53
protonated molecular ions   2
psilocin (PCN)   170–1
psilocybin (PCBN)   170–1
purity of standards   105
pyrethroid metabolites   219

quadrupole/linear ion trap (QLIT)   16
quadrupole/time of flight (QTOF)   16
quadrupoles (Q)   2, 8, 12
quantification   43–70
 matrices   44–51
 pesticide analysis   232–6
 therapeutic drugs   140–2
 therapeutic drugs (table)   136–7

radiofrequency-only daughter-ion-scan mode (RFD)   118
RDX (explosive)   264–5
reconstructed ion current (RIC)   100
recovery   29
recovery, validation   64, 86
reflectrons   10–11
regression plots   75–6, 78–9
regulatory bodies   102–4
relative standard deviations (RSDs)   81–2
repeatability, validation   80

reproducibility
 MS-MS spectra   123–4
 precision   80
 standardisation   114
 time   114
response function   60–2
reversed-phase liquid chromatography (RP-LC)
 chromatography   51
 peptide analysis   244–5
 pesticides   223–4
 sample preparation   45
risperidone   50
road traffic offenders   179
robustness   86–7, 114–15
rofecoxib   25
Rohypnol (FBN)   170–1
ruggedness *see* robustness

samples
 chromatography   36
 cleanliness   34
 dilution   36–7
 handling   44–5
 long-term stability   85
 pesticides   225–7
 preparation   45–50, 125–6
 quantification   44–51
 separation   25
 therapeutic drug analysis   132–4
sampling vessels   86
scopolamine   174
screening
 drugs of abuse   176–9
 therapeutic drugs   139–40
 therapeutic drugs (table)   136–7
selected-ion monitoring (SIM)
 fragmentation   54–6
 identification   99
 ion trap   9
 pesticides   235
 quadrupole   8
 time-of-flight   11
selected-ion retrieval (SIR)   8
selected-reaction monitoring (SRM)   14, 31–3, 119–20
selectivity   12, 14, 17, 63–4, 73–5
'self CI' effect   9
semi-synthetic opiates   152–5
Semtex (explosive)   264
sensitivity, validation 63–4
separation systems for therapeutic drugs 138–9
sheathless interface 246–7
signal processing
 interference 73–5

ratio 60
software 119
systematic toxicological analysis 118
signal-to-noise ratio (S/N) 4, 2, 14
lower limit of quantification 83–4
sildenafil (Viagra) 174–5, 177
single-stage mass spectrometry (MS) 99, 113–18
small molecules, doping control 195–206
'snorting' 164
'snow' 163
Society of Forensic Toxicologists (SOFT) 98
solid-phase extraction (SPE) 38, 45
pesticide sample preparation 223, 225
therapeutic drugs 134
sonic spray ionisation (SSI) 1, 3–5
specificity 63–4, 73–5
stability, validation 64, 85–6
stable isotope labelling 252–3
standard deviation 84
standard operating procedures (SOPs) 103
standards, purity 105
stanozolol 201, 203
statistical decision rules 107
steel, cross contamination 50
stepwise process for method development 24
steroids
conjugates 203–5
doping control 199–205
esters 201–3
stimulants 206
street drugs 152
Substance Abuse and Mental Health Services Administration (SAMHSA) 150

substance P derivatives (SPDs) 248–9
surviving ion technique 14, 56, 232
sulphates, steroid conjugates 204
synthetic anabolic steroids 200–1
systematic toxicological analysis (STA) 111–30, 217, 247

Tabernanthe iboga 170
tacrolimus 31–4
tandem mass spectrometry (MS-MS) 12–16
identification criteria 99
method development 23
molecular identification 118–24
see also LC-MS-MS analysis
tear gas sprays 257–8
Temgesic *see* buprenorphine
tetrabutylammonium (TBA) 224
11-nor-9-carboxyl-Δ^9-tetrahydro-cannabinol (THC-COOH) 60–2
Δ^9-tetrahydrocannabinol (THC) (cannabis) 44, 164–7, 206
tetrahydrogestrinone (THG) 201–2
textile fibres 260, 262
thaw/freeze stability 85
The International Association of Forensic Toxicologists (TIAFT) 103
thebaine 155
therapeutic drug monitoring (TDM) 43
therapeutic drugs 131–48
thermospray ionisation (TSI) 1
thiazides 195–6
time-of-flight (TOF)
ionisation 10–11
mass spectrometry 54
peptide analysis 243–4

systematic toxicological analysis 117–18
tipranavir 74
tolerance windows 104
total ion current (TIC) 2, 100
toxicoproteomics 251–3
trenbolone 201
triacetone triperoxide (TATP) 266
triamcinolone 205
triazines 220
3,5,6-trichloro-2-pyridinol (TCPY) 226, 229, 232, 233–7
D_3-trimiparine 144
2,4,6,-trinitrotoluene (TNT) 264, 266–8
triple quadrupole instrument 23
two-dimensional liquid chromatography 38, 245–6

upper limit of quantification (ULOQ) 84–5
urine analysis 44, 136–7, 161–3
US Food and Drug Administration (FDA) 38, 103
UV-diode array detection (DAD) 111

validation 71–95
methods 71–95
parameters 72–88
preliminary method 38–9
vessels, sampling 86
Viagra 174–5, 177

warfare agents 260–3
World Anti-Doping Agency (WADA) 102

xenobiotics 73

zero samples 74
zolpidem 172–3, 175